Fiber Optic
Communications

THIRD EDITION

Fiber Optic Communications

Joseph C. Palais

Professor of Electrical Engineering
Arizona State University

 PRENTICE HALL, Englewood Cliffs, New Jersey 07632

Library of Congress Cataloging-in-Publication Data

Palais, Joseph C.
    Fiber optic communications / Joseph C. Palais. -- 3rd ed.

        p.   cm.
    Includes bibliographical references (p.   ) and index.
    ISBN 0-13-473554-4
    1. Optical communications.   2. Fiber optics.   I. Title.
TK5103.59.P34 1992
621.381'75--dc20   91-28735
                        CIP

Acquisitions editor: Peter Janzow
Production edtior: Irwin Zucker
Copy editor: Michael Schwartz
Prepress buyer: Linda Behrens
Manufacturing buyer: David Dickey
Supplements editor: Alice Dworkin
Cover design: Lundgren Graphics
Editorial assistant: Phyllis Morgan

© 1992, 1988, 19784 by Prentice-Hall, Inc.
A Simon & Schuster Company
Englewood Cliffs, New Jersey 07632

Printed in the United States of America

10 9 8 7 6 5 4

ISBN 0-13-473554-4

PRENTICE-HALL INTERNATIONAL (UK) LIMITED, *London*
PRENTICE-HALL OF AUSTRALIA PTY. LIMITED, *Sydney*
PRENTICE-HALL CANADA INC., *Toronto*
PRENTICE-HALL HISPANOAMERICANA, S.A., *Mexico*
PRENTICE-HALL OF INDIA PRIVATE LIMITED, *New Delhi*
PRENTICE-HALL OF JAPAN, INC., *Tokyo*
SIMON & SCHUSTER ASIA PTE. LTD., *Singapore*
EDITORA PRENTICE-HALL DO BRASIL, LTDA., *Rio de Janeiro*

Dedicated to Sandra, Michael, and Barbara

Contents

4
INTEGRATED OPTIC
WAVEGUIDES   79

5
OPTIC FIBER WAVEGUIDES   101

6
LIGHT SOURCES   141

7
LIGHT DETECTORS   167

Preface

Fiber optic communications developed very quickly after the first low-loss fibers were produced in 1970. Operational fiber systems are now common, and new installations and applications appear continually. This growth is expected to continue for many years.

Although still evolving, fiber technology has matured sufficiently so that many books have been written on the subject. Many of them are quite detailed in terms of theoretical and mathematical content, and the beginner may find the level difficult.

This text is intended to be less difficult, while still bringing to the reader the information necessary to understand the design, operation, and capabilities of fiber systems. Important theoretical and mathematical results are given without the accompanying lengthy proofs. However, results are explained in physical terms when possible and appropriate,

and extensive tables and figures are used to make those results readily usable. To provide a realistic view, numerical values are given for the range of typical device parameters.

When the first edition of this book appeared in 1984, fibers had already crisscrossed much of the United States and many other countries to deliver telephone messages between the major exchanges. By 1988, when the second edition was published, the land-based long-distance fiber telephone network was nearly complete, and submarine fiber telephone cables were being installed beneath the major oceans. In addition, fiber optic local-area networks (LANs) were in development. As this third edition goes to press, over 10 million kilometers of fiber have been installed worldwide, numerous submarine fiber cables covering the Atlantic and Pacific oceans and many other smaller seas are operational, and installation of fiber LANs is increasing. In addition, numerous tests have been completed for bringing fiber directly to all homes, holding the promise for expanded services to the individual subscriber. This will involve huge amounts of fiber, tremendous numbers of associated components, and a significantly large work force trained to implement it. Some projections foresee the goal being met sometime early in the next century.

It is because of the continued growth and evolution of the fiber industry that this new edition is necessary. The fundamentals have not changed, but there are many new components and techniques available for applying fibers to communications problems. I have attempted to work these new ideas into the previous edition with as little disruption as possible.

This is an introductory book. No background in fiber optics, or optic communications, is presumed. Only the simplest concepts from algebra and trigonometry are invoked in explaining the characteristics of fiber systems. Appropriate background material on optics, electronics, and communications is introduced in the text as needed.

This book is based on a set of notes I developed and used for numerous short courses on fiber optic communications. Participants in these courses had training ranging from 2 years of technical school through the Ph.D. level. Jobs varied from designer to board chairman. Attendees included personnel from industry, government, and academia. Individual backgrounds were in chemistry, physics, and many areas of engineering. In addition to the short course presentations, I have taught this material to hundreds of electrical engineering students at the senior and first-year graduate level.

The professionals benefiting from this book include practicing design engineers concerned with the selection and application of components and with the design and evaluation of systems. Knowledge of the entire system is useful for the device designer as well. Others involved in fiber optics, such as high-level engineering decision makers, project managers, technicians, marketing and sales personnel, and teachers, can also obtain valuable information from the material presented.

The organization of the book is as follows. Block diagrams of entire fiber optic systems are presented at the outset. This identifies the components of fiber systems, providing motivation for their individual study in succeeding chapters. Chapters 2 and 3 contain a review of important results from the fields of optics and wave travel. This basic information is needed for an understanding of fiber optic devices and systems. Chapter 4, on integrated optics, introduces the technology of combining optic components onto a single substrate. The integrated-optic waveguide provides an excellent, simplified model for propagation of light in a fiber. Chapters 5–9 present the main devices encountered in a fiber optic system. These are the fiber, light source, light detector, couplers, and distribution networks. System considerations appear in Chapters 10–12, where modulation formats, the effects of noise on message quality, and system design are covered. The last chapter includes examples of operational systems. In this chapter, the design information developed throughout the book is applied to realistic problems.

I expect the reader who has mastered this material to be able to design and specify systems and to choose and evaluate system components such as fibers, light sources, detectors, and couplers. Commercially available subsystems, such as complete transmitters and receivers, will also be amenable to evaluation by the techniques presented in this book.

This is a new and fully revised version

of the second edition of *Fiber Optic Communications*. Major advances in the fiber industry since publication of the second edition have been incorporated. Because the fundamentals of the technology have remained the same, the number of changes is moderate. Nonetheless, the changes and additions are significant. A few of the added topics are optical solitons, fiber and semiconductor optical amplifiers, optical isolators, improved laser diodes, laser residual intensity noise calculations, new transmission standards, wavelength-dependent couplers, and advances in optical integrated circuits for fiber networks. The present version of the text reflects the changes in the industry.

Because numerous colleges adopted the initial text for an undergraduate course in fiber optics, a homework problem set was inserted in the second edition. Over 50 new problems have been added and several of the old ones have been updated for this new edition, making this text's use in the classroom even more desirable. Some problems are merely "plug-in"–type questions intended to give the student practice and confidence in understanding of the material presented. Other problems take more thought and may even require finding and reviewing material from other sources. Answers to most problems appear at the end of the book, and a new solutions manual is available for instructors.

The bibliography and periodical listing appearing at the end of the text provides a resource for further reading. More than 25 new books have been added for this edition.

I find that the first seven chapters can be covered in a one-semester course. This introduces all the major system components to the students, allowing those who have mastered the material to productively enter into the fiber industry. The last five chapters, on more advanced topics, can be covered in a second term. To simplify the mathematics and reach a wider audience, many of the results presented in the text are not fully derived. Instructors of well-prepared students, such as seniors in electrical engineering programs, may wish to fill in the derivations to deepen student understanding.

FIBER OPTIC DESIGN SOFTWARE

Most technical workers and students have access to a personal computer. Even the simplest of these computers has the potential for enhancing the learning experience and aiding in the application of the concepts described in a technical book such as this one. For these reasons, Dean R. Johnson and I prepared a set of computer programs to illustrate various fiber optic phenomena and design considerations. Many of the programs are extensions of the material presented in this book.

Programs included illustrate the emission patterns of light sources, receiver bandwidth and noise calculations, mode charts and field plots for the dielectric slab waveguide, Gaussian beam patterns, system design, OTDR measurements, the fiber optic gyroscope, and others.

The name of the software package is *Fiber Optics Design Software*. It can be run with any IBM® compatible BASIC program, using either color or black-and-white monitors. The complete software package is available from Kern International, Inc., 190 Duck Hill Road, Duxbury, MA 02332, (617) 934-2452.

An additional software package was developed by one of my students for the design of digital networks. This interactive software analysis tool provides complete automatic design synthesis based on user-supplied system performance criteria. It has many of the features of an expert system. For further details contact C. S. Bergstrom, 501 South Oak, Chandler, AZ 85226.

ACKNOWLEDGMENTS

I am grateful to students in my regular electrical engineering fiber-optic classes at Arizona State University, and to the attendees at the many short courses in fiber optics that I have been privileged to present. The former proof-read the early version of the manuscript and made many valuable suggestions that I incorporated into the final result. The latter impressed upon me the type of information desired by typical workers whose jobs involve applying fiber optic technology. I also thank my wife and children (now grown) for accepting my absences from home without complaint during the preparation and revisions of this book.

Fiber Optic Communications

Chapter 1

Fiber Optic Communications Systems

In this chapter we define the subject of fiber optic communications and explain our approach to this subject. We review the many advantages over alternative technologies and discuss significant applications. Because the reader may have no previous fiber optics experience, this book presents the fundamentals of several subjects on which the technology is based. These include fibers, optics, communications, optic communications, and, finally, complete fiber optic communications systems. The outlines of a complete system are shown here. Later, properties of each part and the dependence between parts is described. Finally, details of the design of practical systems are presented.

1-1 HISTORICAL PERSPECTIVE

Light has always been with us. Communications using light occurred early in our development when human beings first communicated by using hand signals. This is obviously a form of optic communications. It does not work in darkness. During the day, the sun is the source of light for this system. The information is carried from the sender to the re-

1

ceiver on the sun's radiation. Hand motion modifies, or modulates, the light. The eye is the message-detecting device, and the brain processes this message. Information transfer for such a system is slow, the transmission distance is limited, and the chances for error are great.

A later optic system, useful for longer transmission paths, was the smoke signal. The message was sent by varying the pattern of smoke rising from a fire. This pattern was again carried to the receiving party by sunlight. This system required that a coding method be developed and learned by the communicator and receiver of the message. This is comparable to modern digital systems that use pulse codes.

In 1880 Alexander Graham Bell invented a light communication system, the *photophone*.[1] That used sunlight reflected from a thin voice-modulated mirror to carry conversation. At the receiver the modulated sunlight fell on a photoconducting selenium cell, which converted the message to electrical current. A telephone receiver completed the system. The photophone never achieved commercial success, although it worked rather well.

The advent of lamps allowed the construction of simple optic communications systems such as blinker lights for ship-to-ship and ship-to-shore links, automobile turn signals, and traffic lights. In fact, any type of indicator lamp is basically an optic communications system.

All the systems just described have low information capacities. A major breakthrough that led to high capacity optic communications was the invention of the laser, in 1960. The laser provided a narrow-band source of optic radiation suitable for use as a carrier of information. Lasers are comparable to the radio frequency sources used for conventional electronics communications. Unguided optic communications systems (nonfiber) were developed shortly after the discovery of the laser. Communication over light beams traveling through the atmosphere was easily accomplished. The disadvantages of these systems include dependence on a clear atmosphere, the need for a line-of-sight path between transmitter and receiver, and the possibility of eye damage to persons who unknowingly look into the beam. Although somewhat limited in their use, these early applications caused an interest in optic systems that would guide the light beam and thus overcome those disadvantages. In addition, guided beams could bend around corners and could be buried in the ground. The early work on atmospheric laser systems provided much of the fundamental theory and many of the actual components required for communications over fibers. Ironically, it is now known that laser sources are not required for all fiber systems. In many cases the less narrow band light-emitting diode is suitable. (The choice of the proper light source is a matter we discuss in this book.)

In the 1960s the key element in a practical fiber system was missing; that is, an efficient fiber. Although light could be guided by a glass fiber, those available glass fibers attenuated light by far too large an amount. Glass produced by the ancient Egyptians was opaque. The artisans of Venice fabricated glass of much greater purity in the middle ages. Venetian glass was moderately transparent, but still orders of magnitude too lossy for modern long-distance communications. It was not until 1970 that the first truly low-loss fiber was developed and fiber optic communications became practical. This occurred just 100 years after John Tyndall, a British physicist, demonstrated to the Royal Society that light can be guided along a curved stream of water. Guiding of light by a glass fiber and by a stream of water are evidence of the same phenomenon (total internal reflection).

1-2 THE BASIC COMMUNICATIONS SYSTEM

A basic communications system consists of a transmitter a receiver, and an information channel, arranged as in Fig. 1-1. At the transmitter the message is generated and put into a form suitable for transfer over the information channel. The information travels from the transmitter to the receiver over this channel. Information channels can be divided into two categories: unguided channel and guided channel. The atmosphere is an example of an unguided channel over which waves can propagate. Systems using atmospheric channels include commercial radio and television broadcasts and microwave relay links. Guided channels include a variety of conducting transmission structures. A few of these, illustrated in Fig. 1-2, are the two-wire line, coaxial cable, and rectangular waveguide. Guided lines cost more to manufacture, install, and service than do atmospheric channels. Guided channels have the advantages of privacy, no weather dependence, and the ability to convey messages within, under, and around physical structures. Fiber waveguides have these advantages and others. We enumerate them later in this chapter. At the receiver the message is extracted from the information channel and put into its final form.

A more detailed, but still quite general block diagram appears in Fig. 1-3. A brief discussion of each block in this figure gives us

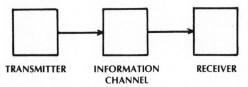

TRANSMITTER   INFORMATION   RECEIVER
                  CHANNEL

Figure 1-1   The basic communications system.

a good feel for the main elements of a communications system. Our descriptions of these elements emphasize those suitable for fiber systems, although the diagram itself is applicable to other types of communications links. Many of the concise descriptions given in this section are expanded later. At present we present an overview of the subject and lay the foundations for further discussions.

Message Origin

The message origin may take several physical forms. Quite often it is a transducer that converts a nonelectrical message into an electrical signal. Common examples include microphones for converting sound waves into currents and video (TV) cameras for converting images into currents. In some cases, such as data transfer between computers or parts of a computer, the message is already in electrical form. This situation also arises when a fiber link comprises a portion of some larger system. Examples include fibers used in the ground portion of a satellite communications system and fibers used in relaying cable television signals. In any case, the information must

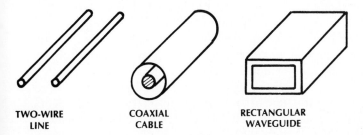

TWO-WIRE   COAXIAL   RECTANGULAR
LINE   CABLE   WAVEGUIDE

Figure 1-2   Some conducting transmission lines.

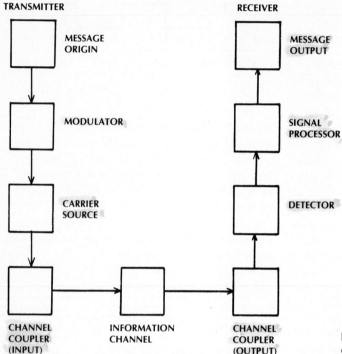

Figure 1-3 Details of a generalized communications system.

be in electrical form before transmission for either electronic or optic communications.

Modulator

The modulator has two main functions. First, it converts the electrical message into the proper format. Second, it impresses this signal onto the wave generated by the carrier source. Two distinct categories of modulation format exist: *analog* and *digital*. An analog signal is continuous and reproduces the form of the original message quite faithfully. For example, suppose a sound wave containing a single tone is to be transmitted. The electrical current produced when a microphone picks up this wave has the same shape as the wave itself. This relationship is illustrated in Fig. 1-4. In this case, the modulator need not change

the format of the signal. It may be appropriate to amplify this signal, however, so that the signal will be strong enough to drive the carrier source.

Digital modulation involves transmitting information in discrete form. This is illustrated in Fig. 1-5. The signal is either on or off. The *on* state represents a digital 1 and the *off* state represents a digital 0. These states are the *binary digits* (or *bits*) of the digital system. The data rate is the number of bits per second (bps) transmitted. The sequence of on or off pulses may be a coded version of an analog message. An analog-to-digital converter develops the digital sequence from the analog message. The reverse process occurs at the receiver, where the digital signal is returned to its analog form. To impress a digital signal onto a carrier, the modulator need only turn the source on or off at the appropriate

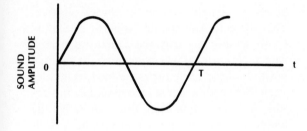

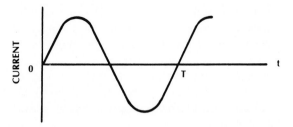

Figure 1-4 Analog modulation.

times. The ease of constructing digital modulators makes this format very attractive for fiber systems.

The choice of format must be made very early in the design of any system. Other considerations and comparisons between analog and digital systems are discussed in more detail later in this chapter and in succeeding chapters.

At the end of this chapter is a list of the decisions a designer faces in the construction of a complete system. We explain the items in this table throughout the book. Later we are more specific in describing the available choices and add a list of advantages, disadvantages, and primary applications appropriate to each choice.

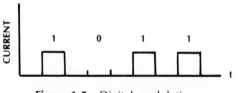

Figure 1-5 Digital modulation.

Carrier Source

The carrier source generates the wave on which the information is transmitted. This wave is called the *carrier*. The carrier is produced by an electronic oscillator in radio-frequency communications systems. For fiber optic systems, a *laser diode* (LD) or a *light-emitting diode* (LED) is used. These two devices can correctly be called optic oscillators. Ideally, these light sources provide stable, single-frequency waves with sufficient power for long-distance propagation. Actual laser diodes and light-emitting diodes differ somewhat from this ideal. They emit a range of frequencies and generally radiate only a few milliwatts of average power. This power is sufficient in many cases, because receivers are so sensitive. However, transmission losses continually decrease the power level along the fiber, so the lack of sufficient source power limits the length of any communications link. The lack of truly single-frequency source also degrades the system. This degradation limits the amount of information that can be carried along a given path length.

LEDs and laser diodes are small, light, and consume only moderate amounts of power. They are relatively easy to modulate; that is, to impress the information on their radiation. Both of these devices are operated by passing current through them. The amount of power that they radiate can be made proportional to this current. In this way, the optic output power takes the shape of the input current coming from the modulator. The results of analog and digital modulation of the carrier are indicated in Fig. 1-6. It should be emphasized that the information being transmitted is contained in the variation of the optic power. This is called *intensity modulation* (IM). Although the signal current shown in Fig. 1-4 has both positive and negative parts, the output

power of a light emitter is always positive. This characteristic should be noted in Fig. 1-6. To achieve linearity the actual modulation current in an analog system must be entirely positive. A dc bias current, added to the desired information signal, accomplishes this result, as pictured in Fig. 1-6. Similarly, the modulating current for a digital system is always positive. Because a laser diode does not turn on (that is, it does not radiate) until some threshold current is applied, the modulation current may include a dc offset equal to this threshold value. The presence of a binary 1 drives the current beyond threshold and makes the diode emit light. A binary 0 leaves the current at threshold, where no radiation occurs. An LED does not have a threshold and turns on whenever positive current flows through it.

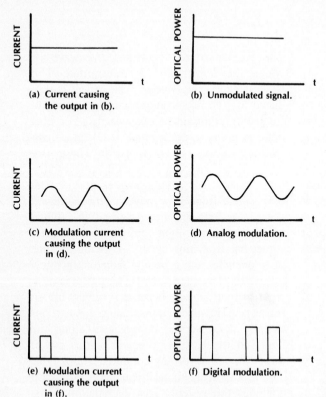

(a) Current causing the output in (b).

(b) Unmodulated signal.

(c) Modulation current causing the output in (d).

(d) Analog modulation.

(e) Modulation current causing the output in (f).

(f) Digital modulation.

Figure 1-6   Analog and digital modulation of an optical carrier.

Laser diodes and LEDs have been constructed that radiate at frequencies where glass fibers are efficient transmitters of light; that is, where fibers have low attenuation. This is indeed fortunate, because suitable sources emitting at arbitrary frequencies are difficult to obtain. Without this match between source frequency and fiber low-loss region, fiber option communications would not exist.

Channel Coupler (Input)

Next we consider the coupler, which feeds power into the information channel. In a radio or television broadcasting system, this element is an antenna. The antenna transfers the signals from the transmitter onto the information channel, in this case the atmosphere. In a guided system using wires, such as a telephone link, the coupler need only be a simple connector for attaching the transmitter to the transmission line being used as the information channel. For an atmospheric optic system, the channel coupler is a lens used for collimating the light emitted by the source and directing this light toward the receiver. In our fiber system, the coupler must efficiently transfer the modulated light beam from the source to the optic fiber. Unfortunately, it is not easy to accomplish this transfer without relatively large reductions in power or somewhat complicated coupler designs. One difficulty arises because of the small size of conventional fibers, which have diameters on the order of 50 millionths of a meter. However, the large loss basically occurs because light sources emit over a large angular extent. Fibers can only capture light within more limited angles. This matter is illustrated in Fig. 1-7. The simplest type of coupler is shown. The light emitter is merely butted against the fiber. As indicated, even if the fiber is big enough to intercept all the light rays emitted by the

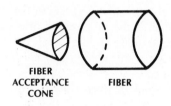

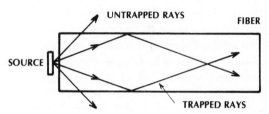

Figure 1-7   Coupling of light to the fiber.

source, the light will not be entirely collected because of the difference between the radiation and acceptance cone angles. More efficient, but also more complex, couplers can be constructed. The channel coupler is an important part of the design of a fiber system because of the possibility of high losses.

Numerical evaluation of expected efficiencies and the design of improved couplers are considered later in this book.

Information Channel

The information channel refers to the path between the transmitter and receiver. In fiber optic communications, a glass (or plastic) fiber is the channel. Desirable characteristics of the information channel include low attenuation and large light acceptance cone angle. Low attenuation and efficient light collection are particularly necessary for transmission over long path lengths. Although highly sensitive receivers are available, the power delivered to the receiver must be above some limiting value to obtain the desired message with appropriate clarity.

Another important property of the information channel is the propagation time of the waves traveling along it. In general, the travel time depends on the light frequency and on the path taken by the light rays. A signal propagating along a fiber normally contains a range of optic frequencies (because optic sources emit a range of frequencies) and divides its power along several ray paths. This results in a distortion of the propagating signal. In a digital system, this distortion appears as a spreading and deforming of the on pulses as indicated in Fig. 1-8. The spreading increases with the distance traveled. Eventually, the spreading is so great that adjacent pulses begin to overlap (see Fig. 1-8) and become unrecognizable as separate bits of information. Errors in the transmission now result. To keep this from occurring, pulses must be transmitted less frequently. This, of course, limits the rate at which the pulses can be sent. The wave velocity dependence on frequency and path results in a limitation on the information rate, whether the modulation is digital or analog.

The requirements for large light acceptance angle and low signal distortion are contradictory. Practical fibers represent a design compromise between these two qualities. For systems having moderate path lengths and information rates, fibers with suitable values of acceptance angle and signal distortion can be obtained. Other interesting qualities of fibers are presented later in this chapter.

Channel Coupler (Output)

In an atmospheric electronic communications system, an antenna collects the signal from the information channel and routes it to the rest of the receiver. In the fiber system, the output coupler merely directs the light emerging from the fiber onto the light detector. This light is radiated in a pattern identical to the fiber's acceptance cone. Because common photodetectors have large surface areas and large acceptance angles, light can be efficiently coupled from the fiber by a simple butt connection, as indicated in Fig. 1-9.

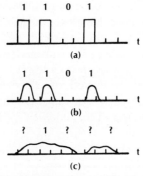

Figure 1-8 Spreading of optical pulses. (a) Original pulse train. (b) After traveling some distance, the pulses have widened. (c) Further travel causes adjacent pulses to merge and to spill over into time slots where 0s were meant to be. Numerous errors will now occur in the detection of this signal.

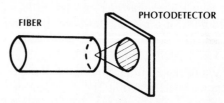

Figure 1-9 Coupling from a fiber to a photodetector is very efficient. The detector can accept most of the light radiated by the fiber.

Detector

The information being transmitted must now be taken off the carrier wave. In an electronic system, this is the process of *demodulating* the signal and is performed by the proper electronic circuit. In the fiber system, the optic wave is converted into an electric current by a photodetector. Semiconductor photodiodes of various designs are most commonly used. The current developed by these detectors is proportional to the power in the incident optic wave. Because the information is contained in the optic power variation, the detector output current contains this information. This current is a replica of that used to drive the carrier light source. The relationship between the signals at various points in the system is illustrated in Fig. 1-10 for an analog signal. The current generated by the transducer at the message origin is sketched in Fig. 1-10(a). This is the information signal we wish to transmit. The modulator adds a constant bias to this current 1-10(b) and applies the result to the light carrier. The carrier power waveform in Fig. 1-10(c) now contains the desired information. The signal is attenuated as it propagates through the fiber, as illustrated by the diminished optic power shown in Fig. 1-10(d). This figure is drawn assuming negligible waveform distortion owing to travel along the fiber. The detector converts the optic waveshape to electrical form, as shown in Fig. 1-10(e). To complete the transmission, the detector output current is filtered to remove the constant bias and amplified if needed. These last two functions take place in the signal-processor block of our system. The result, shown in Fig. 1-10(f), is the desired information waveshape. A similar set of figures could be drawn for a digital system. The result would show the replication of the input pulse sequence at the detector output.

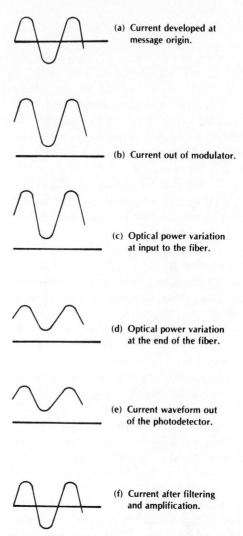

(a) Current developed at message origin.

(b) Current out of modulator.

(c) Optical power variation at input to the fiber.

(d) Optical power variation at the end of the fiber.

(e) Current waveform out of the photodetector.

(f) Current after filtering and amplification.

Figure 1-10 Signals at various points in an analog system.

Important properties of photodetectors include small size, economy, long life, low power consumption, high sensitivity to optic signals, and fast response to quick variations in the optic power. Fortunately, light detectors having these characteristics are readily available.

Signal Processor

For analog transmission, the signal processor includes amplification and filtering of the signal. In addition to filtering of the constant bias, any other undesired frequencies should be blocked from further travel. An ideal filter passes all frequencies contained in the transmitted information and rejects all others. This improves the clarity of the intended transmission. Proper filtering maximizes the ratio of signal power to unwanted power. Random fluctuations in the received signal are referred to as *noise*. Noise is present in all communications systems. We will learn how to evaluate the amount of noise in a fiber system and how to design fiber systems to meet the *signal-to-noise ratio* (SNR) requirements for a given application.

For a digital system, the processor may include decision circuits in addition to amplifiers and filters. The decision circuit decides if a binary 1 or 0 was received during the time slot of any individual bit. Because of unavoidable noise, there will always be some probability of error in this process. The *bit-error rate* (BER) should be very small for quality communications. The digital signal processor must also decode the incoming sequence of 0s and 1s if the original message was analog. This is done by a digital-to-analog converter, which re-creates the original electrical form of the information. If the communications were between machines, then the digital form might be suitable for use without digital-to-analog conversion.

Message Output

We are concerned with two different situations at this point. In one case, the message is presented to a person, who either hears or views the information. To achieve this, the electrical signal must be transformed into a sound wave or a visual image. Suitable transducers for accomplishing this transformation are the loudspeaker for audio messages and the cathode ray tube similar to that used in a television set for pictures.

In the second case, the electrical form of the message emerging from the signal processor is directly usable. This situation occurs, for example, when computers or other machines are connected through a fiber system. It also occurs when the fiber system is only a part of a larger network, as in a fiber link between telephone exchanges or a fiber trunk line carrying a number of television programs. In these last two systems, the processing includes distribution of the electrical signals to the proper destinations. The message output device is simply an electrical connector from the signal processor to the succeeding system.

We are only peripherally interested in signal processing circuits and message output devices in this book, because these components are the same as those already developed for systems that do not use optics.

Some Numbers

Up to now there has been a noticeable lack of numbers associated with our discussions. This omission must be corrected if we hope to understand and design communications systems.

Units appearing frequently in this book are listed in Table 1-1 for convenient reference. This book uses the MKSC (meter-kilogram-second-coulomb) system as much as possible. In practice, fiber lengths and diameters are almost always expressed in metric form. Table 1-2 summarizes a few of the physical constants important to our study of fiber optics.

The frequency unit, the hertz, is equivalent to one cycle of oscillation per second.

Fiber Optic Communications Systems **11**

TABLE 1-1. Units

Unit	Symbol	Measure
meter	m	length
kilogram	kg	mass
second	s	time
coulomb	C	charge
joule	J	energy
watt	W	power
hertz	Hz	frequency
newton	N	force
ampere	A	current
kelvin	K	temperature
degree Celsius	°C	temperature
farad	F	capacitance
ohm	Ω	resistance
volt	V	voltage

TABLE 1-2. Constants

Description	Value	Symbol
Velocity of light	3×10^8 m/s	c
Planck constant	6.626×10^{-34} J $\times$ s	h
Electron charge	-1.6×10^{-19} C	$-e$
Boltzmann constant	1.38×10^{-23} J/K	k

The time between successive peaks of an oscillation is called the *period* and is given by the reciprocal of the wave frequency. That is, seconds per cycle is the reciprocal of the cycles per second. If f is the wave frequency and T is its period, then $T = 1/f$. Figure 1-11 illustrates this point. In fiber optic communications, we encounter frequencies from a few hertz to beyond 10^{14} Hz. We also deal with

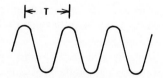

Figure 1-11 A wave having period T seconds. The corresponding frequency is $f = 1/T$.

lengths from a millionth of a meter (10^{-6} m) to tens of kilometers. It is therefore convenient to learn the standard prefixes for very large and very small quantities. Some common ones are listed in Table 1-3.

The wavelengths of light are on the order of 10^{-6} m $= 1$ μm. This length is a micrometer. An even smaller unit of length is the nanometer, which is 10^{-9} m. Thus, 1000 nm equals 1 μm.

Next, we will discuss some of the important numerical characteristics of common analog and digital systems. Table 1-4 summarizes the bandwidth requirements of several analog systems. Telephone links need only transmit messages with frequencies up to 4000 Hz, because most of the energy in normal speech is contained in the frequencies below this value. Messages are intelligible and

TABLE 1-3. Prefixes

Prefix	Symbol	Multiplication Factor
giga	G	10^9
mega	M	10^6
kilo	k	10^3
centi	c	10^{-2}
milli	m	10^{-3}
micro	μ	10^{-6}
nano	n	10^{-9}
pico	p	10^{-12}
femto	f	10^{-15}

TABLE 1-4. Common Analog Systems

Message Type	Bandwidth	Comments
Voice	4 kHz	single telephone channel
Music	10 kHz	AM radio broadcasting station
Music	200 kHz	FM radio broadcasting station
Television	6 MHz	television broadcasting station

individual voices are quite recognizable with this bandwidth. A channel with a larger bandwidth would reproduce sounds with higher quality, but this is not necessary in practical telephone circuits. Bandwidth reductions below 4 kHz are possible if some degradation in speech quality is allowable. For most voice transmission examples in this book we will assume the 4-kHz bandwidth used in commercial telephone systems. The range of frequencies up to 4 kHz is called the *baseband* of the voice message.

Commercial AM (*amplitude modulation*) broadcasting stations transmit messages from 100 to 5000 Hz. The AM format requires a bandwidth of twice the highest modulation frequency, so that AM stations have 10-kHz bandwidths, and their carrier frequencies are separated by 10 kHz. High-quality music reproduction requires transmission of modulating frequencies up to 15 kHz (a particularly responsive ear can detect even faster oscillations, approaching 20 kHz). FM (*frequency modulation*) broadcasting stations transmit information between 50 and 15,000 Hz. The FM format requires a 200-kHz bandwidth to accomplish this result.

Because video signals contain more information than sound signals, they require additional transmissions bandwidths. Commercial television channels have a 6-MHz bandwidth, including both the picture and the sound. The highest video frequency actually transmitted is near 4.2 MHz. The range of frequencies occupied by the TV signal (up to 6 MHz) is the baseband of the television message.

When analog signals are transmitted digitally, the bit rate depends on the rate at which the analog signal is sampled and the coding scheme. According to the *sampling theorem*, an analog signal can be accurately transmitted if sampled at a rate of at least twice the highest frequency contained in that

signal. For this reason the standard 4-kHz telephone channel is sampled 8000 times a second.[2] The coding procedure uses 8 bits to describe the amplitude of each sample so that a total of 64,000 bps are transmitted for a single telephone message. By sending pulses at a rate higher than 64 kbps, several messages can be simultaneously transmitted. The different messages are combined (*multiplexed*) onto a single information channel by interleaving their data bits at the transmitter. The messages are separated (*demultiplexed*) at the receiver. The multiplexing-demultiplexing functions would have to be added to the block diagram of Fig. 1-3. Table 1-5 shows the hierarchy for the asynchronous digital telephone network used in the United States. Shown are the information rates, their designation, and the corresponding number of voice channels. As an example, the basic block is the T1 (transmission at level 1) system. This level carries 24 voice messages. When digital signaling is used, the designation DS-1 (digital signal at level 1) is appropriate. The T2 level is formed by combining 4 T1 systems so that $4(24) = 96$ messages can be carried. Similarly, all the other levels above the first are combinations of lower-level systems. If we look closely at the data rates for each level, then we find that more data bits are being sent than are required for the messages alone. For example, the T3 system should require a rate of $672(64,000) = 43$ Mbps. The actual rate of 44.736 Mbps includes synchronization· and signaling pulses.

Fiber capabilities are so great that systems with even greater capacities than those shown in Table 1-5 can be constructed. For example, systems have been designed with individual fiber line rates of 1.2, 1.7, and 2.3 Gbps. Multigigabit line rates (and tens of thousands of voice channels) are well within the realm of possibility using fibers.

A newer transmission standard has been

TABLE 1-5. Digital Transmission Rates of the Telephone System in the United States

Number of Voice Channels	Transmission Designation	Signaling Designation	Data Rate
1			64 kbps
24	T1	DS-1	1.5444 Mbps
48 (2 T1 systems)	T1C	DS-1C	3.152 Mbps
96 (4 T1 systems)	T2	DS-2	6.312 Mbps
672 (7 T2 systems)	T3	DS-3	44.736 Mbps
1344 (2 T3 systems)	T3C	DS-3C	91.053 Mbps
4032 (6 T3 systems)	T4	DS-4	274.175 Mbps
6048 (9 T3 systems)	—	—	405 Mbps
8064 (12 T3 systems)	—	—	565 Mbps

developed intended for world-wide use. Its name is SONET, which stands for synchronous optical network. This standardized synchronous system will allow greater flexibility in adding new services to existing SONET installations. The basic SONET transmission rate is OC-1 (optical carrier at level 1) at 51.8 Mbps. The electrical equivalent is STS-1 (synchronous transport signal at level 1). Higher level SONET rates are shown in Table 1-6. Introduction of the SONET standard is expected to increase the rate at which fiber optic applications are introduced.

The data rate necessary for digital transmission of a commercial television broadcast is easily determined. The analog signal has a bandwidth of 6 MHz. Sampling at twice this rate, and encoding by 8 bits per sample, requires a data rate of $2(6)(8) = 96$ Mbps. If several of these signals were multiplexed onto

TABLE 1-6. SONET Transmission Rates

Transmission (Electrical)	Designation (Optical)	Data Rate (Mbps)
STS-1	OC-1	51.84
STS-3c	OC-3	155.52
STS-12	OC-12	622.08
STS-24	OC-24	1244.16
STS-48	OC-48	2488.28

a single fiber, the data rate would be several hundred megabits per second. Because the video information is contained in a bandwidth that is less than 6 MHz, the 96-Mbps rate can be reduced. For example, allowing a video bandwidth of 4.5 MHz, sampling at twice this rate, and encoding by using 9 bits per sample yields a data rate of 81 Mbps. An accompanying sound track covering 15 kHz, sampled at 30 kbps, and encoded by using 8 bits per sample requires a rate of 240 kbps. The total signaling rate for this system would be 81.24 Mbps. This signal can be transmitted easily over the standard DS-3C telephone line, which operates at a 91.053-Mbps data rate.

The relationship between bandwidth and message type has now been established for common music, voice, and video communication networks. Transmission involving data, such as from a computer or a workstation, requires bandwidths that depend on the desired rate of information transfer. One popular local-area network, Ethernet, operates at 10 Mbps. It can be implemented by using co-ax or fiber transmission lines. For high-speed data transfer, the fiber distributed-data interface (FDDI) has been specified to operate at a data rate of 100 Mbps. An even faster LAN, the high performance parallel interface (HPPI), operates at 800 Mbps. There is a steady evolution toward higher transmission rates as data transfer becomes more important to business and industry. Fibers seem to be the ideal transmission medium for these applications because of their huge data handling capabilities.

Remember that the bandwidths and rates being discussed are characteristics of the message and do not depend on the type of transmission used. Optic and radio-frequency systems require the same bandwidths and data rates to convey the same messages.

At this point we wish to give the reader some feeling for the ease (or difficulty) of designing, constructing, and testing a fiber system of specified data rate. As expected, the difficulties increase as the data rate becomes larger. The specific classifications to follow are arbitrary, but useful:

Fiber systems operating below 100 kbps have low transmission rates. Such systems can be readily, and cheaply, constructed from available optic and electronic components. Rates from 100 kbps to 10 Mbps are only somewhat more costly and difficult to implement. This is a moderate range of information rates. From 10 Mbps to just over 100 Mbps, improved circuits, light emitters, and light detectors must be used. Despite the expense and difficulties, systems in this range are common, as witnessed by the telephone systems operating at these high rates. The range from a few hundred megabites per second to 1000 Mbps (1 Gbps) is very high and requires added expense and care. Optic components able to emit and detect at such great speeds are costly, and the electronic circuits that interface with them are difficult to construct. Components and systems operating beyond 1 Gbps are possible. Such extremely high data rates are only encountered in very large and sophisticated systems. The amount of information transferred at such rates is probably greater than the amount most of us ever will have to deal with.

For an analog link, the quality of the signal transmission is expressed by the ratio of the signal power S to the noise power N. Noise is present in all receivers so that the signal-to-noise ratio is never infinite. A good, clear television picture requires a signal-to-noise ratio better than 10^4. For values below this, the picture becomes fuzzy; that is, the resolution and contrast are degraded by the noise. Acceptable music and voice signals also require high signal-to-noise ratios for good reception.

In a digital system, a transmitted 1 may be interpreted as a 0 by the receiver or a transmitted 0 may be sensed as a 1. This is caused by system noise. The quality of a digital system is given by its bit-error rate (BER). A BER of 10^{-9} means only 1 bit is misread out of every 1 billion bits sent. A rate of 10^{-9} (or better) is available on standard digital telephone lines, which transmit data as well as audio messages. Data need this degree of precision. Speech can be delivered with an error rate several orders of magnitude larger than 10^{-9} before a listener detects a decline in the quality of reception.

A strong optic signal must appear at the receiver if large signal-to-noise ratios, or low bit-error rates, are to be achieved. The numbers we have reviewed in this section are used to evaluate the devices and systems we encounter in the remainder of this book.

Computing Power Levels in Decibels

A major part of systems design involves keeping an account of the optic power along the communications link. This account is usually necessary to ensure that the wave incident on the detector has sufficient strength to be clearly and correctly recognized. In other instances the received power may even be too large for the receiver. The designer must be certain this does not occur. The *decibel* (dB) is a convenient measure of the relative power levels in a communications system. If the power is P_1 watts at one point in the system and P_2 watts at some point farther along the link, then P_2/P_1 is the fraction of the power transmitted between the two locations. In other words, P_2/P_1 is the efficiency of transmission between the two points. This ratio, expressed in decibels, is

$$dB = 10 \log_{10} \frac{P_2}{P_1} \qquad (1\text{-}1)$$

P_2 and P_1 must be in the same units; for example, both in watts or both in milliwatts. Logarithms of numbers less than one are negative, so the decibel result is negative if P_2 is less than P_1. This is the case when the system has losses. If P_2 is larger than P_1 (as occurs when an amplifier is placed between the two locations), the decibel value is positive. If the dB value relating P_1 and P_2 is known, then P_2 can be found in terms of P_1 by what follows:

$$P_2 = P_1 10^{dB/10}$$

The logarithmic scale is handy because of the ease with which the total change in power level is found when several elements are cascaded. Consider the three-element system sketched in Fig. 1-12. The three blocks could represent a coupler from a light source to a fiber, the fiber itself, and a connector. The output power is determined by multiplying the efficiencies of each block, as seen from the expression

$$\frac{P_4}{P_1} = \frac{P_4}{P_3} \times \frac{P_3}{P_2} \times \frac{P_2}{P_1}$$

$$\frac{P_4}{P_1} = \frac{P_4}{P_3} \times \frac{P_3}{P_2} \times \frac{P_2}{P_1}$$

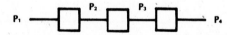

Figure 1-12   Power levels in a cascaded system.

The corresponding loss, expressed in decibels, is

$$dB = 10 \log_{10} \frac{P_4}{P_1}$$

$$= 10 \log_{10}\left(\frac{P_4}{P_3} \times \frac{P_3}{P_2} \times \frac{P_2}{P_1}\right)$$

If we use the property that the logarithm of a product of terms is equal to the sum of the logarithms of these terms, then we find

$$dB = 10 \log_{10} \frac{P_4}{P_3} + 10 \log_{10} \frac{P_3}{P_2}$$

$$+ 10 \log_{10} \frac{P_2}{P_1} \qquad (1\text{-}2)$$

That is, *the total efficiency (in decibels) is just the sum of the efficiencies (in decibels) of the individual cascaded elements*. This illustrates the major advantage of the decibel scale.

Equation (1-1) can be evaluated by using the logarithm function available on inexpensive hand calculators. For convenience, plots of the decibel scale are given in Figs. 1-13 and 1-14. These figures show the decibel equivalent for both power gains and losses. For power gains ($P_2/P_1 > 1$), read the positive decibel values; for losses ($P_2/P_1 < 1$), read the negative decibel values. Sometimes the negative sign is omitted when it is otherwise clear that a loss is intended. For example,

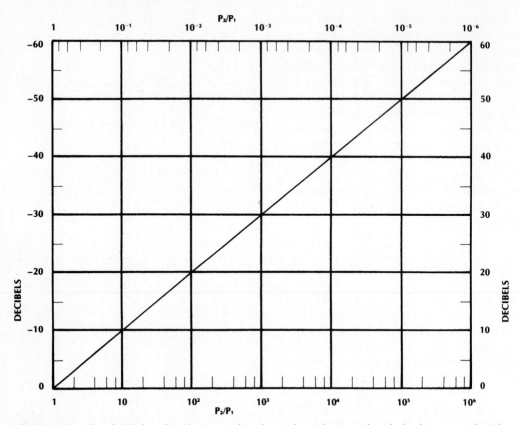

Figure 1-13 The decibel scale. The vertical scale on the right is read with the bottom scale. The vertical scale on the left is read with the top scale. These rules are easily remembered if we realize that a power ratio greater than 1 corresponds to a positive decibel level and that a power ratio less than 1 corresponds to a negative decibel level.

a power change of −3 dB might be called a 3-dB loss. Figure 1-14 is an expanded scale useful when small losses or gains are being computed.

Example 1-1

Suppose that the three elements in Fig. 1-12 have losses of −11, −6, and −3 dB, respectively. Find the total loss of the combination. Find the output power if the input power is 5 mW.

Solution:

According to Eq. (1-2), the total loss is

just the sum −11 − 6 − 3 = −20 dB. Figure 1-13 shows that −20 dB is equivalent to a power ratio of 0.01. The received power is then 5(0.01) = 0.05 mW.

Example 1-2

A system has −23 dB of loss. Compute its efficiency.

Solution:

When Eq. (1-1) is solved for the power ratio, we obtain

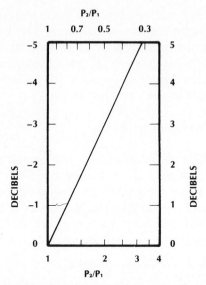

Figure 1-14 Expanded decibel scale. The rules for interpretating this chart are the same as those for Fig. 1-13.

$$\frac{P_2}{P_1} = 10^{dB/10}$$

A calculator with the capability of raising numbers to both positive and negative powers is handy for this calculation. In the present problem

$$\frac{P_2}{P_1} = 10^{-2.3} = 0.005$$

so the efficiency of power transmission is 0.5%. We might also obtain the power ratio corresponding to −23 dB by referring to Fig. 1-13, where we find that $P_2/P_1 = 0.005$. If Figure 1-13 does not have enough resolution, then we can use the expanded scale of Fig. 1-14 in the following way. Note that −23 dB = −20 dB −3 dB. Figure 1-13 shows a loss of 0.01 for −20 dB, and Fig. 1-14 yields a loss of 0.5 for −3 dB. The total

loss, which is the product of these two fractional losses, is then $0.01(0.5) = 0.005$.

So far, the decibel scale has been used to denote relative power levels. We can also use decibels to express the absolute power. To do this, we compare the power to a fixed reference value. A convenient reference level for fiber systems is the milliwatt. The value of power in decibels relative to 1 milliwatt is denoted by the term dBm. It is found from Eq. (1-1) by setting $P_1 = 1$ mW and writing P_2 in milliwatts. That is,

$$dBm = 10 \log_{10} P_2$$

Use Figs. 1-13 and 1-14 for dBm calculations. Just substitute the value of P_2 in milliwatts for P_2/P_1, and then read the vertical scale in dBm.

Example 1-3

A light-emitting diode radiates 2 mW. Compute the dBm value of this radiated power. This power travels through a group of components having a combined loss of 23 dB. Compute the output power.

Solution:

The emitted power, in milliwatts, is 2. Figure 1-14 shows that this power ratio corresponds to 3 dBm. This power is reduced by the loss of 23 dB. That is, the output power is 23 dB less than the input power. The output power is then $3 - 23 = -20$ dBm. From Fig. 1-13 we see that the corresponding power ratio is 0.01 *referred to a milliwatt*, so that the output power is just 0.01 mW.

A microwatt is another common reference level. The value of power in decibels relative to 1 microwatt is

$$dB\mu = 10 \log_{10} P_2$$

where P_2 is in microwatts. Some commercial optic power meters display their readings directly in dBm or dBμ. Others indicate the light power directly in watts. Some have both systems of units available.

1-3 NATURE OF LIGHT

Although light pervades human existence, its fundamental nature remains at least a partial mystery. We know how to quantify light phenomena and make predictions based on this knowledge, and we know how to use and control light for our own convenience. Yet, light is often interpreted in different ways to explain different experiments and observations: Sometimes light behaves as a wave. Sometimes light behaves as a particle.

Wave Nature of Light

Many light phenomena can be explained if we realize that light is an electromagnetic wave having a very high oscillation frequency and a very short wavelength. The frequencies of the *electromagnetic spectrum* are shown in Fig. 1-15. The free-space wavelength and the com-

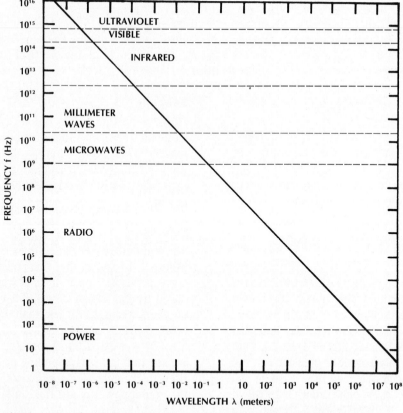

Figure 1-15   The electromagnetic spectrum. The names associated with various frequency regions are shown. Frequency and wavelength are related by $f = c/\lambda$, where $c = 3 \times 10^8$ m/s.

mon names for the various frequency ranges are indicated in the figure. We use the term *optic* (as well as the term *light*) to refer to frequencies in the *infrared, visible,* and *ultraviolet* portions of the spectrum. We do this because so many of the same analyses, techniques, and devices are applicable to these ranges.

The range of frequencies (or wavelengths) that primarily interests us here is shown in Fig. 1-16. Visible wavelengths extend from 0.4 μm (which we distinguish as the color blue) to 0.7 μm (which appears as red). Silica glass fibers are not very good transmitters of light in the visible region. They attenuate the waves to such an extent that only short transmission links are practical. Losses in the ultraviolet are even greater. In the infrared, however, there are three regions in which glass is relatively efficient and occur at wavelengths close to 0.85, 1.3 and 1.55 μm. These three regions are often referred to as the three fiber *transmission windows*. (More details on fiber losses appear in Chapter 5.)

Although light waves have much higher frequencies than do radio waves, they both obey the same laws and share many characteristics. All electromagnetic waves have electric and magnetic fields associated with them, and they all travel very quickly.

In empty space (usually referred to as *free space*), electromagnetic waves travel at a velocity of 3×10^8 m/s. This velocity, indicated by c, is appropriate for wave travel in the atmosphere. In solid media, the wave velocity differs. Its value depends on the material and on the geometry of any waveguiding structure that may be present. The wavelength of a light beam is given by

$$\lambda = v/f \qquad (1\text{-}3)$$

where v is the beam velocity and f is its frequency. The frequency is determined by the emitting source and does not change when the light travels from one material to another. Instead, the velocity difference causes a change in wavelength according to Eq. (1-3). Unless otherwise specified, whenever we refer to a particular wavelength in this book we mean its free space value.

As an example, consider radiation at 0.8 μm. By using Eq. (1-3), with $v = c$, yields a frequency of 3.75×10^{14} Hz. This is a fast oscillation indeed. The period of this oscillation (the reciprocal of its frequency) is then 2.67×10^{-15} s, an extremely short time span. We should also note that the wavelengths of optic beams are on the order of 1 μm near the visible spectrum. Optic wavelengths are so small that most devices used in a fiber system have dimensions of many wavelengths. This is unlike the situation at lower frequencies, where device sizes can be a wavelength or less.

The wave nature of light is used to analyze how optic beams travel through fibers.

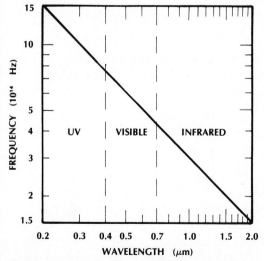

Figure 1-16 Part of the optical spectrum. The names associated with the indicated wavelength ranges are shown. The vertical dashed lines separate the visible from the ultraviolet (UV) and from the infrared.

Results of such analyses show the conditions necessary for light to be guided by a fiber. These analyses show the velocities at which the waves travel. We will study the specifics of this situation in later chapters.

Particle Nature of Light

So far we have described light as a wave. Sometimes light behaves unlike a wave and behaves as though it were made up of very small particles called *photons*. The energy of a single photon is

$$W_p = hf \qquad (1\text{-}4)$$

where $h = 6.626 \times 10^{-34}$ J $\times$ s and is called *Planck's constant*. The energy determined from Eq. (1-4) has units of joules. It is impossible to break a wave into divisions smaller than the photon. Ordinarily, beams of light contain huge numbers of photons. The following example illustrates this.

Example 1-4

Find the number of photons incident on a detector in 1 s if the optic power is 1 μW and the wavelength is 0.8 μm.

Solution:

From Eqs. (1-3) and (1-4) the energy of a single 0.8-μm photon is

$$W_p = hf = hc/\lambda = 2.48 \times 10^{-19} \text{ J}$$

Because power is the rate at which energy is delivered, we can write the total energy as

$$W = Pt$$

Multiplying the power of 1 μW times the 1-s time interval yields an energy of 1 μJ. The number of photons required to make up 1 μJ is

$$\frac{W}{W_p} = \frac{10^{-6} \text{ J}}{2.48 \times 10^{-19} \text{ J/photon}}$$

$$= 4.03 \times 10^{12} \text{ photons}$$

In Example 1-4, if we reduced our observation time to as little as 1 ns, we would still receive over 4000 photons. The most sensitive receivers can detect the presence of radiation when only a few photons arrive.

A convenient unit of energy is the *electron volt* (eV). It is the kinetic energy acquired by an electron when it is accelerated by 1 V of potential difference. The relationship between electron volts and joules is

$$1 \text{ eV} = 1.6 \times 10^{-19} \text{ J}$$

The 0.8-μm photon treated in Example 1-4 has an energy in electron volts given by

$$\frac{2.48 \times 10^{-19} \text{ J}}{1.6 \times 10^{-19} \text{ J/eV}} = 1.55 \text{ eV}$$

A graph showing photon energies (in joules and electron volts) and their corresponding wavelengths appears in Fig. 1-17.

Particle theory explains generation of light by sources, such as light-emitting diodes, lasers, and laser diodes. It also explains detection of light by conversion of optic radiation to electrical current.

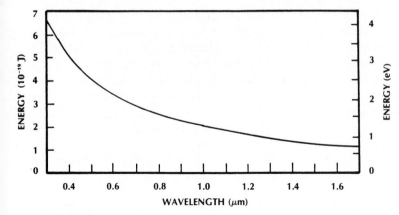

Figure 1-17 Photon energy.

1-4 ADVANTAGES OF FIBERS

We are now ready to discuss the advantages of optic fibers. Before doing so, let us mention a few words of caution. Fiber systems are not perfect. They have technical and economic limitations. For any desired system the relative merits of guided channel versus unguided channel and metallic conductor versus fiber, must be evaluated. The following discussion of desirable fiber properties can be useful in that evaluation.

The basic material for glass fibers is silicon dioxide, which is plentiful. Some optic fibers are made of transparent plastic, another readily available material. Costs are often the most important consideration in a system. Comparisons between fiber and metallic cables must be done with care. There are many fiber cables available, some of which are cheaper than their wire equivalents. The savings may become particularly apparent when the comparison is made on the basis of cost per unit of information transfer. For example, a valid comparison for a telephone link would be on the basis of cost per meter per telephone channel, rather than just cost per meter. This

consideration arises because fibers have greater information-carrying capacities than do metallic channels.

Economic comparisons should also include the costs of installation, operation, and maintenance. Some generalities about these concerns are worth presenting. For long paths, fiber cables are cheaper to transport and easier to install than metal cables. This is because fibers are smaller and lighter. (A light guide would have to be light weight, correct?) One cable design has a fiber of diameter 125 μm enclosed in a plastic sheath with an outer diameter of 2.5 mm. The weight of this cable is 6 kg/km; the loss is 5 dB/km. Let us compare this cable with the RG-19/U coaxial cable, which has an attenuation of 22.6 dB/km when carrying a 100-MHz signal.[3] Its outer diameter is 28.4 mm, and its weight is 1110 kg/km. Smaller and lighter coaxial cables are available, but they have higher losses than the RG-19/U. The significant size and weight advantages of fiber cables are apparent from this example. There are no great differences in the operation of fiber or metallic systems. The costs here should be the same. Maintenance of fiber cables does differ, however. If a line is broken, either as a result of an accident or a system modification, splices must be made or new connectors must be attached. These operations require more time and skill for fibers than for wires. As a result, maintenance costs should be considered when designing a system in which many changes are likely to be made.

Fibers and fiber cables have turned out to be surprisingly strong and flexible. Some fibers are so slender that they do not break when wrapped around curves of only a few centimeters radius. Fibers are often stored and transported while tightly wrapped around spools having this small curvature. Fiber flexibility is attractive for installations containing many turns along the transmission path. For a large radius bend, fibers guide light with negligible loss. There is some loss at a very tight bend, however. When a fiber is protected, for example, by encasing it in a plastic sheath, it is difficult to bend the cable into a radius small enough to break the fiber. Fibers embedded in cables do not break easily.

The addition of a plastic sheath increases the tensile strength of a fiber transmission line. Steel rods can be placed inside the plastic cable to add further strength, if needed. Another strengthening material is Kevlar, a synthetic polymer fiber with great tensile strength. Despite the apparent fragile nature of glass, optic fiber cables are very rugged and serviceable.

Techniques have been developed for the production of fibers with very low transmission losses. Many fiber designs exist, but an attenuation of 4 dB/km is typical of commercial glass fibers when operated at a wavelength around 0.82 μm. According to Fig. 1-14 this represents a transmission efficiency of 40% for a 1-km length. This degree of transparency could not be achieved before 1970. Now, fibers with losses of only a few tenths of a dB/km are available for use around 1.3 μm and 1.55 μm. Very long communications links can be constructed because of the availability of low-loss fibers. Repeaters, needed to amplify weak signals, can be located at large intervals. The losses of wire transmission lines increase rapidly with frequency, as indicated in Fig. 1-18 for the RG-19/U coaxial cable. At high frequencies, link lengths and repeater spacings would be significantly smaller for wire systems than for fiber systems.

One of the most important advantages of fibers is their ability to carry large amounts of information and to do so in either digital or analog form. For example, a single fiber of the type developed for telephone service can propagate data at the T3 rate, 44.7 Mbps. This fiber transmits 672 voice channels. Fibers with even greater capacities are available. Although pulse spreading (see Fig. 1-8) limits the maximum rate, fiber capabilities meet the requirements of most data-handling systems and exceed the capabilities of conducting cables.

In the analog format, modulation rates of hundreds of megahertz, or more, can propagate along fibers. As with the digital systems, the rate is limited by distortion of the optic signal. A representative plot showing how the signal changes with modulation frequency appears in Fig. 1-18. In this figure we see a 4-dB loss when modulation frequencies are low. At 500 MHz, the loss increases by 3 dB. We say that this length of fiber has a *3-dB bandwidth* (which we will denote by the symbol $f_{\text{3-dB}}$) of 500 MHz. Above this frequency, the modulation is further attenuated. The high-frequency attenuation requires some explanation. It is not caused by any added power losses, such as absorption in the fiber. In fact, the transmission efficiency of the fiber remains at 4 dB regardless of the modulation rate. Figure 1-19 illustrates the problem that arises at high modulation frequencies. The in-

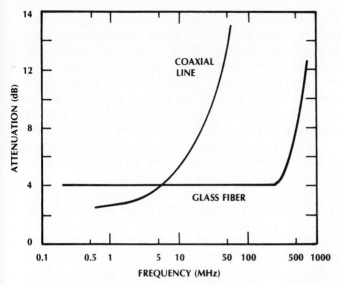

Figure 1-18 Effective attenuation of a 1-km length of coaxial line and glass fiber. The 3-dB bandwidth of the fiber is 500 MHz. (Coaxial line data from manufacturer's literature, Alpha Wire Corporation, Elizabeth, N.J.)

formation being transmitted is contained in the time *variation* of the optic power. As the modulation frequency increases, the signal distortion causes a loss in the amplitude of this variation. This occurs owing to spreading of the regions of peak power into the adjacent minima. The result is lower peak power and higher minimum power. At low frequencies this effect is negligible because the spread is small compared to the separation between adjacent peaks and nulls. At high frequencies the spread is significant compared to this spacing,

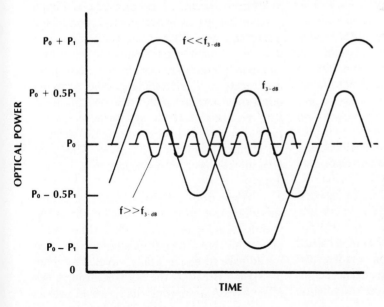

Figure 1-19 Optical power at the output of a fiber cable for various modulation frequencies. The 3-dB bandwidth is $f_{3\text{-dB}}$, the peak variation in the optical power is half of that at low modulation frequencies.

so the power variation diminishes greatly. The optic power is still efficiently transmitted (at a 4-dB loss in this example), but the information has been lost. The losses shown in Fig. 1-18 for the coaxial cable are more easily interpreted. They represent the actual efficiencies of power transmission. The relative superiority of the glass fiber at higher information rates is apparent.

A dramatic comparison can be made between a standard wire telephone cable and a fiber cable. The metal cable contains 900 twisted-wire pairs, and its diameter is 70 mm. Each pair carries 24 voice channels (T1 rate), so the cable capacity is 21,600 calls. One fiber cable developed for telephone applications has a 12.7-mm diameter and contains 144 fibers, each operating at the T3 rate (672 channels). This cable has a total capacity of 96,768 calls. The fiber cable has nearly 4.5 times as much capacity as the wire cable and has a cross-sectional area that is 30 times less. Still more remarkable comparisons can be made if operation at a higher rate (say 565 Mbps or 2.3 Gbps) is assumed.

Optic fibers, glass or plastic, are insulators. No electric currents flow through them, either owing to the transmitted signal or owing to external radiation striking the fiber. In addition, the optic wave within the fiber is trapped, so none leaks out during transmission to interfere with signals in other fibers. Conversely, light cannot couple into the fiber from its side. We conclude that a fiber is well protected from interference and coupling with other communications channels, whether they are electrical or optic.

For those reasons noted, fibers have excellent rejection of *radio-frequency interference* (RFI) and of *electromagnetic interference* (EMI). RFI refers to interference caused by radio and television stations, radar, and other signals originating in electronic equipment. EMI includes those sources of radiation and those caused by natural phenomena (such as lightning) or caused unintentionally (such as sparking). If not rejected, then these undesired signals could increase the system noise level beyond acceptable limits. A fiber excels at rejecting externally caused background noise. The ability of a fiber to isolate itself from its environment allows us to pack numerous fibers together in a cable to transmit many channels of information along a single path. No crosstalk occurs.

Because fibers are insulators, they will not pick up or propagate electromagnetic pulses (EMP) caused by nuclear explosions, which can induce millions of volts in a conducting transmission line. The voltage pulse can travel many miles along the wire and eventually (because of its strength) destroy the electronics at the end of the path.

The insulating nature of a fiber has several other practical consequences. In an environment in which high-voltage lines are present, a wire communications link could possibly short circuit the lines by falling across them, causing considerable damage. Sparking occurring in this process could ignite combustible gases in the area. This problem disappears with fibers. Another advantage is that optic coupling eliminates the need for a common ground between a fiber transmitter and receiver. Troublesome ground loops are not formed. Additionally, it is possible to repair the fiber while the system is on without the possibility of short circuiting the electronics at the transmitter or receiver. This problem might occur when repairing a metallic cable.

Fibers offer a degree of security and privacy. Because fibers do not radiate the energy within them, it is difficult for an intruder to detect the signal being transmitted. The fiber would have to be physically violated to obtain the signal. Breaking the fiber, or fusing a new fiber to the transmitting fiber, would provide access to the optic beam. During such modification of the link, the power reaching the receiver would drop. A sensitive receiver can measure this loss, providing knowledge that an intrusion is occurring. To improve the success of detection, the system would have to be monitored continuously.

Electronic communications systems include processing of information before it is delivered to the information channel and after it reaches the receiving terminal. Fiber systems require processing that is very similar. This allows incorporation of fibers into systems originally conceived for wire transmission with only moderate modifications. Fiber compatibility with the basic structure of the telephone system is a good example. It is even possible to make an optic system transparent to the user. This means the user need not be aware that the electrical signal has been converted to optic form, transmitted as a beam of light, and then returned to its electrical form. Users simply supply electrical inputs and receive electrical outputs, just as they do for all-electronic systems. Compatibility with electronic systems also means that persons trained in electronic communications can transfer their skills to fiber communications rather easily.

Corrosion caused by water or chemicals is less severe for glass than for the copper it replaces. However, water must not penetrate the glass. For submerged applications, fibers are encapsulated within cables, which protect them from the water.

Glass fibers themselves can withstand extreme temperatures before deteriorating. Temperatures approaching 800°C leave glass fiber unaffected. Other components of a fiber system are much more sensitive to increases in temperature. Plastic cable sheathing can melt, leaving the fiber unprotected and possibly distorting the fiber. This distortion would increase the fiber loss. Fiber cables with an operating range of $-25°$ to $+65°C$ are commercially available at moderate costs. Large temperature variations also cause expansions and contractions within a cable that upset the critical alignments required for low-loss connections.

Fibers are available in long lengths, reducing the need for numerous splices. A common length is 1 km, but continuous strands of several kilometers have been produced.

Conscience requires that something be noted about one of the disadvantages of optic fibers. A concern is the optic connector: its cost is high, its loss is also high, and its installation is time consuming. The reasons for this situation are well understood. For a good connection the two fibers must be aligned very precisely. Metal connectors having the precision needed for a loss of just under 1 dB when the fibers are repeatedly coupled and uncoupled are costly to manufacture. Designers would like to have connectors available with even lower losses, say, 0.1 dB. This is difficult to achieve, although techniques are available for permanently splicing fibers with this loss. Inexpensive plastic connectors are available with typical losses of 2 dB or more. These are sufficient for some applications.

1-5 APPLICATIONS OF FIBER OPTIC COMMUNICATIONS

In the first few years following discovery of the laser, applications developed so slowly that the laser was described as a solution in search of a problem to solve. No such comments were heard regarding the emergence of fiber optics as a practical technology. Introduction of fibers into operational systems proceeded quickly in comparison with the time engineering innovations usually require for acceptance. The first large-scale applications were for telephone links. The pressures to expand service and the suitability of fibers for voice communications combined to hasten the design and test of operational telephone equipment. Telephone experience demonstrated the reliability and practicality of fiber communications. It also provided devices and system design methods that could be used in other applications.

A few fiber applications are described in this section. The list is not exhaustive and is merely indicative of areas in which fiber optics is successful. The characteristics cited are not the ultimate limits of fiber performance but are typical achievements. Furthermore, many details of system construction and performance are left for later chapters. These details will be better understood after absorbing the foundation material in the intervening chapters.

The small size and large information carrying capacity of optic fibers make them attractive as alternatives to conventional copper twisted-pair cables in telephone systems. In one of the first installed systems, fiber trunk lines connected telephone offices in Chicago. The offices were spaced 1 and 2.4 km apart. Operating at the T3 rate, each of the 24 fibers in the cable had a capacity to carry 672 voice messages.

Continuous passive links (no repeaters) more than 100 km long have been demonstrated, allowing construction of intercity trunk lines. Repeaters increase permissible path lengths by boosting weakened signals and restoring their shape. With repeaters messages can be sent over thousands of kilometers of fiber. Because of low attenuation, separation between repeaters in a fiber system can be greater than in a coaxial link. The savings in installation and maintenance costs can be considerable when large repeater spacings are feasible.

Because repeaters can be spaced very far apart in fiber systems, underwater fiber links can be designed to span the oceans. One such system, TAT-8, covers about 6000 km between the east coast of the United States and Europe.[4] With repeater spacings of 50 km, just over 100 repeaters are necessary. Two fiber pairs, each operating at 295.6 Mbps and with special coding methods, provide a total capacity of 40,000 voice channels. Future underwater systems will use lower-loss fibers to increase the spacing between repeaters. The low weight of fiber cables, compared to coaxial lines, gives them distinct advantages for submerged cable applications because of the relative ease of transporting and laying the fibers.

The "wired city" refers to a community in which each home has electronic access to a large number of information services. When the connections are optic, the term "fibered city" is more accurate. Such a community was established in Japan under the experimental Hi-OVIS (Higashi-Ikoma Optical Visual Information System) program. Hi-OVIS also stands for Highly Interactive Optical Visual Information System. The system consists of a center, subcenter, and home terminals linked by optic transmission lines. The lines connect computers and video equipment. Each home terminal has a TV set, camera, microphone, and keyboard. Two-way interactive communications is obtained. Terminals were connected in 158 private homes initially. Subscriber services include direct reception of TV programs, which provides superior pictures and sound compared to conventional antenna reception. TV programs of a very localized nature are broadcast; for example, programs involving the local police and fire departments and local shopping information are broadcast. A video request service is available where the subscriber can request that a particular video cassette be played from a central storage location. Home study courses are available. Still pictures giving information about local events, medical facilities, train timetables, and the like, are available. Services such as those described require large bandwidths only available by using *broadband* transmission systems. Fiber optic communications provides the needed bandwidths.

The Hi-OVIS system pioneered in two areas: reduction to practice of several types of voice and video fiber communications and development of greatly expanded services available from the home. Other communities throughout the world have expanded the coverage and scope of the fibered city. There are tens of communities where (in the late 1980s and early 1990s) fiber-to-the-home (FTTH) is being tested on a limited basis by connecting anywhere from around 50 to more than 1000 homes. Results of these tests will determine the future directions of fiber-to-the-home installations. Answers to the questions of consumer desire for the added services and consumer willingness to pay for the added services will be just as important as solutions to the technical problems.

Metallic communications links installed along electrified railway tracks suffer from electromagnetic interference from the electricity powering the vehicles. Because of fiber rejection of EMI, signals traveling through fibers laid along the track do not degrade. Optic communications are compatible with electrified railways. Wire systems are not. Similarly, fibers can be placed near high-voltage power lines without adverse effects, whereas wire systems would be noisy. Fibers can even pass unaffected through areas where electrical power is generated or through power substations. Optic cables can be suspended directly from power line towers, or poles, if clearance space permits and if the load can be tolerated.

An alternative is to include the fiber within one of the cables housing a metallic conductor; that is, to embed it in a wire cable. An earth conductor for lightning protection is often located at the top of a group of high-power transmission lines. This is a good location for an embedded fiber.

Applications that are primarily video include broadcast television, cable television (CATV), remote monitoring, and surveillance. The broadcast television industry uses fiber transmission for short links; for example, studio to transmitter, or live event to equipment van, or live event directly to studio. In the coverage of live events, the low weight of a fiber cable allows considerable range of movement for the regular TV cameras and for TV minicameras. In these short-range broadcast applications only one channel is required, so the signals are modulated in analog form and transmitted in the baseband. Bandwidths of 6 MHz are sufficient. For longer transmission paths, where repeaters may be used, or where the signals will be transferred to commercial telephone lines, digital modulation is preferred.

Cable television systems collect and distribute a large number of color channels. The distances covered range from a few tens of meters to several kilometers. CATV systems obtain their signals from various sources. These include satellite earth stations, microwave links, antennas picking up broadcasts from nearby transmitters, and local studios where programming originates. All these sources can be connected to the central distribution location (the CATV *headend*) by fibers. For CATV it is common to use frequency-modulated (FM) video pictures occupying about 20 MHz of bandwidth. FM improves signal-to-noise ratios and yields systems with greater distortion tolerance, so the increase in bandwidth (from 6 to 20 MHz) is acceptable. Multiple channels are accommodated by placing a separate fiber in the cable for each channel or by *frequency-division multiplexing* (FDM). In FDM each channel is modulated onto a different radio-frequency subcarrier before being applied to the optic source. In this way several channels are transmitted simultaneously along a single fiber. In a four-channel FM system, the subcarriers can be located at 30, 50, 70, and 90 MHz. At the receiver, the four channels are separated by filters and then demodulated to recover the baseband signals. Trunk capacity is further increased if several fibers, each carrying several multiplexed channels, are placed in the cable. Fibers are useful in all parts of CATV distribution systems. We have only described fiber transmission from the source to the CATV headend. Fibers can be used throughout the video distribution network, including the final link into the subscriber's home.

Fiber optic video transmission successfully competes with coaxial cable for surveillance and remote monitoring systems. EMI rejection and low susceptibility to lightning damage are important in these applications. Specific examples are surveillance of power-generating stations, critical control points along a railroad right-of-way, parking lots, and the perimeter of military installations. Railroad car identification numbers can be remotely read. Black-and-white pictures are normally acceptable for these applications. The transmission lengths are less than 5 km, although railways may require paths up to 20 km. Because each camera transmits a single channel, baseband analog modulation is acceptable. That is, the baseband signal directly intensity modulates the optic carrier source. Longer and more complex systems may use FM for improved signal quality and FDM for multiple channels. Oneway, or *simplex*, transmission is often sufficient. If not, a second fiber can send messages back to the camera location. *Full-duplex*, where signals can propagate simultaneously in both directions along a single fiber, can also be accomplished, although the equipment is more complex.

Fiber systems are particularly suited for transmission of digital data such as that generated by computers. Interconnections can be made between the central processing unit (CPU) and peripherals, between CPU and memory, and between CPUs. A good example is the connection of several hundred cathode ray tube (CRT) terminals, located throughout a high-rise, to a processor located on one of the floors. Low weight, small size, and the security resulting from a nonradiating transmission line make fibers attractive for data transfer over any distance.

When the connected equipment is all within one room, the transmission distances are so small that very good error rates (10^{-12} or better) are achieved. Data rates on the order of 200 Mbps are easily obtained for these *intraroom* applications. *Intersite* installations are connections between equipment located in different rooms or in different buildings, or even in different cities. A *local-area network* (LAN) distributes information to several stations within a limited region (for example, all the stations are within one building). A variety of network topologies are available for local-area networks by using fiber transmission.

Fiber optic transmission of control data is useful in areas where high voltages are found. Such an environment exists when laser induced fusion experiments are performed. Microprocessors, which control the firing sequence of the lasers and laser amplifiers, are linked by fibers to eliminate interference that the large voltages would create on metallic conductors.

Military applications of fiber optics abound. They include communications, command and control links on ships and aircraft, data links for satellite earth stations, and transmission lines for tactical command post communications. The important fiber characteristics are low weight, small size, EMI rejection, and no signal radiation. On aircraft and ships the reduced shock, fire, and spark hazards are significant assets. The high resistance to corrosion justifies the use of fibers at sea either aboard ship or in the ocean. For field applications, lightweight fibers speed cable deployment.

Tactical communications range from short-distance links (connecting field shelters) to long-haul links (60-km path lengths can occur). A novel application is the fiber-guided missile. The fiber is payed out while the missile is in flight. Sensors on the missile transmit video information through the fiber to a ground control van. Commands are transferred back to the missile from the van, again along the fiber.

Although not strictly communications networks, fiber sensors still represent an important fiber optics applications. Fiber sensors have been used to measure temperature, pressure, rotary and linear position, and liquid levels. In some of these devices fibers have a dual purpose. The sensor itself depends on some fiber property, and the collected information is transmitted along the fiber to the

readout location. We will briefly describe two sensor applications: the fiber gyroscope and the fiber hydrophone.

Gyroscopes measure rotational motion. Until the development of the ring laser optic gyroscope, all practical devices were mechanical spinning gyroscopes. Optic gyroscopes have the advantage of no moving parts. The laser ring suffers from a *lock-in* phenomenon at low rotation rates, so that slow rates cannot be detected without complicating the system. The fiber gyroscope does not exhibit the lock-in problem. The basic sensor is a long fiber coil with an optic signal (from a single source) traveling through it in both directions.[5] The phase difference of the counter-propagating beams is measured. If the coil is stationary, then this difference is zero. If the coil is rotating, then the phase difference is a measure of the rotation rate.

Hydrophones are used to measure acoustic disturbances in water. A conceptually simple design is shown in Fig. 1-20. The fiber is not continuous but has a break in it. At the break, one of the fibers is fixed and the other is attached to a speaker diaphragm. A sound wave vibrates the diaphragm and displaces the movable fiber. The coupling efficiency changes according to the amplitude and frequency of the displacement. The power delivered to the receiver is then a measure of the

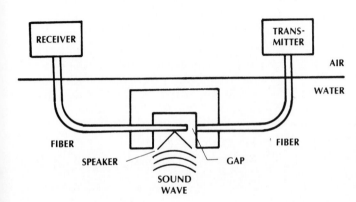

Figure 1-20 Fiber hydrophone. The fiber on the left is displaced when an acoustic wave is present, changing the amount of light coupled across the gap. The change in light intensity is measured by the receiver.

TABLE 1-7.　Fiber Applications

Voice
  Telephone trunk
      Interoffice
      Intercity
      Transoceanic
  Subscriber service
      Fiber-to-the home
      Broadband service
  Near power plants
  Along power lines
  Along electric railways
  Field communications

Video
  Broadcast TV
      Live events
      TV minicameras
  CATV
      Source-to-headend trunk lines
      Distribution
      Subscriber taps
  Surveillance
  Remote monitoring
  Fiber-guided missile
  Fiber-to-the-home

Data
  Computers
      CPU to peripherals
      CPU to CPU
  Interoffice data links
  Local-area networks
  Fiber-to-the-home
  Aircraft wiring
  Ship wiring
  Satellite ground stations

Sensors
  Gyroscope
  Hydrophone
  Position
  Temperature
  Electric and magnetic fields

frequency and amplitude of the acoustic wave. In this system the fiber acts as the sensor and as the transmission channel for the information. Many other fiber hydrophone designs have been successfully tested.

Table 1-7 lists fiber applications we have been describing. Subdivisions into the four categories of voice, video, data, and sensors have been made, although in some cases the system belongs in more than one category. The fibered city is an example in which fibers carry voice and video (and possibly data). Nonetheless, most systems fit into just one of these major areas. System architects should review other designs in the same major category. In the course of completing their own designs, the architects should apply strategies that have already proven successful.

1-6　SUMMARY

The serious student now possesses some fairly general knowledge about fiber optics communications systems: What they are, what they do, and the advantages they may have over wired alternatives. This knowledge includes the configuration for a point-to-point link and the major devices in that link. The rest of this book involves the detailed study of these devices (their design and operating characteristics) and how they fit together to meet desired performance specifications. We have not yet given the tools for choosing system architectures and components, but the reader is in a position to appreciate the array of decisions that system designers face. Some of these decisions are

1. Fiber or metal cable
2. Simplex, half-duplex, or full-duplex configuration

3. Form of modulation
4. Multiplexing strategy
5. Wavelength of operation
6. Choice of light source
7. Specification of fiber
8. Specification of cable
9. Choice of connectors and splicing techniques
10. Choice of detectors

We will briefly discuss each of these topics next.

The advantages of fibers have been presented earlier in this chapter. Metallic cables may still be superior in any particular system. Conducting lines are readily available from suppliers in most major cities, they are easier to splice, connectors are easily attached, and taps are relatively simple and inexpensive. Costs of fiber and metal cables must be evaluated for the desired application. The decision to use fiber or metal links is sometimes made before the system designers are assigned to the job. Similarly, the system designer may not have a free hand with respect to other choices in the list. For example, compatibility with existing systems might dictate a particular source, wavelength, or fiber.

Point-to-point communications in only one direction is a simplex link. Two-way communications is achieved by simultaneous transmission along a single fiber (full-duplex). A simpler (but possibly costlier) solution uses two fibers inside a cable, one for each direction of information travel. This is *half-duplex transmission*. Fiber designs can accommodate simplex, half-duplex, and full-duplex transmission.

The modulation format (analog or digital) must be decided on early in the system design. When the information is already in digital form, digital transmission is the most likely choice. When the information is generated in analog form (telephone voice messages or pictures from a video camera, for example), the decision may be difficult. For single channels over short paths, analog baseband signals will reach the receiver with adequate strength and form. Signal distortion owing to transmission will be negligible, so that the expense and complexity of analog-to-digital conversion would not be warranted. For long paths, particularly if repeaters are necessary, conversion to digital may be desirable. Digital repeaters are simpler than analog repeaters, and digital transmission yields higher-quality received signals. A disadvantage of the digital form is the increased bandwidth required for its transmission.

A multiplexing scheme must be chosen if more than one channel is to be transmitted. Multiplexing formats for simultaneous transmission of several channels on a single fiber exist for both analog and digital modulation. Alternatively, individual channels can propagate along separate fibers, all encased within a single cable. This strategy works but is costly and does not fully use the broadband capabilities of fibers. You may have already realized that installation of a fiber cable of which the full information carrying capacity is not used can be practical. The system can be upgraded by adding new transmitting and receiving facilities without changing the previously installed fiber cable.

The choices of wavelength may be grouped as operation in the visible spectrum (0.4 to 0.7 μm), operation in the near infrared (close to 0.85 μm), or operation at long wavelengths (1.1 to 1.6 μm). In the visible spectrum, fiber losses are moderately high, so only short links are practical. In some special circumstances the information is generated directly on a visible laser beam and the purpose of the link is to transmit this information without conversion to another wavelength. In this case, the fiber with lowest loss at the required wavelength is chosen. Near 0.85 μm, glass attenuation is moderately low and light sources and detectors are highly developed. The first generation of fiber networks was designed in this region. It will continue to be desirable for inexpensive fiber links. Better transmission efficiency occurs at the longer wavelengths. In addition, signal distortion owing to transmission is lower in the long wavelength region. For these reasons, the longer wavelengths (particularly 1.3 and 1.55 μm) are attractive for long paths and large information rates. Sources and detectors for these wavelengths became practical in the early 1980s, long after the development of such devices for the near infrared. The result was an early cost, availability, and reliability advantage for the near-infrared components.

The major light sources available are the light-emitting diode and the laser diode. The LED is cheaper and requires simpler circuitry. The laser diode provides an output carrier that has a narrower output spectrum than an LED. The LD emission is more nearly single frequency, or *coherent*. Long systems with greater information-bearing capacity can be constructed with carriers having narrower spectra. Laser diodes can also be modulated at higher rates than LEDs. Component and circuit costs, reliability, and lifetime are considerations involved in the choice of a suitable light source. The packaging of the source is also important. A construction permitting easy attachment of the fiber is desirable.

A wide selection of fibers exists. The differences among them involve size, material (glass, plastic, or plastic-cladded glass), ease of coupling light into them, attenuation, and information-bearing capacity (related to the transmission distortion of the signal). Structural variations are divided into *step-index* (SI) and *graded-index* (GRIN) waveguides.

Propagation characteristics include *single-mode* and *multimode* wave travel. We will define and explain these terms later. For the moment, we merely want to plant the idea that all fibers are not created equal. Different fibers exist for different purposes.

The cable, which encases and protects the fiber, can be specified separately from the fiber. At least this is true in principle. When a system must withstand a particularly harsh environment, it is necessary to design a cable capable of surviving the rigors of the application. A transoceanic link, for example, requires a customized cable. In simpler circumstances it is most economical to specify a standard cabled fiber if one can be found that meets system requirements. Cable variations include single-fiber or multifiber bundle, and light-duty or heavy-duty cable. The multifiber bundle is available for multiple channel transmission or for redundant transmission of a single message. Sometimes it makes sense to install a multifiber cable when only one, or a few, of the fibers are needed. The other fibers can be used later when more information channels become necessary. The purpose of the cabling is to protect the fiber from abrasion and provide crush resistance. Cables contain strength members for relief from tensile stresses. This relief becomes important when the cable must be pulled through ducts or when the cable must support its own weight during installation or operation.

Our previous discussion of connectors indicated their potential high cost and high loss. When designing a system, the complete link loss is calculated to determine if sufficient signal strength will be available for the required clarity of reception. For this reason, the losses of all connectors and splices must be known. This implies that the designer will choose specific connectors and splicing techniques and will certify that the loss values used in the system analysis are correct. In addition to having low loss, a connector should be rugged. It should provide the same loss with repeated connection and disconnection, and it should be easy to install.

The designer must choose an appropriate photodetector for converting the optic signal back into electrical form. The small size and low operating power of semiconductor photodiodes make them the preferred detectors for fiber systems. There is a wide range of diodes from which to choose. Most important, the diode must be highly responsive at the wavelength of the light source. Other considerations include the response time, simplicity of the required receiver circuit, noise characteristics, and ease of attachment to the output end of the fiber. Various packaging arrangements exist to facilitate this connection. The designer must also decide whether a detector with internal amplification is needed. The *avalanche photodiode* is such a device. It is more expensive and requires a more complex circuit than a detector without internal signal gain, but the resulting receiver has improved sensitivity.

Transmitter and receiver circuits must be designed. If they are going to be purchased, then specifications must be written. Some basic circuits will be shown when modulation and detection is discussed. In long-distance systems, repeaters may be needed. They are undesirable because they add to the initial system expense and complexity and increase maintenance costs. Providing power to them at remote locations can also be a problem. If repeaters are required, then the system designer will determine the number required and their spacing.

We are now ready to proceed with a study of the many details that will help in making the decisions outlined in this summary.

PROBLEMS

1-1. Compute and plot the fraction of power transmitted as a function of the transmission-line loss in decibels. Do this on log paper for a loss range of 0–50 dB.

1-2. One milliwatt of optical power enters a fiber. Compute and plot the output power as a function of fiber loss for losses in the range 0–50 dB.

1-3. Two 1-km fibers are spliced together. Each fiber has a 5-dB loss and the splice adds 1 dB of loss. If the power entering is 2 mW, then how much power is delivered to the end of this combined transmission line?

1-4. A receiver requires an input power of 10 nW. If all the system losses add up to 50 dB, then how much power is required from the source?

1-5. How much does 1 mile of RG-19/U coaxial cable weigh in pounds?

1-6. RG-19/U is used at 100 MHz with an input power of 10 mW. The receiver sensitivity is 1 μW. Compute the maximum length of the communications link. Repeat if a fiber with loss of 5 dB/km replaces the coaxial cable. The fiber system's input power and receiver sensitivity are the same as those of the coaxial cable system.

1-7. The telephone transmission rate at the T3 level is 44.7 Mbps. Each telephone message occupies 64,000 bps. How many simultaneous messages could be sent along this system. In the actual system only 672 message channels are used. The additional pulses are used for other functions such as synchronization.

1-8. Estimate how many light pulses can be transmitted per second by a manually operated blinker light system. What do you conclude about the information capacity of this manual system when compared with the capacity of modern fiber optic telephone links?

1-9. A fiber telephone cable contains 144 fibers, each capable of carrying 672 voice messages. A conducting telephone cable contains 900 copper twisted pairs and each pair can carry 24 messages. Compare the capacities of the fiber and conducting cables. How many of the conducting cables are required to equal the capacity of the fiber cable? Repeat the calculations if each fiber operates at the DS-4 signaling rate.

1-10. The 144-fiber telephone cable has a 12.7-mm diameter. A 900-pair copper cable has about a 70-mm diameter. Compute the ratio of the cross-sectional area of these two types of transmission lines.

1-11. Make a table with following frequencies in the first column: 10, 60, 10^3, 2×10^4, 10^6, 10^9, 10^{10}, and 10^{14} Hz. In the second column write the corresponding wavelengths in meters. In the third column write the name of the corresponding region of the electromagnetic spectrum.

1-12. Compute the frequencies at the edges of the visible spectrum. Also compute the bandwidth of the visible spectrum (that is, the difference between the highest and lowest visible frequencies).

1-13. Compute the energy of a photon at 0.6, 0.82, and 1.3 μm. Which has more energy, a visible or an infrared photon?

1-14. There are 10^{10} photons per second incident on a photodetector at wavelength 0.8 μm. Compute the power incident on the detector. If this detector converts light to current at a rate of 0.65 mA/mW, what current is produced?

1-15. How many photons are arriving per second at a receiver if the power is 1 nW at wavelength 1.3 μm.

1-16. Assume that a digital system can be operated at a data rate that has a value equal to 1% of the carrier frequency. Compute the allowed bit rates by using carriers having frequencies of 10 kHz, 1 MHz, 100 MHz, and 10 GHz and wavelength 1.0 μm. This problem emphasizes how the system capacity can increase with increasing carrier frequency (one of the principal advantages of optical transmission over radio frequencies).

1-17. Draw 30 cycles of a sinusoid oscillating at 10^6 Hz. Draw this sinusoid if it is modulated by a square wave whose repetition rate is 10^5 pulses per second (pps). Repeat for repetition rate of 5×10^5 pps. What problems occur as the repetition rate approaches the carrier frequency?

1-18. How many voice channels can be modulated onto a carrier at wavelength 1.06 μm? Assume a system bandwidth equal to 1% of the carrier frequency.

1-19. Suggest a fiber optic application other than one already mentioned in the text. Draw the system block diagram. List the characteristics and requirements of your system (for example, give the information bandwidth or data rate, length of line, etc.).

1-20. Assume that there is a phone in every home on earth. If these were to transmit simultaneously over one transmission line by using frequency-division multiplexing, what is the minimum bandwidth required? Could a single optical beam carry this multiplexed signal? (Assume that 10 billion homes need to transmit.)

1-21. In Problem 1-20 assume digital modulation, time-division multiplexing, and 64,000 bps for each voice message. What data rate is required to transmit the multiplexed signal? Could a single optical beam carry this signal?

1-22. The power incident on a detector of light is 100 nW. (a) Determine the number of photons per second incident on the detector if the wavelength is 800 nm. (b) Repeat the calculation if the wavelength is 1550 nm. (c) Which wavelength requires the most photons to produce the 100 nW of power?

1-23. To operate properly, a fiber optic receiver requires -34 dBm power. The system losses total 31 dB from the light source to the receiver. How much power does the light source emit (in mW)?

1-24. A T3 system has a 10^{-9} error rate (one error for every 10^9 bits transmitted). Compute the number of errors per minute.

1-25. A cable contains 144 single-mode fibers, each operating at 2.3 Gbps. How many digitized voice messages can be transmitted simultaneously along this cable?

REFERENCES

1. Forrest M. Mims III. "Alexander Graham Bell and the Photophone: The Centennial of the Invention of Light-Wave Communications, 1880–1980," *Opt. News* 6, no. 1 (1980): 8–16.

2. Mischa Schwartz. *Information, Transmission, Modulation, and Noise,* 3rd ed. New York: McGraw-Hill, 1980, pp. 138–40, 157–58.

3. Manufacturer's literature. Elizabeth, N.J.: Alpha Wire Corporation.

4. Peter K. Runge and Patrick R. Trischitta. "The SL Undersea Lightwave System." *J. Lightwave Technol.* 2, no. 6 (Dec. 1984): 744–53.

5. Thomas G. Giallorenzi, Joseph A. Bucaro, Anthony Dandridge, G. H. Sigel, Jr., James H. Cole, Scott C. Rashleigh, and Richard G. Priest. "Optical Fiber Sensor Technology," *IEEE J. Quantum Electron.* 18, no. 4 (April 1982): 626–65.

Chapter 2

Optics Review

This chapter contains basic concepts of classical optics that apply to fiber communications. We call it a review because many students have studied optics in a high school or college physics course. The present material simply consolidates a few fundamentals of rays, waves, and lenses. For those unfamiliar with classical optics, this chapter is an introduction to some useful topics. The subjects of this chapter are ray theory and the focusing, collimating, imaging, and light-collecting properties of lenses. These subjects apply to the problems of coupling light from sources into fibers and coupling light from one fiber to another.

2-1 RAY THEORY AND APPLICATIONS

A number of optic phenomena (particularly those associated with lenses) are adequately explained by considering light as narrow rays. The theory based on this approach is called *geometrical optics*. These rays obey a few simple rules.

1. In a vacuum, rays travel at a velocity of $c = 3 \times 10^8$ m/s. In any other medium, rays travel at a slower speed, given by

$$v = \frac{c}{n} \qquad (2\text{-}1)$$

The factor n is the *index of refraction* (or *refractive index*) of the medium. For air and gases, the ray velocity is very close to c, so that $n \cong 1$. At optic frequencies, the refractive index of water is 1.33. Glass has many compositions, each with a slightly different ray velocity. An approximate refractive index of 1.5 is representative for the silica glasses used in fibers, while more precise values for these glasses lie between 1.45 and 1.48. Table 2-1 lists the refractive index for several materials. Because the index varies with a number of parameters (such as temperature and wavelength), the values given in the table are not exact under all circumstances. These numbers are close enough to the actual values to allow us to perform meaningful calculations and make useful predictions, however.

TABLE 2-1   Index of Refraction for Some Materials

Material	Index of Refraction
Air	1.0
Carbon dioxide	1.0
Water	1.33
Ethyl alcohol	1.36
Magnesium fluoride	1.38
Fused silica	1.46
Polymethyl methacrylate polymer	1.49
Glass	$\cong 1.5$
Sodium chloride	1.54
Polystyrene	1.59
Calcite	1.6
Sapphire	1.8
Zinc sulfide	2.3
Indium phosphide	3.21
Gallium arsenide	3.35
Silicon	3.5
Indium gallium arsenide phosphide	3.51
Aluminum gallium arsenide	3.6
Germanium	4.0

2. Rays travel in straight paths unless deflected by some change in the medium.

3. At a plane boundary between two media, a ray is reflected at an angle equal to the angle of incidence, as illustrated in Fig. 2-1. Note that the angles are measured with respect to the boundary normal; that is, the direction perpendicular to the surface. This is the conventional notation in optic work. Referring to the drawing, it is clear that

$$\theta_r = \theta_i \qquad (2\text{-}2)$$

where θ_i is the angle of incidence and θ_r is the angle of reflection.

4. If any power crosses the boundary, the transmitted ray direction is given by *Snell's law*[1]

$$\frac{\sin \theta_t}{\sin \theta_i} = \frac{n_1}{n_2} \qquad (2\text{-}3)$$

where θ_t is the angle of transmission and n_1 and n_2 are the refractive indices of the incident and transmission regions, respectively.

The only angles having physical significance in the preceding paragraphs are those lying between $0°$ and $90°$. The trigonometric sine function is plotted in

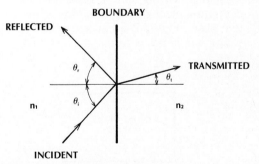

Figure 2-1   Incident, reflected, and transmitted rays at a boundary between two media.

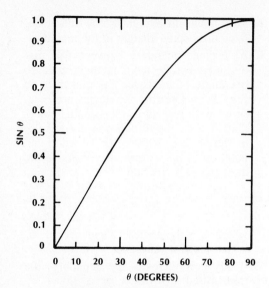

Figure 2-2 The sine function.

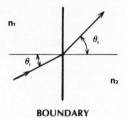

BOUNDARY

Figure 2-4 When $n_1 > n_2$, the ray is bent away from the normal and toward the boundary surface.

Fig. 2-2 over this range. If n_1 is less than n_2, then Snell's law predicts that sin $\theta_t <$ sin θ_i. Figure 2-2 indicates that smaller angles have lower sine values, so $\theta_t < \theta_i$ in this example. The angle of transmission is less than the angle of incidence. It is helpful when tracing rays from one medium to another to remember this result. We can summarize it as follows: The transmitted ray is bent toward the normal when traveling from a medium having a low refractive index into a medium with a higher refractive index. Figure 2-3 illustrates this situation for a ray entering a glass fiber from air.

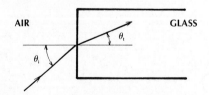

Figure 2-3 Bending of a light ray as it enters a glass fiber.

If $n_1 > n_2$, Snell's law yields sin $\theta_t >$ sin θ_i. So $\theta_t > \theta_i$ and the ray is bent away from the normal, as shown in Fig. 2-4. What follows is the result to remember: The transmitted ray is bent away from the normal when traveling from a medium having a high refractive index into a medium with a lower refractive index.

Example 2-1

A light ray proceeds from air ($n_1 = 1$) into glass ($n_2 = 1.5$). Find the transmission angles when $\theta_i = 0°$ (the incident ray is normal to the boundary) and when $\theta_i = 15°$.

Solution:

When the incident angle is 0, then sin $\theta_i = 0$. Snell's law yields sin $\theta_t = 0$ and finally $\theta_t = 0°$ itself. The ray is undeflected. When $\theta_i = 15°$, however, sin $\theta_t = (1/1.5)$ sin $15° = 0.17$ and $\theta_t = 9.94°$. As expected, the ray is redirected toward the normal.

Example 2-2

The last ray in Example 2-1 now travels from glass back into air. Assume that this second boundary is parallel to the first one. The new incident angle is 9.94°, as can be determined from Fig.

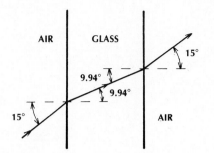

Figure 2-5 A ray is undeflected after transversing a parallel glass plate.

2-5. Find the direction of the transmitted ray.

Solution:
From Snell's law, sin θ_t = 1.5 sin 9.94° = 0.259. Then, θ_t = 15°. The transmitted ray is bent away from the normal. In addition, by combining these last two examples, we find that a ray incident on a parallel plate of glass will suffer no net deflection. This will always be true. As illustrated in Fig. 2-5, the ray enters the glass (at 15°), is deflected toward the normal (at 9.94°), and then is deflected back parallel to its original direction (at 15°). The translation (sideways displacement) of this ray will be negligible if the thickness of the glass is small.

2-2 LENSES

Fibers can be tested by sending visible beams of light through them. The simplest tests are those for continuity (checking for a break in the fiber) and more moderate physical damage causing only small losses. A continuity check can be made by observing whether any light emerges from the end of the fiber. Cracks or inhomogeneities in a bare fiber (one that is not inside a cable) can be located by observing the scattered light they produce. Gas lasers are convenient for these tests. Because their output beams have diameters on the order of a millimeter (and fibers are much smaller), a lens is used to focus the light onto the fiber end face, as illustrated in Fig. 2-6. To simplify our discussion, we will consider only *thin lenses*. A lens is thin when its thickness is so small that the translation of a ray passing through it is negligible. In other words, rays enter and leave at (approximately) the same distance from the lens axis. We will assume initially that our lenses are ideal, have no absorption or reflection losses, and produce no aberrations. We can add in these complications later if they become important.

In Fig. 2-6, a parallel beam of light (called a *collimated beam*) is focused to a point. This beam is traveling parallel to the lens axis. The incident light is made up of a number of parallel rays. Only the two outermost rays are drawn in the figure. All the rays converge to the position shown, known as the *focal point*. It lies a distance *f* (called the *focal length*) away from the lens. The plane that passes through the focal point and is perpendicular to the axis of the lens is the *focal plane*. The lens itself has two spherical surfaces. Think of the lens as being constructed by connecting the caps of two solid glass spheres. The radii (or curvatures) of these spheres are R_1 and R_2. The lens has diameter D and refractive index n. Its focal length is found from[2]

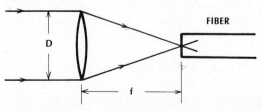

Figure 2-6 Focusing a light beam onto a fiber.

$$\frac{1}{f} = (n - 1)\left(\frac{1}{R_1} + \frac{1}{R_2}\right) \qquad (2\text{-}4)$$

The ratio f/D is called the *f-number* of the lens.

A long focal length is easily obtained by having large curvatures (R_1 and R_2); that is, by having fairly flat surfaces for the lens. A small focal length lens is harder to design because the curvatures have to be small, resulting in a small lens. A limiting case illustrates the problem. Consider a lens that is a complete sphere of glass. If the lens diameter is fixed, this design yields the smallest possible focal length. In this example, the radius of the sphere is the lens curvature, and the lens diameter is twice the curvature. That is, $R_1 = R_2 = D/2$. By using these values in Eq. (2-4) yields

$$f = \frac{D}{4(n - 1)}$$

For $n = 1.5$ we conclude that the lens f number is $f/D = 0.5$, showing that a small focal length requires a small lens diameter. A lens such as this would have severe spherical aberrations, causing the focused spot in Fig. 2-6 to smear out considerably. Most lenses have f numbers greater than 0.5 to correct this problem, making it even more difficult to obtain small focal lengths. When a lens is used to couple light from a gas laser beam into a fiber, aberrations may be unimportant. This is because the fiber diameter, although small, is not infinitesimal. For coupling, the beam does not have to be focused to a point. It must only be reduced to a size smaller than the fiber core.

Parallel rays of light that are incident at some angle relative to the lens axis are focused in the focal plane, as shown in Fig. 2-7. The position of the focused point is determined by the intersection of the central ray

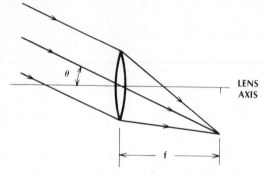

Figure 2-7 Focusing of an off-axis beam.

with the focal plane. The *central ray* (a ray directed toward the center of the lens) is undeviated by a thin lens because the ray enters and leaves at surfaces that are nearly parallel. In Section 2-1 we illustrated that a ray incident on a plate of glass with parallel sides suffered no net deflection.

A thin lens can collimate a beam that emerges from a point, as shown in Fig. 2-8. If the light source is located at the focal point, then the transmitted beam travels parallel to the lens axis. If the source lies anywhere else in the focal plane, then the transmitted beam

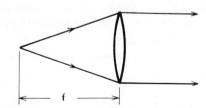

Figure 2-8 Collimating a diverging beam.

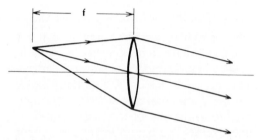

Figure 2-9 Collimating an off-axis point source.

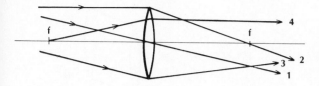

Figure 2-10 Ray paths through a thin lens. The numbers refer to the rules, listed in the text, which they illustrate.

will again be collimated, but its direction will differ. As shown in Fig. 2-9, this beam will travel in the direction of the ray connecting the source to the center of the lens. As previously mentioned, this ray is undeviated.

The rules for tracing rays through a thin lens are illustrated in Fig. 2-10. They are

1. Rays traveling through the center of the lens are undeviated.
2. Incident rays traveling parallel to the lens axis pass through the focal point after emerging from the lens.
3. An incident ray traveling parallel to a central ray intersects that ray in the focal plane after transmission through the lens.
4. An incident ray passing through the focal point travels parallel to the lens axis after it emerges from the lens.

These rules will enable you to trace rays for focusing, collimating, and imaging by using thin lenses.

Cylindrical lenses have surfaces that are portions of cylinders, as drawn in Fig. 2-11. This lens only deflects rays in one direction (in Fig. 2-11 this is vertically). The cylindrical lens is a one-dimensional version of the spher-

ical lens. In fact, Eq. (2-4) is true for the cylindrical lens, where R_1 and R_2 are now the curvatures of the cylindrical faces. The focal length f locates a line (the *focal line*) that is parallel to the axes of the cylindrical surfaces, a distance f from the lens, and passes through the lens axis (see Fig. 2-11). Light from a line source located along the focal line will be collimated by the lens. The ray paths are drawn in Fig. 2-12. Similarly, a collimated beam of rays entering the lens and traveling parallel to the lens axis will be focused to a line a focal length away from the lens.

It is interesting to consider the effects of a cylindrical lens on a point source located along the focal line. The light emerging from the lens will be collimated vertically but will continue to expand horizontally. Figure 2-13 illustrates this effect. The point to remember is that cylindrical lenses act like spherical lenses in one direction and have no effect in the orthogonal direction. This property is useful for fiber optics, because the light emitted from laser diodes and LEDs is often radiated nonsymmetrically. That is, the emitted light spreads more quickly in one direction (verti-

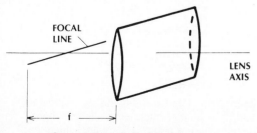

Figure 2-11 Cylindrical lens.

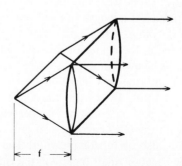

Figure 2-12 Collimating a line source.

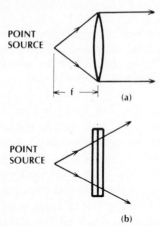

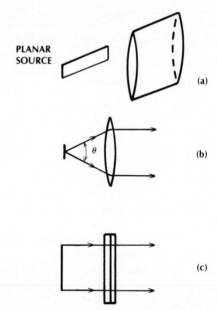

Figure 2-13 Point source and cylindrical lens. (a) Side view, showing collimation. (b) Top view.

cally) than the other (horizontally). A cylindrical lens can make the beam spreading more symmetrical by reducing the larger of the two divergence angles. This possibility is sketched in Fig. 2-14. The source emission has negligible beam spread in one of the directions.

The *graded-index rod lens,* or GRIN rod lens, is a modern development that has been applied to fiber systems in a number of ways.[3] The graded-index rod has a refractive index that decreases with distance from its axis. This causes light rays to travel in sinusoidal paths (see Fig. 2-15). (Section 5-2 contains a more extensive discussion of ray travel in a graded-index rod.) The length of one complete cycle is called the lens *pitch P*. Notice what will happen if a length of rod is cut equal to a *quarter pitch*. The light from a point source located at the center of this rod will be collimated as shown in Fig. 2-15(b). Collimated light entering this lens will be focused as in Fig. 2-15(c). Evidently, the GRIN rod has focusing and collimating properties in common with the classical spherical lens. The GRIN rod is also useful for imaging. The rod is desirable because small focal lengths can be obtained, permitting construction of short,

Figure 2-14 (a) Light from an unsymmetrical source is collimated by a cylindrical lens. (b) Side view. (c) Top view.

solid optic structures. For example, the light emitted from the end of a fiber can be collimated by a conventional lens (Fig. 2-16(a)) or by a rod lens (Fig. 2-16(b)). By using the spherical lens an air gap exists between the fiber and the lens. The rod lens needs no gap.

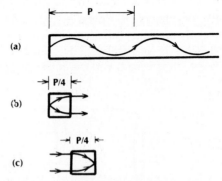

Figure 2-15 Graded-index rod. (a) Typical ray path. (b) A quarter-pitch lens collimates light emerging from a point. (c) A quarter-pitch lens focuses a collimated light beam.

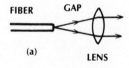

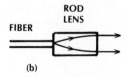

(a)

(b)

Figure 2-16 Collimating the light radiated from a fiber using (a) a spherical lens and (b) a GRIN rod lens.

The fiber can be cemented to the rod, yielding a continuous and solid mechanical structure. The rod collimator would be easier to assemble, align, and maintain than the spherical lens collimator.

2-3 IMAGING

Imaging by a thin lens is illustrated in Fig. 2-17. The object is an arrow with height O, located a distance d_o from the lens. The image is an arrow with height I, a distance d_i from the lens. The position of the image is found by tracing the rays emitted from the tip of the object. One ray passes through the center of the lens. According to tracing rule 1, it is undeviated. The second ray travels parallel to the lens axis, so it passes through the focal point when it emerges from the lens, following rule

2. The intersection of these two rays defines the focused image point corresponding to the tip of the object. In general, the intersection of any two rays that emanate from the same point determines the image position of that point. Notice how a single lens inverts the object.

The positions of the object and focused image are related by the *thin-lens equation*, [4]

$$\frac{1}{d_o} + \frac{1}{d_i} = \frac{1}{f} \qquad (2\text{-}5)$$

The *magnification M* is the ratio of the size of the image to that of the object. It is given by

$$M = \frac{d_i}{d_o} \qquad (2\text{-}6)$$

The magnification may be greater than, equal to, or less than unity.

Example 2-3
Find the object and image distances if the magnification is unity.

Solution:
If $M = 1$, then $d_i = d_o$. In the thin-lens equation, then,

$$\frac{1}{d_o} + \frac{1}{d_o} = \frac{1}{f}$$

or, $d_o = 2f$. Finally, d_i also equals $2f$.

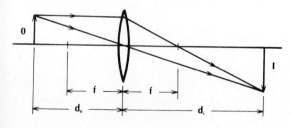

Figure 2-17 Image formation by a thin lens.

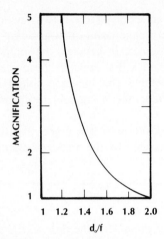

Figure 2-18 Magnification as a function of object position.

Equations (2-5) and (2-6) can be combined to show the direct relationship between magnification and object distance. The result is

$$M = \frac{1}{(d_o/f) - 1} \qquad (2\text{-}7)$$

This equation is plotted in Fig. 2-18. For magnifications greater than unity, the range of object positions is

$$1 < \frac{d_o}{f} < 2 \qquad (2\text{-}8)$$

When using lenses to couple light into fibers it is important to keep track of the angles at which the light rays travel. Referring to

Fig. 2-19, we note the angular light spread (α_o) at the object and the resulting angular spread (α_i) at the image for an object point on the lens axis. Figure 2-19 is the same as Fig. 2-17, except now we are considering angular changes rather than size changes. The thin-lens equation still predicts the image position in terms of the object distance and the lens focal length. A little trigonometry shows that

$$\frac{\tan(\alpha_i/2)}{\tan(\alpha_o/2)} = \frac{1}{M} \qquad (2\text{-}9)$$

A plot of the tangent function appears in Fig. 2-20. Note that the tangent of an angle is equal to the angle itself when that angle is small and expressed in radians. This approximation is very good (less than 4% error) up to 20° (0.35 rad). Conversion between radians and degrees is made by using the relationship 1 rad = 57.3°. Assuming small angles, we can replace the tangent functions in Eq. (2-9) with the angles themselves, yielding

$$\frac{\alpha_i}{\alpha_o} = \frac{1}{M} \qquad (2\text{-}10)$$

This equation can be used when the cone half angles are less than 20°, that is, for full angular beam spreads (α_i or α_o) as much as 40°. Although Eq. (2-10) was developed with α_i and α_o measured in radians, it is also correct when α_i and α_o are expressed in degrees.

We may conclude from Eq. (2-10) that an increase in object size owing to magni-

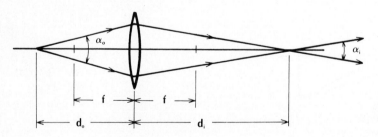

Figure 2-19 Angular changes owing to imaging.

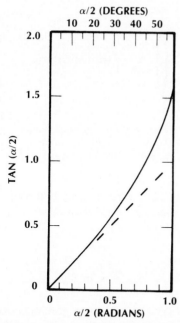

α/2 (DEGREES)

Figure 2-20 The tangent function is shown by the solid line. The dashed line represents the approximation $\tan(\alpha/2) = \alpha/2$.

fication is accompanied by a decrease of the beam spread. The imaging lens tends to collimate the rays of light radiating from the object. Because laser diodes and LEDs radiate over large angles, and fibers accept rays only over small angles, lenses can increase the efficiency of coupling between sources and fibers.

Example 2-4

A source radiates light uniformly over a region having a 40° full cone angle. The source is a square, planar radiator measuring 20 μm on a side. Design a lens system that will decrease the beam spread to a 10° cone. Determine the image size.

Solution:

The imaging system is the same as that in Fig. 2-19. In this example, $\alpha_o = 40°$

and $\alpha_i = 10°$. We may use the approximate result in Eq. (2-10) because $\alpha_o/2$ is only 20°. Equation (2-10) shows that the magnification is 4, making the planar image 4(20) = 80 μm on a side. For a magnification of 4, Eq. (2-7) or Fig. 2-18 shows that $d_o/f = 1.25$. If we choose a lens having focal length 10 cm, then $d_o = 12.5$ cm. Finally, Eq. (2-6) yields the image distance $d_i = Md_o = 50$ cm.

In Example 2-4 a more compact arrangement will be obtained if we choose a much smaller focal length. With $f = 1$ mm, the object and image distances become just 1.25 and 5 mm, respectively. An increase in source size from 20 to 80 μm would be acceptable when coupling light into a fiber having a core diameter of 100 μm or more.

Coupling from a source larger than the fiber presents a bigger problem. If we attempt to demagnify the source size ($M < 1$), then the angular spread will increase as predicted by Eq. (2-10). The fiber might not accept rays over this expanded range.

2-4 NUMERICAL APERTURE

An important characteristic of an optic system is its ability to collect light incident over a wide range of angles. Figure 2-21 illustrates an optic receiver consisting of a lens and a photodetector. The lens is much larger than the detector surface, so it intercepts more rays than the detector would by itself. The lens focuses this light onto the detector. Together, the lens and detector make an efficient collection system. It is easy to locate the position on the detector where the light is focused by using a ray tracing rule 1. Simply extend the incident ray passing through the center of the

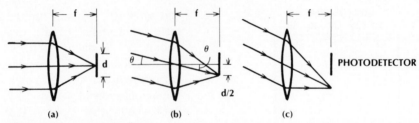

Figure 2-21 An optical receiver with the photodetector placed in the focal plane of the lens. (a) Light is incident parallel to the lens axis. (b) The light rays are at the extreme angle for reception (acceptance angle). (c) The incident rays are beyond the system acceptance angle.

lens until it hits the detector, as illustrated in the figure. By applying this rule to Fig. 2-21 it is apparent that rays incident at large angles will not strike the detector and will be lost.

By referring to the figure the *maximum acceptance angle* is determined from

$$\tan \theta = \frac{d}{2f} \qquad (2\text{-}11)$$

where d is the diameter of the circular photodetector surface and f is the lens focal length. Because of the circular symmetry of the receiver, it will detect light incident within a cone having half angle θ. The *numerical aperture* (NA) is defined to be[5]

$$NA = n_0 \sin \theta \qquad (2\text{-}12)$$

where n_0 is the refractive index of the material between the lens and the photodetector, and θ is the maximum acceptance angle. For the receiver in Fig. 2-21, θ is given by Eq. (2-11).

Example 2-5

A receiver has a 10-cm focal length, a 1-cm photodetector diameter, and air between the lens and detector. Compute the receiver's NA.

Solution:

Because $d/2f$ is so small, we can use the small-angle approximation $\sin \theta \simeq$ $\tan \theta$. (The error is less than 6% if θ is less than 20°.) In this case, with $n_0 = 1$, Eq. (2-12) yields

$$NA = \sin \theta = \frac{d}{2f} = 0.05$$

This corresponds to an acceptance angle θ of 2.87°. The full cone angle is twice this value, 5.74°.

The definition of numerical aperture given in Eq. (2-12) applies to all light-collecting systems, including optic fibers. The collection cone for a fiber is sketched in Fig. 2-22. Light rays incident at angles outside this cone will not propagate along the fiber but instead will attenuate rapidly. The numerical aperture is usually measured with air in front of the fiber, so $n_0 = 1$ in Eq. (2-12) and

$$NA = \sin \theta \qquad (2\text{-}13)$$

A plot of this equation appears in Fig. 2-23. A low NA indicates a small acceptance angle. Because of this, coupling to a low-NA fiber is more difficult (mechanical alignment is more sensitive) and less efficient (some of the rays are outside the acceptance angle) than coupling to a high NA fiber. Lenses can be used as we illustrated in Section 2-3 to reduce the beam spread and, consequently, to improve the coupling efficiency. Typically, fibers

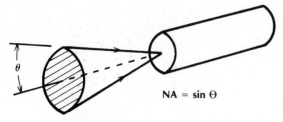

$$NA = \sin\Theta$$

Figure 2-22 The fiber will only accept light rays incident within a cone having half-angle θ.

for long path lengths are designed to have numerical apertures from about 0.1 to 0.3. The low NA does make coupling efficiency tend to be poor but improves the fiber's bandwidth (as we will show in Chapter 5). Plastic, rather than glass, fibers are available for short path lengths. These fibers are restricted to short lengths because of the high attenuation in plastic materials. Plastic fibers are designed to have high numerical apertures (typically, 0.4–0.5) to improve coupling efficiency, partially offsetting the high propagation losses.

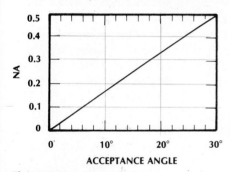

Figure 2-23 Numerical aperture and acceptance angle. NA = sin θ.

$$\Theta = NA \ \sin^{-1} \ Air$$

2-5 DIFFRACTION

In some experiments, geometrical optics (ray theory) correctly predicts the gross results but does not agree with the fine details of the observation. In other experiments, even the gross behavior is predicted erroneously. In these instances, a more complete theory based

on the wave nature of light is needed to explain the observed phenomena. This theory is called *diffraction* or *physical optics*. We can say that diffraction is the deviation from the predictions of geometrical optics. A few important examples requiring diffraction analysis follow.

First, let us define a few terms. The plane perpendicular to the direction of wave travel is called the *transverse plane*. Throughout this course we will be dealing with distributions of the light power in the transverse plane. In doing so, we will use the words *power* and *intensity* of the light beam interchangeably. A *uniform* beam is one whose intensity is the same at all points in the transverse plane.

Figure 2-6 shows a lens focusing a collimated uniform light beam. Diffraction theory and careful experimentation show the beam does not converge to a point but instead reduces to a central spot of light surrounded by rings of steadily diminishing intensity. The central spot has diameter[6]

$$d = 2.44\lambda f/D \qquad (2\text{-}14)$$

where λ is the wavelength, f is the focal length, and D is the lens diameter. Figure 2-24 illustrates the situation. The central spot is normally pretty small. For example, if $f = 2D$ and the wavelength is 1 μm, then Eq. (2-14) predicts a spot diameter of 4.88 μm. This may be negligible in some applications, meaning that the geometrical optics treatment is sufficient. On the other hand, suppose that

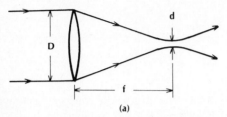

Figure 2-24   (a) Focusing a uniform light beam according to diffraction theory and experiment. (b) Light distribution in the focal plane.

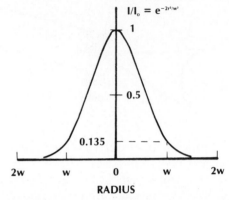

Figure 2-26   The normalized Gaussian intensity distribution.

this beam and lens are used to couple light into a fiber having a diameter less than 4 μm (as in Fig. 2-25) or into a glass film with a thickness less than 4 μm. The coupling efficiency will be low, because the focused spot is larger than the fiber (or the film). Clearly, diffraction theory is required to explain the results of these experiments.

Actual light sources often produce nonuniform beams. The intensities vary across the transverse plane. A particularly important *transverse pattern* is the *Gaussian distribution*. This is the familiar bell curve drawn in Fig. 2-26. For simplicity, the figure is *normalized* such that the peak of the curve is unity. Most gas lasers, and some specially designed laser diodes, radiate in this pattern. Very small fibers (having diameters of a few micrometers) also have their light distributed in this manner. The Gaussian intensity distribution is given mathematically by

$e = $ *decaying exopential*

$$I = I_o e^{-2r^2/w^2} \qquad (2\text{-}15)$$

$e = 2.718$

Here, $e = 2.718$ is the base of the natural logarithm. Because $e^0 = 1$, I_o is the inten-

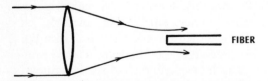

Figure 2-25   Focusing a beam onto a small fiber may result in inefficient coupling.

sity at the center of the beam ($r = 0$). When viewed by the eye, this pattern appears to be a circle of light. The edges of the circle are not sharp, but, instead, the light intensity drops off gradually. An accepted definition of the radius of the spot is the distance at which the beam intensity has dropped to $1/e^2 = 0.135$ times its peak value I_o. This radius is called the *spot size*. For the beam described by Eq. (2-15), the spot size is just w.

Focusing a Gaussian light beam with a lens (as in Fig. 2-27) yields a distribution of light in the focal plane that is also Gaussian shaped. There are no surrounding rings like those that appear when focusing a uniform beam. The spot size in the focal plane is

$$w_o = \frac{\lambda f}{\pi w} \qquad (2\text{-}16)$$

and the intensity distribution is $I = I_o' \exp(-2r^2/w_o^2)$. The size of the focused Gaussian spot is not much different from the size of the central spot obtained by focusing a uniform beam. This can be seen by writing Eq. (2-16) in the same form as Eq. (2-14). We do this by defining the focused spot diameter as $d = 2w_o$ and the incident beam spot diameter as $D = 2w$. Making these replacements in

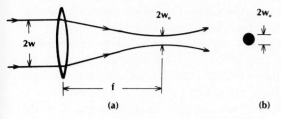

Figure 2-27 (a) Focusing a Gaussian light beam. (b) The spot appearing in the focal plane.

Eq. (2-16) yields $d = 4\lambda f/\pi D = 1.27\ \lambda f/D$, which is comparable to Eq. (2-14). We conclude that the shape of the incident beam does not greatly change the degree to which light can be concentrated.

Next, we see what corrections to ray theory are needed when collimating a beam. Referring to Fig. 2-28, a small source is located at the focal point of a lens. Ray theory predicts a parallel beam of light will emerge from the lens. If the light distribution is Gaussian, then the beam just to the right of the lens is given by $I = I_o \exp(-2r^2/w^2)$. Diffraction theory agrees with the geometrical prediction of collimation if we restrict our observation to regions close to the lens. For longer distances, diffraction theory shows that the beam diverges at a constant full angle given by

$$\theta = \frac{2\lambda}{\pi w} \qquad (2\text{-}17)$$

where θ is in radians. Experiments verify this result. The radiated field pattern is $I = I_o' \exp(-2r^2/w_o^2)$ where $w_o = \lambda z/\pi w$.

Example 2-6

Consider a Gaussian beam whose spot size is 1 mm when collimated. The wavelength is 0.82 μm. Compute the divergence angle. Also find the spot size at 10 m, 1 km, and 10 km.

Solution:

The divergence angle is

$$\theta = \frac{2(0.82\times10^{-6})}{\pi(10^{-3})} = 0.55 \times 10^{-3}\ \text{rad}$$

or $\theta = 0.032°$. The spot size at 10 m is

$$w_o = \frac{(0.82 \times 10^{-6})10}{\pi(10^{-3})} = 2.6 \times 10^{-3}\ \text{m}$$

$$= 2.6\ \text{mm}$$

At 1 km the spot size is 260 mm and at 10 km is 2.6 m.

The preceding equations and Example 2-6 illustrate several interesting results. Equa-

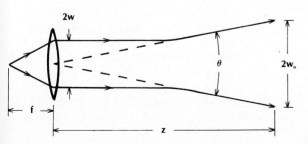

Figure 2-28 Collimating a Gaussian beam.

tion (2-17) shows that very small divergence angles are obtained when the spot size is much larger than the wavelength. Optic wavelengths are so small that this condition is easily achieved. Figure 2-28 is the optic analog of a radio-frequency transmitting antenna. In fact, Eq. (2-17) applies qualitatively to an antenna whose largest dimension is on the order of $2w$. In general, the divergence of a beam radiated at any wavelength is inversely proportional to the size of the radiator as measured in wavelengths. Transmitters that emit narrow beams are many wavelengths long. At radio frequencies such an antenna must be very large. We conclude that optic transmission provides narrow, highly directed beams.

An atmospheric communication system is drawn in Fig. 2-29. Because of divergence over a long path, the beam at the receiver may be quite large; in fact, a lot larger than the receiving lens itself. Much of the transmitted power will be lost in this case. Although atmospheric systems perform acceptably over short paths, the desirability of more efficient power transfer over long distances is apparent. This need prompted investigations into guided schemes, such as the optic fiber. Loss dependence on weather poses another problem for atmospheric systems. Poor weather conditions decrease system performance. The weather problem could be overcome by sending the beam down an evacuated pipe. This would be acceptable if the beam were truly collimated. Because it is not, the expanding beam would

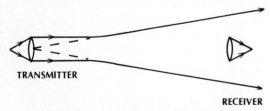

Figure 2-29 Atmospheric transmission link.

strike the sides of the pipe, undergoing losses by absorption, scattering, and imperfect reflection. By using the numbers generated in Example 2-6, a 1-km pipe would need a radius of more than 260 mm to keep the light from contacting its sides. A pipe this large is clearly unacceptable. A short length of tubing (a few centimeters) is practical for protecting a narrow optic beam from its surroundings because the beam will not expand much over a very short path.

2-6 SUMMARY

This summary contains two lists. One list denotes major points learned from the discussions presented. The other list contains new topics suggested by these discussions. First, the list of what we learned:

1. Rays travel through a medium at a speed determined by the material's refractive index n.
2. Rays are deflected according to Snell's law when crossing boundaries.
3. Lenses can focus and collimate light beams. They can also form magnified images with accompanying changes in ray angles.
4. The GRIN rod lens can perform the same functions as a classical spherical lens. Its compact structure and small focal length make it attractive for fiber systems.
5. Optic systems, including fibers, accept light only over a limited range of incident angles. The numerical aperture is a measure of this characteristic.
6. Diffraction tells us that light cannot be focused to an infinitesimally small point and cannot be perfectly collimated. The first of these results applies to attempts

at coupling light into very small fibers by focusing, and the second applies to construction of unguided optic communications systems.

7. The Gaussian intensity distribution often occurs in laser and fiber systems. We should be familiar with this pattern.

8. Atmospheric optic communications systems are practical. Over short unobstructed paths they may be preferable to guided systems. For long paths in which waveguides cannot be installed (for example, communications between satellites), they may provide a viable system. Generally, however, many more applications exist for fiber systems than for unguided systems.

Here is the list of the new topics. These subjects are addressed in succeeding chapters.

1. The propagation properties of light beams within an optic fiber. The velocity given in Eq. (2-1) applies to the speed of light in an unrestricted medium.

2. How much light is transmitted and reflected at a boundary.

3. How to design a lens system for efficient coupling from sources to fibers and between fibers. How to compute the resulting efficiencies.

4. How to design with GRIN rod lenses. What specific devices can profitably include these lenses.

5. Why fibers do not accept rays at all angles. How the numerical aperture of a fiber is computed. How coupling efficiency depends on NA.

These topics are important in the design and evaluation of components, such as couplers, connectors, and multiplexers. These topics will also help us understand how fibers guide beams of light.

PROBLEMS

2-1. Image a point source of light with a single lens of focal length f. The point source radiates within a cone having a full angle α_o. Compute the angular light spread (α_i) at the image in terms of the image and object distances and the angular light spread of the source. If $\alpha_o = 40°$ and the magnification is 5, compute α_i.

2-2. Plot numerical aperture versus acceptance angle over the range $0 \leq \text{NA} \leq 0.7$. Assume that the refractive index of the surrounding material is 1.0 in the calculation.

2-3. Plot magnification versus the normalized object distance d_o / f.

2-4. Let the focal length of an imaging lens be 20 mm. Plot the object distance versus the image distance.

2-5. A uniform collimated beam is focused by a lens whose focal length is 20 mm and whose diameter is 10 mm. The wavelength is 0.8 μm. Compute the focused spot size.

2-6. A collimated Gaussian beam has a spot size of 1 mm and wavelength of 0.8 μm. Compute the focused spot size when focused by a lens whose focal length is 20 mm.

2-7. Plot the normalized intensity of a Gaussian beam versus distance from the beam axis if the spot size is 1 mm.

2-8. Compute the divergence angle of a Gaussian beam of wavelength 0.8 μm and spot size 1 mm. If this beam is aimed at the moon, what is its spot size on the moon's surface? What is its spot size at distances of 1 km and 10 km?

2-9. A 6,000-km undersea glass fiber telephone line crosses the Atlantic ocean connecting the United States and France. (a) How long does it take for a message to traverse this link? (b) How long does it take for a message to travel from the United States to France by using a satellite link? The satellite is stationed about 22,000 miles above the earth between the United States and France. (c) Will two people having a conversation across these two different links notice the travel delays?

2-10. A beam of light is incident on a plane boundary between two dielectrics. The incident ray angle is at 10° to the boundary normal and the transmitted beam is at 12°. Which of the two media has the higher refractive index?

REFERENCES

1. George Shortley and Dudley Williams. *Elements of Physics,* 5th ed. Englewood Cliffs, N.J.: Prentice Hall, 1971, pp. 748–50.
2. Ibid. p. 778.
3. Teji Uchida, Moatoaki Furukawa, Ichiro Kitano, Ken Koizumi, and Hiroyoshi Matsumura. "Optical Characteristics of a Light-Focusing Fiber Guide and Its Applications" *IEEE J. Quantum Electron* 6, no. 10 (Oct. 1970): 606–12.
4. Shortley and Williams. *Elements of Physics.* p. 778.
5. Jurgen R. Meyer-Arendt. *Introduction to Classical and Modern Optics.* Englewood Cliffs, N.J.: Prentice Hall, 1972, pp. 136–37.
6. Shortley and Williams, *Elements of Physics.* p. 813.

Chapter 3

Lightwave Fundamentals

Wave propagation is important in fiber optics. In this chapter we present fundamental aspects of wave travel that are particularly valuable. For some reason, the prospect of studying electromagnetic waves frightens many people. Indeed, expositions of electromagnetic theory are often quite formidable. The discussion to follow is as cheerful and painless as possible. Important results are explained, but the lengthy derivations required to develop them are omitted. Mathematics is minimized. The specific concepts developed are velocity, power, dispersion, polarization, interference, and reflections at boundaries. All of these relate directly to fiber optics.

3-1 ELECTROMAGNETIC WAVES

Light consists of an electric field and a magnetic field that oscillate at very high rates, on the order of 10^{14} Hz. These fields travel in wavelike fashion at very high speeds. A picture of an electromagnetic wave[1] traveling along the z direction appears in Fig. 3-1. The electric field is plotted at three times, showing the progress of the wave. At any fixed location, the field amplitude varies at the optic frequency. The amplitude repeats itself after one period of the oscillation. The wave repeats itself in space, at a fixed time, after a distance λ. This distance is the *wavelength*. Its reciprocal, $1/\lambda$, is the *wave number*.

The electric field for the wave sketched in Fig. 3-1 can be written as

$$E = E_o \sin (\omega t - kz) \qquad (3-1)$$

where E_o is the peak amplitude, $\omega = 2\pi f$ rad/s, and f is the frequency in hertz. The factor ω is called the *radian frequency*. The

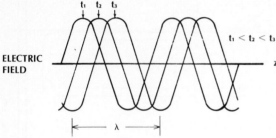

ELECTRIC
FIELD

Figure 3-1 Electric field for a wave traveling in the *z* direction. The field is drawn at three different times to illustrate the motion of the wave in the direction of travel.

term *k* is the *propagation factor*. It is given by

$$k = \frac{\omega}{v} \qquad (3\text{-}2)$$

where v is the phase velocity of the wave. The factor $\omega t - kz$ is the *phase* of the wave, while kz is the *phase shift* owing to travel over length z. A *plane wave* is one whose phase is the same over a planar surface. In the present example, the phase is the same over any plane defined by a fixed value of z, so that Eq. (3-1) represents a plane wave. If time is held constant, then Eq. (3-1) shows the sinusoidal spatial variation of the field. For example, if $t = 0$, then $E = E_o \sin(-kz) = -E_o \sin kz$. On the other hand, if the position is fixed, then Eq. (3-1) shows the sinusoidal time variation of the field. Taking the fixed position as the origin, $z = 0$, yields $E = E_o \sin \omega t$, illustrating this point.

In terms of the refractive index n, the velocity is $v = c/n$, so that

$$k = \frac{\omega n}{c} \qquad (3\text{-}3)$$

The propagation constant in free space will be denoted by k_0. Since $n = 1$ in free space,

$$k_0 = \frac{\omega}{c} \qquad (3\text{-}4)$$

Combining Eqs. (3-3) and (3-4) the propaga-

tion constant in any medium can be given in terms of the free-space propagation value by

$$k = k_0 n \qquad (3\text{-}5)$$

According to Eq. (1-3), $\lambda = v/f$. Substituting this into Eq. (3-2) yields

$$k = \frac{2\pi}{\lambda} \qquad (3\text{-}6)$$

This equation relates the propagation constant in a medium to the wavelength in that medium. The free-space wavelength is $\lambda_0 = c/f$, and the wavelength in any medium is $\lambda = v/f$, so that

$$\frac{\lambda_0}{\lambda} = \frac{c}{v} = n \qquad (3\text{-}7)$$

The wavelength in a medium is shorter than in free space, because the refractive index is greater than unity.

The power in an optic beam is proportional to the light *intensity* (defined as the square of the electric field). Intensity is proportional to *irradiance,* the power density. The units of irradiance are watts per square meter. In Section 2-5 we investigated the intensity variation of a particular light distribution, the Gaussian beam. Sometimes intensity is used to describe the total power in a wave. This use, although not accurate, is common.

If a wave does not lose energy as it propagates, then Eq.(3-1) and Fig. 3-1 provide appropriate descriptions. If attenuation is important, than the equation and the figure must be modified. The corrected equation is

$$E = E_o e^{-\alpha z} \sin (\omega t - kz) \qquad (3-8)$$

where ω and k have the same meaning as in Eq. (3-1). The term α is the *attenuation coefficient*. Its value determines the rate at which the electric field diminishes as it travels through the lossy medium. Although the decay is exponential, the attenuation coefficient is so small for quality fibers that there is little attenuation (maybe just a few decibels), even over long paths. In a lossy medium, the field appears as shown in Fig. 3-2. The dashed line on the figure is a curve of the factor $\exp(-\alpha z)$, describing the loss in Eq. (3-8).

The intensity of a light beam is proportional to the square of its electric field. Therefore, the power in the beam corresponding to Eq. (3-8) diminishes as $\exp(-2\alpha z)$. For a path of length L, the ratio of the output power to the input power is $\exp(-2\alpha L)$. The power reduction in decibels is thus

$$dB = 10 \log_{10} \exp(-2\alpha L)$$

This will turn out to be a negative number for propagation through a lossy medium. From this last expression, we can find the relationship between the attenuation coefficient and the power change in dB/km. The result is

$$dB/km = -8.685\alpha$$

where α is given in units of km^{-1}.

3-2 DISPERSION, PULSE DISTORTION, AND INFORMATION RATE

Up to this point we have been assuming that optic sources in fiber systems emit light at a single wavelength (or equivalently, at a single frequency). This is never true in practice. Real sources produce radiation over a range of wavelengths. This range is the source *line width,* or *spectral width*. The smaller the line width, the more *coherent* the source. A perfectly coherent source emits light at a single wavelength. Thus, it has zero line width and is perfectly *monochromatic.* Typical line widths of common sources are listed in Table 3-1. The conversion between spectral width in wavelengths $\Delta\lambda$ and bandwidth in frequency Δf is

$$\frac{\Delta f}{f} = \frac{\Delta \lambda}{\lambda} \qquad (3-9)$$

TABLE 3-1.   Typical Source Spectral Widths

Source	Linewidth ($\Delta\lambda$) (nm)
Light-emitting diode	20–100
Laser diode	1–5
Nd:YAG laser	0.1
HeNe laser	0.002

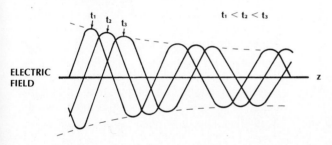

ELECTRIC
FIELD

Figure 3-2   Attenuation of a traveling wave.

where f is the center frequency, λ is the center wavelength, and Δf is the range of frequencies radiated. This conversion is simply the mathematical statement that the fractional emission width is the same whether computed on the basis of wavelength spread or frequency spread.

Figure 3-3 illustrates some of the preceding points. It is a plot of the wavelength distribution of power radiated by a representative LED. The wavelength, or frequency, content of a signal is called its *spectrum*. For the LED in the figure, the center wavelength is 820 nm (0.82) μm. The linewidth is normally taken to be the width to the half-power points; so, in this example, $\Delta\lambda = 30$ nm (805–835 nm). The fractional bandwidth is $30/820 = 0.037$, or 3.7%.

According to Table 3-1, laser diodes are more coherent than LEDs. The solid state neodymium yttrium-aluminum-garnet laser (Nd:YAG) and the helium-neon gas laser (HeNe) are even better. However, the small size and low power requirements of the LED and LD sources make them the most practical for fiber systems, even though they have much

greater line widths than the other laser emitters.

At this point, this is a natural question to ask: Do we consider a source to have negligible bandwidth (that is, treat it as a perfectly coherent source), or do we need to consider its lack of coherence? In the discussion that follows, we will examine how the source's spectral width limits the information capacity of a fiber system. If the limiting capacity is higher than needed, then the noncoherence can be ignored.

Material Dispersion and Pulse Distortion

In Section 2-1 we related the wave velocity v to the refractive index n by the equation $v = c/n$. For the glasses used in optic fibers, the refractive index varies with wavelength. Therefore, the wave velocity also varies with wavelength. *Dispersion* is the name given to the property of velocity variation with wavelength. When, as in the present example, the velocity variation is caused by some property of the material, the effect is called *material dispersion*. For fibers and other waveguides, dispersion can also be caused by the structures themselves. This case, treated in Section 5-6, is *waveguide dispersion*.

Consider what happens when a real source (nonzero bandwidth) emits a pulse of light into a dispersive glass fiber. The initial pulse consists of a sum of pulses that are identical, except for their wavelengths. This is illustrated in Fig. 3-4 for a few of the source wavelengths. The several pulses travel at different velocities, reaching the end of the fiber at slightly different times. When summed at the output, the slightly displaced pulses add together, yielding an output that is lengthened, or spread, relative to the input signal. This illustrates how dispersion creates pulse

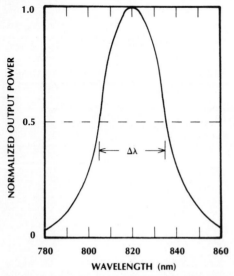

Figure 3-3 Spectrum of a light-emitting diode.

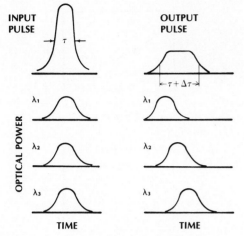

Figure 3-4 Pulse spreading caused by propagation through a dispersive material. The complete pulse contains wavelengths λ_1, λ_2, and λ_3, each traveling at a different speed.

distortion. The farther the pulse travels, the greater the spreading.

Dispersion will also distort an analog signal. Figure 3-5 shows an analog waveshape propagated at three different wavelengths. At the input the three wavelengths vary together in phase with each other, creating a large sig-

nal variation. After travel through the dispersive media, these wavelengths are no longer in phase. When added together they produce a signal variation lower in amplitude than the input signal variation. Dispersion does not change the average power or the modulation frequencies, but it does lower the signal variation. The transmitted information is contained in this variation, so its attenuation is troublesome. We may think of this result as broadening the signal peak (lowering its amplitude) and filling in the valley (raising its level). Excessive broadening will cause loss of the signal variation altogether.

Distortion caused by material (or waveguide) dispersion can be reduced by using sources with smaller bandwidths; that is, by using more coherent emitters. A laser diode has the advantage over an LED in this respect. In principle, dispersive distortion could be reduced by filtering the optic beam at the transmitter or receiver, allowing only a very narrow band of wavelengths to reach the photodetector. This technique has two drawbacks: Filters cannot be constructed with passbands narrow enough to be effective, and a

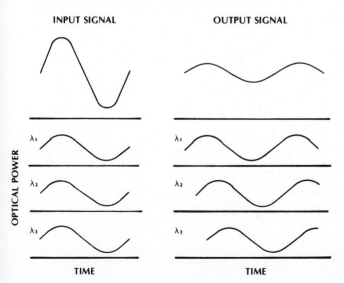

Figure 3-5 Dispersion causes loss in amplitude of an analog signal.

narrow-band filter would greatly reduce the optic power by eliminating the light at the unwanted wavelengths.

Dispersion in glass is easily observed. We have all seen the results of dispersion when a glass prism separates white light into its component colors. This experiment is explained by the wavelength dependence of the refractive index of glass. The incident light rays are bent according to Snell's law [Eq. (2-3)]. The different colors are bent at different angles, because the refractive index is different for each color. The refractive index for pure silicon dioxide (SiO_2) glass used in optic fibers has the wavelength dependence shown in Fig. 3-6. There are several noteworthy characteristics. The refractive index decreases with increasing wavelengths, so the slope of the curve in Fig. 3-6(a) is negative. The magnitude of this slope changes with wavelength. At a particular wavelength (λ_0 in the figure) there is an inflection point on the refractive index curve. The magnitude of the slope is minimum at this wavelength, as indicated in Fig. 3-6(b). Because of this, the slope of curve (b) is 0 at λ_0. The slope of curve b appears in Fig. 3-6(c). This last figure is the second derivative of the refractive index with respect to the wavelength. For pure silica, the refractive index is close to 1.45 and the inflection point is near 1.3 μm. Doping SiO_2 with small amounts of other materials, for example with germanium oxide (GeO_2), shifts the refractive index curves slightly.

Now that we have determined qualitatively how dispersion distorts signals transmitted through glass, we must find how much broadening is introduced and how this relates to the amount of information we can transfer.

Let τ be the time for a pulse to travel a path having length L. Figure 3-7 shows a plot of the travel time per unit length (τ/L) as a function of wavelength. In Fig. 3-7(a), we have the curve for a nondispersive medium,

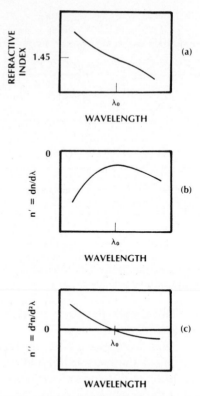

Figure 3-6 (a) Wavelength dependence of the refractive index of SiO_2 glass. (b) Derivative (slope) of the curve in (a). (c) Derivative (slope) of the curve in (b).

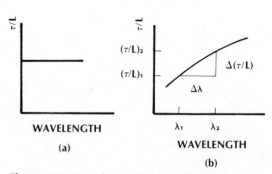

Figure 3-7 Travel time per unit length. (a) Nondispersive and (b) dispersive media.

where the travel time is independent of the wavelength. In Fig. 3-7(b), the travel time does depend on the wavelength, as is appropriate for a dispersive material. Now consider a pulse whose shortest and longest wavelengths are λ_1 and λ_2. We will determine the spreading of this pulse. We may consider the two wavelengths to be the edges of a band emitted by a source. In other words, we can let $\lambda_2 - \lambda_1 = \Delta\lambda$, where $\Delta\lambda$ is the source spectral width. All wavelengths between λ_1 and λ_2 will arrive after the faster of these two wavelengths and before the slower. The travel time per unit length is not directly important. The important quantity is the difference in travel time per unit length for the two extreme wavelengths. We denote this quantity by $\Delta(\tau/L)$. Then, *TOU = Time*

$$\Delta(\tau/L) = (\tau/L)_2 - (\tau/L)_1 \quad (3\text{-}10)$$

where $(\tau/L)_1$ and $(\tau/L)_2$ are the values corresponding to λ_1 and λ_2, respectively. The term $\Delta(\tau/L)$ is the *pulse spread per unit length*, often simply (but imprecisely) called the *pulse spread*. The term $\Delta\tau = \tau_2 - \tau_1$ is the actual pulse spread. Of course, $\Delta\tau = L\,\Delta(\tau/L)$.

In real situations, pulses do not have sharply defined beginnings and endings. They gradually increase to a peak and then diminish similarly, as is illustrated in Fig. 3-4. The value of the duration of a pulse depends on the definition of its starting and stopping points. Various definitions have been used, each based on the time at which the pulse reaches a desired level relative to its peak.

We will use the following definition throughout this text: The pulse duration is the interval from the time the optic power rises to half its peak value to the time it falls to half its peak value. This definition describes the *full duration half-maximum* (FDHM) pulse duration. It is illustrated for the pulses in Fig. 3-4.

As seen from Fig. 3-7(b), the slope of the τ/L curve, denoted by $(\tau/L)'$, is

$$(\tau/L)' = \frac{\Delta(\tau/L)}{\Delta\lambda} \quad (3\text{-}11)$$

or

$$\Delta(\tau/L) = (\tau/L)'\Delta\lambda \quad (3\text{-}12)$$

Analysis shows that

$$(\tau/L)' = -\frac{\lambda}{c}\frac{d^2n}{d\lambda^2} = -\frac{\lambda}{c}n'' \quad (3\text{-}13)$$

We can visualize this term by looking at Fig. 3-6(c), where n'' is sketched. Combining the last two equations yields $\Delta(\tau/L) = -\lambda n''\Delta\lambda/c$, showing how the pulse spread depends on the behavior of the refractive index. It is convenient to define the *material dispersion* as $M = \lambda n''/c$. The pulse spread per unit length can then be written as

$$\Delta(\tau/L) = -\frac{\lambda}{c}n''\Delta\lambda = -M\Delta\lambda \quad (3\text{-}14)$$

The material dispersion is plotted as a function of the free space wavelength in Fig. 3-8 for pure silica. A plot of M would appear quite similar to Fig. 3-6(c) because M is proportional to n''. The units of M are ps/(nm $\times$ km). This is read as picoseconds of pulse spreading per nanometer of source spectral width and per kilometer of path length. Let us also interpret the negative sign in Eq. (3-14). Because $\Delta\lambda$ is always positive, this equation predicts that the pulse spread will be negative when M is positive. This means that $(\tau/L)_1 > (\tau/L)_2$; that is, the travel time for the shorter wavelength (λ_1) is longer than the travel time for the longer wavelength (λ_2). The longer wavelength travels faster. According to Fig. 3-8 this is the case for pure silica at

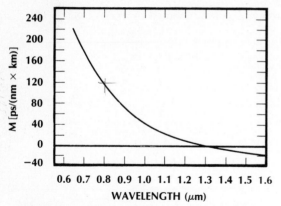

Figure 3-8   Material dispersion for pure silica. [From S. H. Wemple, "Material Dispersion in Optical Fibers," *Applied Optics,* 18, no. 1 (Jan. 1979), p. 33, and Corning Glass Works Product Bulletin 1519 (September 1985). Adapted with permission.]

wavelengths below 1.3 μm where M is positive. When M is negative, the pulse spread is positive and the shorter wavelength travels faster (its travel time is less) than the longer wavelength. This is the case for pure silica at wavelengths above 1.3 μm. For some calculations, only the magnitude of the pulse spread is important. In these cases we will ignore the sign in Eq. (3-14). Later, when we combine material and waveguide dispersion, we will need to account for the sign of the pulse spread.

At 1.3 μm, the material dispersion is zero for pure silica. Pulse spreading owing to material dispersion disappears at this wavelength. Silica-based glasses used in fiber optics have zero material dispersion near 1.3 μm. *Doping* (adding small amounts of other constituents to silica) may change the zero dispersion wavelength by about 0.1 μm.

Example 3-1

Find the amount of pulse spreading in pure silica for an LED operating at 0.82 μm and having a 20-nm spectral width.

The path is 10 km long. Repeat if λ = 1.5 μm and $\Delta\lambda$ = 50 nm.

Solution:

Figure 3-8 shows that $M = 110$ ps/ (nm $\times$ km) at 0.82 μm. From Eq. (3-14)

$$\Delta(\tau/L) = 110(20)$$

$$= 2200 \text{ ps/km} = 2.2 \text{ ns/km}$$

The spread after 10 km is then 22 ns. Changing the wavelength to 1.5 μm results in a material dispersion of 15 ps/ (nm $\times$ km) and Eq. (3-14) then yields $\Delta(\tau/L) = 750$ ps/km. After 10 km the spread is 7.5 ns. The spread is considerably reduced at the longer wavelength, even with the increased source bandwidth.

Example 3-2

Repeat Example 3-1 if the source is a laser diode with a 1-nm spectral width.

Solution:

A decrease in source spectral width by a given factor results in a corresponding decrease in the pulse spread by this same factor, according to Eq. (3-14). The 10-km pulse spreads are then 22/20 = 1.1 ns at 0.82 μm and 7.5/50 = 0.15 ns at 1.5 μm. Use of a more coherent source greatly reduces the amount of material dispersion.

In practice, when operating near the zero dispersion wavelength it is incorrect to completely neglect material pulse spreading. The reasons are that it is unlikely that the light source emits at precisely the zero dispersion wavelength, the emission wavelength will vary with temperature and drive current and thus wander away from the zero point, and a real light source does not emit at a single wavelength but emits over a range of wavelengths. For these reasons a maximum permissible dispersive pulse spread (nonzero) is usually specified for systems designed around 1300 nm. As an example, a maximum spread of 3 ps/km is attainable by using laser diodes whose linewidths are 2 nm or less. The dispersion curve is nearly a straight line between 1200 and 1600 nm. A useful analytic approximation in this range is [2]

$$M = \frac{M_0}{4}\left(\lambda - \frac{\lambda_0^4}{\lambda^3}\right)$$

where the slope M_0 is approximately -0.095 ps/(nm^2 × km) and λ_0 is the zero dispersion wavelength. The minus sign is needed because of the negative slope of the dispersion curve. Some reverse the sign conventions followed in this section, so that their dispersion curves have a positive slope and the minus sign is missing in Eq. (3-14).

Example 3-3

Compute the material dispersion at 1.55 μm if the zero dispersion wavelength is 1.3 μm.

Solution:

It is most straightforward to solve the preceding equation by using the wavelengths expressed in nm. Otherwise, the slope coefficient M_0 would have to be converted into the appropriate units. Thus

$$M = \frac{-0.095}{4}\left(1550 - \frac{1300^4}{1550^3}\right)$$

$$= -18.6 \text{ ps/(nm × km)}$$

a result that checks nicely with the value obtained directly from Fig. 3-8.

Example 3-4

Compute the pulse spread when the light source emits at 1320 nm and has a 2-nm spectral width. The zero dispersion wavelength is 1300 nm.

Solution:

The dispersion turns out to have a magnitude of 1.86 ps/(nm × km), so that Eq. (3-14) yields a spread of $\Delta(\tau/L) = 2 \times 1.86 = 3.72$ ps/km. A 10-km length of this material would produce a pulse spread of only 37.2 ps = 0.0372 ns, considerably smaller than that computed in Examples 3-1 and 3-2 for propagation at wavelengths farther away from the dispersion minimum.

Solitons

Pulse spreading reduces the bandwidth and data capacity of a fiber communications link in the manner described later in this section. Because of this, many techniques for minimizing pulse spreading have been pursued. A few that we already know about are (1) operating at the zero dispersion wavelength and (2) choosing very coherent (small spectral width) light sources. These solutions (often applied together) have been common since the mid 1980s. Improvements now take the form of shifting the fiber's zero dispersion point to wavelengths of lower fiber attenuation and producing more coherent laser sources.

Another technique that shows promise for reducing pulse spreading is the production of *solitons*.[3] A soliton is a pulse that travels along a fiber without changing shape. How can this happen? The actual procedure is fairly complicated, but some insight into soliton propagation can be easily developed. Pulses broaden because dispersion causes some wavelengths emitted by the light source to travel faster than other wavelengths. All we need do is find some property of the fiber that counters this tendency. It turns out that such a property does exist. It is a fiber nonlinearity where the index of refraction depends upon the intensity of the light beam. Since the pulse velocity depends on the index of refraction, it is clear that the intensity of the beam can itself influence the speed of the various wavelengths propagating along the fiber. Usually, this phenomenon is not observed because it is quite small and requires a moderately large amount of optical power before becoming significant.

To form a soliton, the initial pulse must have a particular peak energy and pulse shape. To be specific, the product of pulse energy and pulse width must be a constant. The value of the constant depends on the magnitudes of the dispersion and the nonlinearity. With too little power, the nonlinearity is too weak to be effective in compensating for dispersion. If the power is too great, then the pulse may actually continually change widths as it travels owing to imperfect (and distance dependent) compensation. In addition, the nonlinear compensation is such that solitons are only produced at wavelengths longer than the zero dispersion wavelength in glass fibers. That is, the nonlinearity acts with dispersion to further broaden pulses at the shorter wavelengths and only compensates at the longer ones. We conclude that soliton pulses can be expected in silica fibers only when operating in the 1300- to 1600-nm range.

Although solitons retain pulse widths during propagation, solitons do attenuate just like other waves. It will be imperative for long systems that the optical beam be amplified periodically so that the pulse energy not fall below that required for soliton maintenance.

Various optical amplifiers (to be described in Section 6-7) are candidates for the amplification process.

Soliton widths of a few picoseconds are realizable. The corresponding data rates (the inverse of the soliton widths) are over 10 Gbps. Multigigabit per second systems covering several thousand kilometers with amplifier spacings of several tens of kilometers can be envisioned with soliton pulses. The product of data rate and fiber path length for such systems is far greater than can be achieved by more conventional fiber techniques.

Information Rate

Pulse spreading limits the information capacity of any transmission system in the manner described in what follows. For numerical calculations we will use the spreads generated by material dispersion. The equations developed apply regardless of the cause of the distortion. We will investigate the limits on both analog and digital links. Without long and complex derivations, exact results cannot be obtained. Reasonable limits can be developed based on approximate intuitive analyses. The results obtained will be useful in first-order design and will deepen understanding of the ability of fiber links to carry information.

First, consider a sinusoidally modulated beam of light (like that shown in Fig. 3-5). The modulation frequency is f and the period is $T = 1/f$. Suppose that the source radiates optic wavelengths between λ_1 and λ_2. How much delay between the fastest and slowest wavelength is acceptable? Figure 3-9 shows the received power at λ_1 and λ_2 when the delay is equal to half the modulation period; that is,

$$\Delta\tau = \frac{T}{2} \qquad (3-15)$$

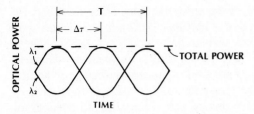

Figure 3-9   Canceling of the modulation when two carrier wavelengths have a delay of half the modulation period. $\Delta\tau = T/2$.

With this amount of delay, the modulation cancels out completely when the two waves are added. Modulated power carried at wavelengths between λ_1 and λ_2 will have delays smaller than $T/2$ and will partially cancel, resulting in a small signal variation at the receiver. If we take Eq. (3-15) as the maximum allowable pulse spread (that is, require that $\Delta\tau \le T/2$), then the modulation frequency is limited by

$$f = \frac{1}{T} \le \frac{1}{2\Delta\tau}$$

The upper frequency determined by this expression turns out to be a good approximation to the 3-dB bandwidth (the modulation frequency at which the signal power diminishes by half). A more analytical approach concludes that $f = 1/(2.27\ \Delta\tau) = 0.44/\Delta\tau$. This result assumes a particular impulse response characteristic (Gaussian), which approximates the behavior of actual fibers. It makes little difference in initial system design which of these similar results are used. In either case, a bandwidth margin should be included to account for the difference between the actual and assumed fiber response. The 3-dB optic bandwidth is now $f_{3\text{-dB}} = (2\Delta\tau)^{-1}$, and the frequency-length limit is

$$f_{3\text{-}dB} \times L = \frac{1}{2\ \Delta(\tau/L)} \quad (3\text{-}16)$$

The attenuation of a transmission medium as a function of modulation frequency appears in Fig. 3-10. The total loss (in decibels) is $L_a + L_f$, where L_a is the fixed loss (mainly owing to absorption and scattering) and L_f is the modulation-frequency dependent loss (owing to pulse spreading). For the Gaussian response, L_f can be modeled by

$$L_f = -10\ \log\left\{ \exp\left[-0.693\left(\frac{f}{f_{3\text{-dB}}}\right)^2 \right] \right\} \quad (3\text{-}17)$$

For $f \ll f_{3\text{-dB}}$, L_f is negligible.

As determined from Eq. (3-17), the loss is 1.5 dB at a frequency equal to $0.71\ f_{3\text{-dB}}$. That is,

$$f_{1.5\text{-dB}} = 0.71\ f_{3\text{-dB}}$$

The 1.5-dB optic bandwidth is important because, as proven later in Section 12-1, it corresponds to the frequency at which the electrical power in the receiver diminishes by half. Thus, *the optic 1.5-dB bandwidth equals the electrical 3-dB bandwidth*. In equation form,

$$f_{1.5\text{-dB}}(\text{optic}) = f_{3\text{-dB}}(\text{electrical})$$
$$= 0.71\ f_{3\text{-dB}}(\text{optic}) \quad (3\text{-}18)$$

Since $f_{3\text{-dB}}(\text{optic}) = (2\Delta\tau)^{-1}$, we conclude that

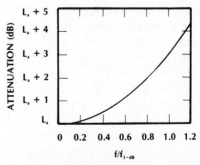

Figure 3-10   Loss dependence on modulation frequency. L_a is the fixed loss.

$$f_{3\text{-}dB}(\text{electrical}) = \frac{0.35}{\Delta\tau}$$

and

$$f_{3\text{-}dB}(\text{electrical}) \times L = \frac{0.35}{\Delta(\tau/L)} \qquad (3\text{-}19)$$

Next, consider a return-to-zero (RZ) digital signal, as illustrated in Fig. 3-11. Each bit is allocated a time T. The data rate is $R = 1/T$ bps. In this format, pulses occupy half of the time slot. The pulse duration is $T/2$. The spectrum (frequency content) of these pulses is sketched on the figure. The RZ signal is adequately transmitted by a system having a bandwidth of $1/T$ Hz because most of the signal power lies below this frequency. We can reach this same conclusion by approximating the RZ signal by a sinusoid. A system passing this sinusoid should transmit the actual pulses without excessive deterioration. As drawn in Fig. 3-11, the approximating sinusoid has frequency $1/T$, verifying the bandwidth requirement.

To be conservative, we will use the electrical 3-dB frequency for the system bandwidth. Applying Eq. (3-19),

$$R_{RZ} = \frac{1}{T} = f_{3\text{-}dB}(\text{electrical}) = \frac{0.35}{\Delta\tau}$$

or

$$R_{RZ} \times L = \frac{0.35}{\Delta(\tau/L)} \qquad (3\text{-}20)$$

We can also obtain this last result by assuming an allowable pulse spread equal to

70% of the pulse duration. Since the RZ pulse duration is half the repetition period, this condition yields $\Delta\tau = 0.7T/2 = 0.35T$. Then, $R = 1/T = 0.35/\Delta\tau$, as before. Adjacent pulses are well separated by requiring that the pulse spread be less than 35% of the time slot. When this is not accomplished, portions of the pulse may spread into the neighboring time slot, producing *intersymbol interference* and increasing the probability of detection errors.

Finally, consider a non-return-to-zero (NRZ) digital signal, as drawn in Fig. 3-12. The time allotted for each bit is T and the data rate is $1/T$. The spectrum of this signal is sketched in the figure. The required transmission bandwidth is $1/2T$, just half that of the RZ system. This follows because the NRZ pulses are twice as long as the RZ pulses, and the bandwidth of a pulse is inversely proportional to the pulse duration. A sinusoid that approximates the NRZ signal is drawn in the figure for the case of alternating ones and zeros. This situation produces the quickest variations and, consequently, the highest frequencies. The approximating sinusoid has period $2T$ and frequency $1/2T$, verifying the bandwidth requirement. We conclude that the maximum allowed data rate is $R = 1/T = 2f$, where f is the system bandwidth. By using the 3-dB electrical bandwidth from Eq. (3-19), we obtain

$$R_{NRZ} = 2f_{3\text{-}dB}(\text{electrical}) = \frac{0.7}{\Delta\tau}$$

or

$$R_{NRZ} \times L = \frac{0.7}{\Delta(\tau/L)} \qquad (3\text{-}21)$$

The allowable pulse spread is 70% of the pulse duration T for the NRZ pulse train.

Example 3-5

For the conditions stated in Examples

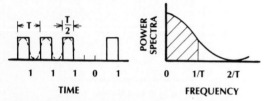

Figure 3-11 Return-to-zero signal and its power spectra. The dashed curve is the approximating sinusoid. The crosshatched region indicates the required transmission bandwidth.

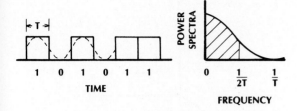

Figure 3-12   Non-return-to-zero signal and its power spectra. The dashed curve is the approximating sinusoid. The crosshatched region indicates the required transmission bandwidth.

3-1 and 3-2, compute the rate-length and frequency-length products.

Solution:

The pulse spreads, taken from the two preceding examples, are summarized in Table 3-2. These data are used in Eqs. (3-16), (3-19), (3-20), and (3-21) to produce the results in the last few columns of the table.

Example 3-6

What are the frequency and data limits for a 10-km link for the sources listed in Table 3-2?

Solution:

We simply divide the frequency-length and rate-length products in the table by 10. For the LED at 0.82 μm, we find

$$f_{3\text{-}dB} = 23 \text{ MHz}$$

$$R_{NRZ} = 32 \text{ Mbps}$$

$$f_{3\text{-}dB}(\text{electrical}) = 16 \text{ MHz}$$

$$R_{RZ} = 16 \text{ Mbps}$$

For the LED at 1.5 μm,

$$f_{3\text{-}dB} = 67 \text{ MHz}$$

$$R_{NRZ} = 94 \text{ Mbps}$$

$$f_{3\text{-}dB}(\text{electrical}) = 47 \text{ MHz}$$

$$R_{RZ} = 47 \text{ Mbps}$$

For the LD at 0.82 μm,

$$f_{3\text{-}dB} = 455 \text{ MHz}$$

$$R_{NRZ} = 637 \text{ Mbps}$$

$$f_{3\text{-}dB}(\text{electrical}) = 320 \text{ MHz}$$

$$R_{RZ} = 320 \text{ Mbps}$$

For the LD at 1.5 μm,

$$f_{3\text{-}dB} = 3.3 \text{ } GHz$$

$$R_{NRZ} = 4.7 \text{ Gbps}$$

$$f_{3\text{-}dB}(\text{electrical}) = 2.33 \text{ GHz}$$

$$R_{RZ} = 2.33 \text{ Gbps}$$

TABLE 3-2.   Information Capacity Examples[a]

Source	λ (μm)	$\Delta\lambda$ (nm)	$\Delta(\tau/L)$ (ns/km)	Optic $f_{3\text{-}dB} \times L$ (GHz × km)	$R_{NRZ} \times L$ (Gbps × km)	Electrical $f_{3\text{-}dB} \times L$ (GHz × km)	$R_{RZ} \times L$ (Gbps × km)
LED	0.82	20	2.2	0.23	0.32	0.16	0.16
LED	1.5	50	0.75	0.67	0.94	0.47	0.47
LD	0.82	1	0.11	4.55	6.4	3.2	3.2
LD	1.5	1	0.015	33.33	46.7	23.3	23.3

[a] Limited by material dispersion in silica.

The results displayed in Table 3-2 dramatically illustrate the advantages of operation at the longer wavelengths and the superiority of the laser diode over the LED for high data rates. Systems using laser diodes in the long-wavelength region are more complex and costly than LED systems in the shorter wavelength range 0.8–0.9 μm, so they are used only when necessary to achieve higher performance. The tabulated rates are fairly high. They will be lower in some systems because of additional pulse spreading caused by modal distortion, as described in Section 5-6.

The information limits in Eqs. (3-16), (3-19), (3-20), and (3-21) apply whether the pulse spread is due to material dispersion or other causes. These results are approximate because of the assumptions made in developing them. They do, however, yield reasonable values for initial system design. They are also important because they show the relationships between pulse spreading and the allowed digital data rates and analog modulation frequencies.

3-3 POLARIZATION

The electric field of a light beam has several directions associated with it. One of these, the direction of travel, has already been discussed with regard to phase shift, wavelength, velocity, and attenuation of the propagating wave. The other direction is that of the electric field vector itself. Figure 3-13 shows the relationship between the vector E and the direction of travel for a simple plane wave. The wave travels in the z direction, and the electric field vector points in the x direction. An electric field that points in just one direction is said to be *linearly polarized*, because it always points along the same single line.

The electric vector is always perpendicu-

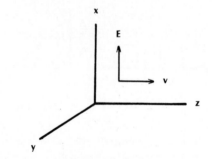

Figure 3-13 Wave traveling in the z direction having its electric field polarized in the x direction.

lar to the direction of travel for a plane wave in an unbounded medium. This being so, the field in Fig. 3-13 could also point in the y direction while traveling in the z direction. The actual direction of polarization is determined by the polarization of the light source and by any polarization-sensitive elements through which the beam passes. It is also possible for two waves to simultaneously travel in the z direction, one polarized in the x direction and one polarized in the y direction. These two waves would be independent of each other because of their orthogonal polarization. The term *mode* refers to the different ways a wave can travel in a given direction. The two independent waves just described are the two plane-wave modes of an unbounded medium. It might occur that other modes are possible, having polarizations in the xy plane at some angle to the x or y axis. Any electric field vector can be decomposed into its x and y components, so that such a field is simply the combination of the two modes already described.

A wave is *unpolarized* if its electric vector varies randomly in direction. Waves in most optic fibers are unpolarized.

In a guided structure, such as an optic fiber, many modes can exist. Polarization is just one of the differences among modes in a waveguide. Modes will be investigated in Chapters 4 and 5. They play an extremely important part in determining the design and capabilities of an optic communications system.

3-4 RESONANT CAVITIES

A radio-frequency oscillator consists of an amplifier, a tuned circuit, and a feedback mechanism. The feedback connects the amplifier output to its input, causing the signal to increase as it periodically passes through the amplifier. A short time after being turned on, a steady state is reached where the system losses (the power extracted from the oscillator as useful output, plus any other losses, such as those caused by heating) are just made up by the gain through the amplifier. After this point, the oscillator maintains a constant output power. The tuned circuit determines the oscillation frequency.

A laser is a very high frequency oscillator. It may correctly be referred to as an *optic oscillator*. Its components have functions paralleling those of lower frequency oscillators. The laser sketched in Fig. 3-14 consists of a cylindrically shaped medium with mirrors attached at each end. The medium provides the amplification. Light is amplified in this material by the mechanism described in Chapter 6. Properties of the medium determine the output frequency and spectral width of the laser.

In this section we are primarily interested in the purpose of the mirrors. The mirrors provide feedback for the light oscillator, reflecting the light back and forth through the amplifying medium. Power exits the laser through one of the mirrors, which is partially transmitting. In some lasers, both mirrors transmit, allowing power to be obtained from both ends of the device. This construction is valuable for laser diodes in fiber systems. Light from one emitting end is coupled to the transmitting fiber, and light from the other end is measured to monitor the source status. Fluctuations in source power are quickly determined, and automatic corrections in the drive circuit return the laser to the required output level.

The two mirrors in Fig. 3-14 form a cavity (called a *Fabry-Perot* resonator) within which two waves exist, one moving to the right and one moving to the left. These waves are drawn at various times in Fig. 3-15 for a cavity of length L. The top figures show the wave moving to the right, and the middle ones show the wave moving to the left. The total field in the cavity is the sum of the two moving waves and is shown in the bottom figure at the times indicated. These drawings illustrate the ways in which electromagnetic waves can interfere with each other. When waves have the same phase, they add *constructively*. This is the condition at times t_1 and t_3 on the figure. The total field is greater than either of its components. When waves are 180° out of phase, as at time t_2, they interfere *destructively*. The total field is zero when waves of equal amplitude interfere destructively. This is an example of the wavelike behavior of light. If we draw the total wave for all periods of time on the same figure, we find a repeating pattern of peaks and nulls. This results in the stationary *standing-wave pattern* shown in Fig. 3-16. At certain points the field is always zero. At other points, the field oscillates within the envelope drawn in the figure. The envelope itself is stationary. This is the same phenomenon that occurs when a string, fixed

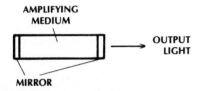

Figure 3-14 A laser consists of an amplifying medium and two end mirrors.

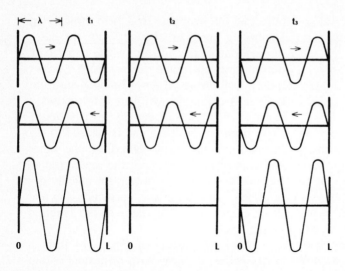

Figure 3-15 Optical waves in a cavity of length L at various times: $t_3 > t_2 > t_1$. The top figures portray the wave moving to the right; the middle ones, the wave moving to the left; the bottom ones, the total wave.

at its ends, is plucked. The vibrations form a standing-wave pattern, with peaks and nulls along the length of the string.

To produce a stationary standing-wave pattern, the cavity must be an integral number of half wavelengths long. That is,

$$L = \frac{m\lambda}{2} \qquad (3\text{-}22)$$

where λ is the wavelength as measured in the material within the cavity, and m is a positive integer. The picture in Fig. 3-15 is drawn for a cavity that is two wavelengths long, as can be seen by counting two complete cycles of the wave over the cavity length. Since $L = 2\lambda$, then $m = 4$ in Eq. (3-22). Only wavelengths satisfying Eq. (3-22) can exist inside the cavity in a steady state. Waves of other lengths, launched into the cavity, inter-

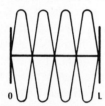

Figure 3-16 Standing-wave pattern in a cavity.

fere destructively with each other as they pass back and forth between the mirrors. These waves attenuate very quickly. We say that the cavity is *resonant* at wavelengths satisfying Eq. (3-22). These are

$$\lambda = \frac{2L}{m} \qquad (3\text{-}23)$$

Equation (3-22) can be developed by reasoning that the phase shift for a wave completing a round trip of the cavity should be an integral number of 2π radians if the pattern is to repeat itself. From Section 3-1, the phase shift is kz where $k = 2\pi/\lambda$ and z is the path length. For a complete round trip, the resonance condition is then $k2L = m2\pi$, leading directly to Eq. (3-22).

According to Eq. (3-23), cavities are resonant at a number of wavelengths or frequencies. The resonant frequencies are found by combining Eqs. (3-23) and (1-3) with the relationship $v = c/n$. The result is

$$f = \frac{mc}{2nL} \qquad (3\text{-}24)$$

where n is the refractive index of the material

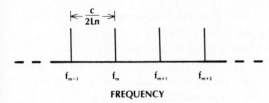

Figure 3-17 Cavity resonant frequencies.

within the cavity. The various resonant frequencies, depicted in Fig. 3-17, are the *longitudinal modes* of the cavity. The spacing between adjacent cavity longitudinal modes is

$$\Delta f_c = \frac{c}{2Ln} \qquad (3\text{-}25)$$

We will need the corresponding free-space wavelength spread $\Delta\lambda_c$. It is found by using the relationship $\Delta f_c / f = \Delta\lambda_c / \lambda_0$, where λ_0 is the free-space value of the mean wavelength and f is the mean frequency. Because $f = c/\lambda_0$ we find

$$\Delta\lambda_c = \frac{\lambda_0^2}{c} \Delta f_c \qquad (3\text{-}26)$$

Example 3-7

Compute the frequency spread and wavelength spread between longitudinal modes for a cavity filled with aluminum gallium arsenide (AlGaAs), which is 0.3 mm long. This structure is typical of an AlGaAs laser diode whose mean (center) wavelength is 0.82 μm and whose refractive index is 3.6.

Solution:

From Eq. (3-25), the mode spacing is

$$\Delta f_c = \frac{3 \times 10^8}{2(0.3 \times 10^{-3})(3.6)}$$

$$= 139 \times 10^9 \text{ Hz}$$

and from Eq. (3-26), the wavelength spread is

$$\Delta\lambda_c = \frac{(0.82 \times 10^{-6})^2(139 \times 10^9)}{3 \times 10^8}$$

$$= 3.11 \times 10^{-10} \text{ m}$$

or

$$\Delta\lambda_c = 0.311 \text{ nm}$$

We have previously stated that laser diodes have spectral widths of 1 to 5 nm. Suppose in Example 3-5 that the width is $\Delta\lambda = 2$ nm. This means that the AlGaAs medium has sufficient amplification for laser oscillation between 819 and 821 nm, as indicated in Fig. 3-18. The cavity will allow only the existence of waves within this range that are resonant. Because the resonances are spaced by 0.311 nm, there will be $\Delta\lambda/\Delta\lambda_c = 2/0.311 \simeq 6$ distinct wavelengths in the output. These six longitudinal modes are illustrated in Fig. 3-18. The individual modes could only have zero width if the mirrors were perfectly reflecting. Since this is not the case in practice, the modes in Fig. 3-18 were drawn slightly broadened. Because pulse distortion caused by material dispersion depends mainly on the spread between the highest and lowest wavelengths emitted by a source, the exact distribution of power among the longitu-

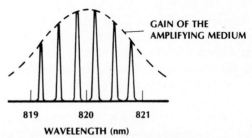

GAIN OF THE
AMPLIFYING MEDIUM

WAVELENGTH (nm)

Figure 3-18 Output power of a laser diode (solid curve), showing six longitudinal modes and a total spectral width of nearly 2 nm.

dinal modes is unimportant. However, if the resonator is designed so that only one longitudinal mode exists, then a significant reduction in source output bandwidth is achieved, and reduced pulse spread can be obtained. Methods exist for producing such *single-longitudinal-mode* lasers. Their added complexity makes them more expensive than multimode laser diodes.

3-5 REFLECTION AT A PLANE BOUNDARY

Problems concerning the amount of light reflected at a boundary between two dielectrics are an important part of the study and practice of optics. These problems are particularly critical in the design and analysis of fiber systems. Reflecting surfaces occur in the situations illustrated in Fig. 3-19. These are

1. The air-to-glass boundary where light is coupled from a source into a fiber
2. The interface between the fiber core and its surrounding layer
3. The two air-glass boundaries where there is an air gap between two fibers being connected

Light reflected at the input and at the connector gap should be small, because these reflections reduce the power being transmitted. Include these losses in calculations of the total system power budget. On the other hand, the internal reflection at the core boundary (point B in Fig. 3-19) should be high to keep the light inside the fiber. We will determine the amounts of reflection in this section.

The simplest computations for reflection loss are those for which the incident beam is traveling normal to the boundary, as in Fig. 3-20. The *reflection coefficient* ρ is the ratio of the reflected electric field to the incident electric field. For normal incidence, it is[4]

$$\rho = \frac{n_1 - n_2}{n_1 + n_2} \qquad (3\text{-}27)$$

where n_1 is the refractive index in the incident region and n_2 is the index in the transmitted region. If $n_2 > n_1$, then the reflection coefficient becomes negative. This indicates a 180° phase shift between the incident and reflected electric fields.

The *reflectance R* is the ratio of the reflected beam intensity to the incident beam intensity. Because the intensity in an optic beam is proportional to the square of its electric field, the reflectance is equal to the square of the reflection coefficient. Thus

$$R = \left(\frac{n_1 - n_2}{n_1 + n_2}\right)^2 \qquad (3\text{-}28)$$

Example 3-8

For an air-to-glass interface, compute the fraction of reflected and transmitted power. Also, compute the transmission loss in decibels. Use 1.5 for the refractive index of glass.

Solution:

From Eq. (3-28),

$$R = \left(\frac{1 - 1.5}{1 + 1.5}\right)^2 = 0.04$$

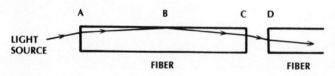

Figure 3-19 Reflecting surfaces in a fiber system. Light rays are reflected at the input (A), at the core interface (B), and at the boundaries of an air gap formed at a connector or splice (C or D).

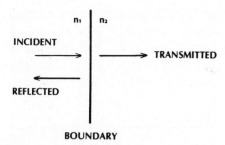

Figure 3-20   A wave incident on a plane boundary between two dielectrics (refractive indices n_1 and n_2) is partially transmitted and partially reflected.

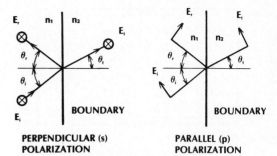

PERPENDICULAR (s)
POLARIZATION

PARALLEL (p)
POLARIZATION

Figure 3-21   Reflection at a boundary is divided into the two polarizations shown. The circled crosses represent vectors pointing into the paper.

so that 4% of the light is reflected. The remainder, 96%, is transmitted. The transmission loss in decibels is then $-10 \log_{10} 0.96 = 0.177$ dB.

Roughly speaking, there will be about a 0.2-dB loss when light enters glass from air. Because of the symmetry of Eq. (3-28), the same loss will occur when light goes in the opposite direction, from glass into air.

In Chapter 2 we studied the relationships among the angles of incidence, reflection, and transmission for arbitrary directions of the incoming wave. Figure 2-1 illustrates the problem. For reference purposes, we define the *plane of incidence* as the plane specified by the normal to the boundary and the direction of travel of the incident wave. In Fig. 2-1 the plane of incidence is the plane of the figure itself. The fraction of light reflected depends on the angle of incidence and on the polarization of the electric field relative to the plane of incidence. We have previously noted that the electric field vector is perpendicular to the direction of travel. The reflection coefficient depends on whether the electric field is polarized perpendicular or parallel to the plane of incidence. We call the perpendicular wave *s* polarization and the parallel wave *p* polarization. Figure 3-21 illustrates the two cases. Any incident field can be decomposed into its *p* and *s*

components. The reflection coefficients for the *p* and *s* cases, known as *Fresnel's laws of reflection,* are[5]

$$\rho_P = \frac{-n_2^2 \cos \theta_i + n_1 \sqrt{(n_2^2 - n_1^2 \sin^2 \theta_i)}}{n_2^2 \cos \theta_i + n_1 \sqrt{(n_2^2 - n_1^2 \sin^2 \theta_i)}}$$

(3-29)

(parallel polarization), and

$$\rho_s = \frac{n_1 \cos \theta_i - \sqrt{(n_2^2 - n_1^2 \sin^2 \theta_i)}}{n_1 \cos \theta_i + \sqrt{(n_2^2 - n_1^2 \sin^2 \theta_i)}}$$

(3-30)

(perpendicular polarization).

Although somewhat formidable in appearance, these equations are easily evaluated when the two indices of refraction, the incident angle, and the polarization are known. The importance of Eqs. (3-29) and (3-30) cannot be understated, because they predict the phenomenon by which dielectric fibers guide light.

The reflectance is found by squaring the magnitudes of the reflection coefficients. That is, $R = |\rho|^2$. Results are shown in Fig. 3-22 for an air-to-glass interface and in Fig. 3-23 for a glass-to-air interface. The general characteristics shown on the figures appear when there are reflections between any two dielectrics. Some interesting, and perhaps unex-

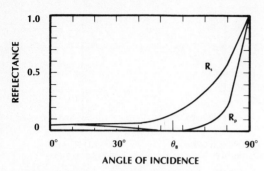

Figure 3-22 Reflectance at an air-to-glass interface. $n_1 = 1.0$, $n_2 = 1.5$.

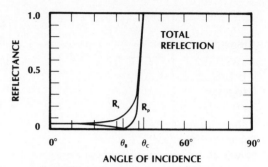

Figure 3-23 Reflectance at a glass-to-air interface. $n_1 = 1.5$, $n_2 = 1.0$.

pected, features can be noted. Three of these features follow:

1. The reflectance does not vary a great deal for incident angles near zero. For the air-glass interface, the reflectance value of 4%, calculated for normal incidence, is a good approximation for angles as large as 20°.
2. The reflectance is zero, meaning full transmission, for certain incident angles and polarization states.
3. The reflectance is unity, indicating total reflection, for a range of incident angles.

First consider the case of zero reflection. Figures 3-22 and 3-23 show that it occurs only for the parallel polarization. The reflection coefficient ρ_p (and thus the reflectance $|\rho_p|^2$ will be zero when the numerator of Eq. (3-29) is zero. This occurs at an incident angle θ_B, called the *Brewster angle*, satisfying the equation

$$\tan \theta_B = \frac{n_2}{n_1} \qquad (3\text{-}31)$$

There is no incident angle that will make ρ_s in Eq. (3-30) zero. The Brewster angle is useful for transmitting a light beam into (or out of) a dielectric without reflection losses. A specific application is shown in Fig. 3-24, where glass windows, at the ends of a helium-neon gas laser tube, are placed at the Brewster angle. The light beam, polarized in the parallel plane, will pass back and forth between the mirrors without reflection losses at the windows.

Example 3-9

Find the Brewster angle for the air-to-glass and glass-to-air interfaces.

Solution:

Using Eq. (3-31) for air-to-glass, $\tan \theta_B = 1.5$, so that $\theta_B = 56.3°$. For glass-to-air, $\tan \theta_B = (1.5)^{-1}$, or $\theta_B = 33.7°$.

Returning now to Fig. 3-19, there is a reflection loss of 0.2 dB at the input to the

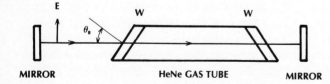

MIRROR   HeNe GAS TUBE   MIRROR

Figure 3-24 Brewster-angled window (W) at the ends of a helium-neon gas laser tube.

fiber. In Chapter 8, where we cover source coupling more thoroughly we will find additional (considerably larger) losses. There is a 0.2-dB loss at each of the two interfaces of the fiber-to-fiber connection. The total reflection loss is 0.4 dB. Further analysis of this connector also appears in Chapter 8. At the interface between the fiber core and its surrounding layer, we have total reflection. Total reflection is so important that we will devote the entire following section to it.

The amount of light reflected when a beam moves from one material to another can be reduced by placing a thin coating layer between them, as shown in Fig. 3-25. If the coating thickness is one quarter of a wavelength (where the wavelength is measured in the middle layer), than the reflectance is

$$R = \frac{[n_1 n_3 - n_2^2]^2}{[n_1 n_3 + n_2^2]^2} \qquad (3\text{-}32)$$

This result shows that the reflectance becomes zero if the index of the coating layer is

$$n_2 = \sqrt{n_1 n_3} \qquad (3\text{-}33)$$

A coating that reduces the reflectance is an *antireflection coating*. Unfortunately, we cannot always find a transparent material with the precise index of refraction required by Eq. (3-33). However, any transparent material with a refractive index lying somewhere between the indices n_1 and n_3 will lower the reflection.

Example 3-10

For an air-to-glass interface, compute the refractive index of the coating required to yield zero reflection. Next, assume the coating material is magnesium fluoride and compute the fractional amount of light reflected and the coating thickness if the illuminating light wavelength is 0.8 μm.

Solution:

From Eq. (3-33) we find that $n_2 = (1.5)^{0.5} = 1.225$ for zero reflection. Unfortunately, there is no suitable material with this index. From Table 2-1, we find that the refractive index of magnesium fluoride is 1.38. By using this for n_2 in Eq. (5-32) yields

$$R = \frac{[1.5 - 1.38^2]^2}{[1.5 + 1.38^2]^2} = 0.014$$

Thus, the reflection has been reduced from 4% without the coating to about 1.4% with it. The wavelength in the magnesium fluoride, computed from Eq. (3-7), is 0.8/1.38 = 0.5797. A quarter wavelength layer has thickness 0.145 μm.

The analyses just presented are correct if the boundary surface is smooth. Reflections from a smooth surface are *specular*. They occur when surface deviations from flatness are small when compared with the wavelength of the incident light. If the surface is rough, then the incident light is scattered over a wide range of directions. This is *diffuse* reflection. Diffuse reflections do not obey Snell's law of refraction or Fresnel's laws of reflection.

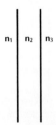

Figure 3-25   Antireflection coating.

3-6 CRITICAL-ANGLE REFLECTIONS

As drawn in Fig. 3-23, there is total reflection for incident angles greater than a particular value, labeled θ_c. θ_c is called the *critical angle*. It is easily determined from Eqs. (3-29) and (3-30) by noting that $|\rho_p| = 1$ and $|\rho_s| = 1$, when $n_2^2 - n_1^2 \sin^2 \theta_i = 0$. The angle satisfying this equation is the critical angle, so that

$$\sin \theta_c = \frac{n_2}{n_1} \qquad (3\text{-}34)$$

Because the sine of an angle is never greater than 1, it is clear from this result that critical angle reflections occur only when $n_1 > n_2$, that is, when a wave travels from a region of higher refractive index into a region of lower index. This explains the occurrence of a critical angle in Fig. 3-23 (glass-to-air boundary), but not in Fig. 3-22 (air-to-glass boundary). We should emphasize that Eq. (3-34) is independent of the wave polarization. It is valid for both parallel and perpendicular directions of the electric field vector.

For angles greater than the critical angle, $\sin \theta_i$ will be greater than $\sin \theta_c$, so that $n_1^2 \sin^2 \theta_i > n_2^2$. The factors under the square root signs in Eqs. (3-29) and (3-30) will now be negative. Because the square root of a negative number is imaginary, both ρ_p and ρ_s take the form

$$|\rho| = \frac{|A - jB|}{|A + jB|}$$

where A and B are real numbers and j indicates an imaginary term. Because the magnitudes of $A - jB$ and $A + jB$ are both $\sqrt{(A^2 + B^2)}$, the magnitude of ρ is unity. The reflectance $R = |\rho|^2$ is then unity for all angles satisfying $\theta_i \geq \theta_c$.

An alternative and instructive development of total reflection involves Snell's law. We will consider a glass-to-air boundary and find the angle of transmission for all incident angles from Eq. (2-3), $\sin \theta_t = (n_1 / n_2) \sin \theta_i$. The result is plotted in Fig. 3-26. As shown in the figure, the transmission angle increases faster than the incident angle. It reaches 90° when $\sin \theta_i = n_2 / n_1$, precisely the critical-angle condition, Eq. (3-34). By referring to Fig. 3-27, the meaning of a 90° transmission angle is clear. The transmitted wave no longer propagates into the second medium. We conclude that all the light must reflect back into the first medium. Perfect reflection at a dielectric-dielectric boundary is called *total internal reflection*.

Critical angles, computed from Eq. (3-34) for several combinations of materials, are listed in Table 3-3. The plastic-plastic boundary in the table is typical of an all-plas-

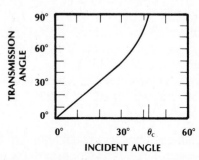

Figure 3-26   Transmission angle for a glass-to-air interface. $n_1 = 1.5$, $n_2 = 1.0$.

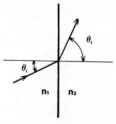

Figure 3-27   As θ_i increases, θ_t approaches 90° if $n_1 > n_2$.

TABLE 3-3. Critical Angles

Boundary	n_1	n_2	θ_c
Glass–air	1.5	1.0	41.8°
Plastic–plastic	1.49	1.39	68.9°
Glass–plastic	1.46	1.4	73.5°
Glass–glass	1.48	1.46	80.6°

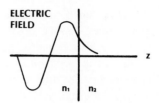

Figure 3-28 A standing wave and an evanescent wave exist on opposite sides of a totally reflecting boundary.

tic fiber whose core and surrounding cladding have different refractive indices. The glass-plastic entry corresponds to a fiber having a glass core surrounded by plastic. The glass-glass boundary is typical of an all-glass fiber in which the core and cladding have slightly different compositions and, thus, slightly different refractive indices. These fibers guide light by totally reflecting the rays that strike their boundaries. The rays must be at, or beyond, the critical angle to be guided without loss, however.

Interference between the incident and reflected waves creates a standing wave in the incident region, as pictured in Fig. 3-28. Although all the power is reflected, a field still exists in the second medium. The amplitude of this field diminishes with distance from the boundary, as indicated in the figure. This result is not inconsistent with total reflection, because no power propagates away from the boundary into the second medium. A field such as this, fading away and carrying no power, is termed *evanescent*. The evanescent electric field decays exponentially according to the expression $e^{-\alpha z}$, where the attenuation factor α has the value

$$\alpha = k_0 \sqrt{(n_1^2 \sin^2 \theta_i - n_2^2)} \qquad (3\text{-}35)$$

and k_0 is the free-space propagation factor. At the critical angle, $\sin \theta_i = n_2/n_1$, making $\alpha = 0$. There is no attenuation of the field. As θ_i increases beyond θ_c, α becomes larger and the fields decay faster. We can now state

the following important conclusion: Rays incident at angles greater than, yet close to, the critical angle produce evanescent waves that decay slowly and penetrate deeply into the second medium, and rays incident far above the critical angle produce waves that disappear after only a short penetration into the second medium.

The reflection coefficient, computed from Eq. (3-29) or (3-30), is a complex quantity, having a magnitude and an angle when $\theta_i > \theta_c$. We have shown that the magnitude is unity. The angle represents the phase shift of the reflected wave relative to the incident wave. Its value varies with the incident angle.

3-7 SUMMARY

This chapter concentrated on developing fundamental ideas about lightwaves that apply directly to fiber optics. The wave concepts of amplitude, phase, wavelength, and polarization should be clear. Pulse distortion owing to material dispersion was studied extensively because of its impact on the information handling capacity of fibers.

Other causes of pulse distortion will be considered in Chapter 5. The dependence of information rate on the spectral width of the optic source indicated the importance of this light-emitter property. Cavity resonance was studied because it determines the longitudinal mode structure appearing in the output spectrum of a laser diode. As we shall see in Chapter 4, resonance also explains the mode structure in a dielectric waveguide. Reflections at dielectric boundaries play a major role in fiber optics. Total internal reflection makes it possible for dielectrics to form waveguides for light rays.

To provide a convenient reference, a few of the most important results of this chapter are now summarized:

1. Pulse spread for material dispersion

$$\Delta(\tau/L) = - M\Delta\lambda$$

2. 3 dB optic bandwidth-length product

$$f_{3\text{-dB}} \times L = \frac{1}{2\Delta(\tau/L)}$$

3. 3 dB electrical bandwidth-length product

$$f_{3\text{-dB}} \times L = \frac{0.35}{\Delta(\tau/L)}$$

4. Rate-length product, RZ

$$R_{\text{RZ}} \times L = \frac{0.35}{\Delta(\tau/L)}$$

5. Rate-length product, NRZ

$$R_{\text{NRZ}} \times L = \frac{0.7}{\Delta(\tau/L)}$$

6. Longitudinal mode separation

$$\Delta f_c = \frac{c}{2Ln}$$

7. Reflectance for normal incidence

$$R = \left(\frac{n_1 - n_2}{n_1 + n_2}\right)^2$$

8. Critical angle for total reflection

$$\sin \theta_c = \frac{n_2}{n_1}$$

PROBLEMS

3-1. Consider a pulse emitted at two discrete optical wavelengths. Will the pulse at the longer or shorter wavelength reach the receiver first (in pure silica)?

3-2. For silica, compute the magnitude of the pulse spread per unit length if the source wavelength is 0.85 μm. The spectral width is 30 nm. Repeat for a spectral width of 2 nm.

3-3. Repeat Problem 3-2 for a source wavelength of 1.55 μm. Assume that the material dispersion is $M = -20$ ps/(nm $\times$ km).

3-4. Use the results of Problems 3-2 and 3-3 to compute the maximum data rates and modulation frequencies. Give your answers for 100 m, 1 km, and 10 km and for RZ and NRZ codes.

3-5. Compute the propagation constant in air and in glass if the free-space wavelength is 0.82 μm.

3-6. Compute the fractional bandwidth of a source of 0.82 μm having a spectral width of 1 nm and 20 nm. Compute the bandwidth in hertz.

3-7. Compute the reflectance at an AlGaAs-to-air boundary (at normal incidence). Compute the transmission loss in decibels.

3-8. Plot the reflectance versus incident angle if $n_1 = 1.48$ and $n_2 = 1.46$. Do this for s and p polarizations.

3-9. Prove Eq. (3-31).

3-10. Use Eq. (3-35) to plot the decaying

wave $e^{-\alpha z}$ versus z. $n_1 = 1.48$, $n_2 = 1.46$, wavelength $= 0.82 \ \mu m$, $0 < z < 4 \ \mu m$, and incident angle $\theta_i = 82°$. Repeat the plot (on the same graph) for incident angles of θ_i equal to 84°, 86°, 88°, and 90°.

3-11. Using the frequency-dependent loss expression Eq. (3-17), show that $f_{1-dB} = 0.58 f_{3-dB}$.

3-12. A source emitting two wavelengths (λ_1 and λ_2) is intensity modulated at frequency f_m. The power at λ_1 is $P_1 = P_{01} + P_{11} \cos(\omega_m t + \phi_1)$. The power at λ_2 is $P_2 = P_{02} + P_{22} \cos(\omega_m t + \phi_2)$. Find an expression for the total power. Now assume that $f_m = 1$ kHz. At the receiver $P_{01} = P_{02} = 2 \ \mu W$, $P_{11} = P_{22} = 1 \ \mu W$. Plot P_1, P_2, and total power (as a function of time) if $\phi_2 - \phi_1 = 0$. Repeat if $\phi_2 - \phi_1 = 1.57$ rad. Repeat if $\phi_2 - \phi_1 = 3.14$ rad. Plot the peak-to-peak value of ac power versus $\phi_2 - \phi_1$.

3-13. A parallel polarized ray is incident at an angle of 85° when traveling from a medium of index 1.48 into a medium having index 1.465. The wavelength is 1300 nm. (a) Compute the reflection coefficient. (b) At what distance from the boundary (in the transmission medium) does the evanescent electric field decay to 10% of its value at the boundary?

3-14. As noted in Section 3-1, the power in a light beam is proportional to the square of its electric field and the electric field for a beam traveling in an attenuating medium can be given by Eq. (3-8). The attenuation coefficient in that equation determines the loss. On the other hand, we more frequently discuss the loss in terms of decibels per kilometer. Show that the power change in dB/km and the attenuation coefficient α are related

by $L\,(dB/km) = -8.685\alpha$, where α is given in the units of km^{-1}.

3-15. If the attenuation coefficient for the wave in Eq. (3-8) has a value of $2 \times 10^{-5} \ cm^{-1}$, compute the power loss in dB for 1 km. Compute the fractional loss for 1 km.

3-16. If the loss in a medium is 0.2 dB/km (as it is for a good fiber operating at a wavelength of 1550 nm), then find the attenuation coefficient.

3-17. The modulation-frequency dependent loss of a fiber was given in Section 3-2. If this loss is measured to be 6 dB at a modulation frequency of 2 GHz, then find the 3-dB electrical bandwidth of the fiber.

3-18. Compute the material dispersion for a wavelength 10 nm lower than the zero dispersion wavelength (assumed to be 1300 nm).

3-19. The material dispersion coefficient was given as -0.095 ps/(nm² × km) in Section 3-2. (a) Convert this to units of s/m³. (b) Repeat the calculation for units of ns/(nm² × km).

3-20. Assuming a maximum allowable pulse spread owing to material dispersion of 3 ps/km and a source spectral width of 2 nm, how far can the operating wavelength be from the zero dispersion wavelength?

3-21. Consider soliton pulses having a pulse width of 20 ps at the transmitter. They propagate in a single mode fiber at a wavelength of 1550 nm with no pulse distortion.
   (a) What is the maximum data rate that can be transmitted using these soliton pulses?
   (b) What will limit the length of fiber over which this data can be transmitted?

REFERENCES

1. Good introductions to electromagnetic waves appear in numerous texts. These include William H. Hayt, Jr. *Engineering Electromagnetics,* 4th ed. New York: McGraw-Hill, 1981.
   John D. Kraus. *Electromagnetics,* 3rd ed. New York: McGraw-Hill, 1984.
   G. G. Skitek and S. V. Marshall. *Electromagnetic Concepts and Applications.* Englewood Cliffs, N.J.: Prentice Hall, 1982.
   David K. Cheng. *Field and Wave Electromagnetics.* Reading, Mass.: Addison-Wesley, 1983.

2. Felix P. Kapron. *Fiber Optics Handbook.* Frederick C. Allard, ed. New York: McGraw-Hill, 1990, pp. 4.32–4.33.

3. Dietrich Marcuse. *Optical Fiber Telecommunications II.* Stewart E. Miller and Ivan P. Kaminow, eds. New York: Academic Press, 1988, pp. 90–98.

4. Kraus. *Electromagnetics.* p. 457.

5. Ibid. pp. 515–518.

Chapter 4

Integrated Optic Waveguides

Integrated optics is the technology of constructing optic devices and networks on substrates.[1] It is similar to the construction of integrated electronic circuits. The terms *integrated optoelectronics* and *integrated photonics* are also used to describe this field. *Photonics* itself refers to any system combining optics and electronics. Integrated optics offers the capability of combining optic and electronic components on a single substrate to produce functional systems or subsystems. Integrated components often have dimensions on the order of the light wavelength. This technology has many of the same advantages of integrated circuits: ruggedness, small size, and potential low cost. Complete optic transmitters, receivers, and repeaters can be designed for interconnection over long distances by optic fibers.

Within an integrated optic network, light is transferred between components by a rectangular dielectric slab waveguide. Because of its role in integrated optics and because the slab is so similar to the optic fiber, we will investigate how light propagates in the slab. The study of wave travel in a slab will help us visualize propagation in fibers. We treat the slab before the fiber because the rectangular structure is so much easier to analyze than the circular fiber geometry.

In addition to the slab waveguide, this chapter briefly covers integrated components and coupling to integrated circuits. We also show a few examples of integrated optic network designs.

4-1 DIELECTRIC SLAB WAVEGUIDE

The dielectric slab waveguide appears in Fig. 4-1. The wave travels primarily in the central layer, which has a refractive index n_1.

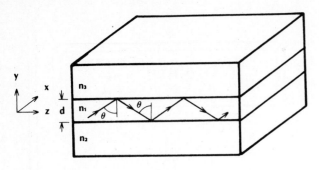

Figure 4-1 Dielectric slab waveguide. $n_1 > n_2$, $n_1 > n_3$.

This layer is so small, often less than a micrometer, that it is referred to as a film. The film is sandwiched between a bottom layer and top layer having indices n_2 and n_3, respectively. Light rays are trapped in the film by total internal reflection. As we found in the preceding chapter, this can occur if both n_2 and n_3 are less than n_1. From Eq. (3-32), the critical angle at the lower boundary is found from

$$\sin \theta_c = \frac{n_2}{n_1} \qquad (4\text{-}1)$$

while the critical angle at the upper boundary is given by

$$\sin \theta_c = \frac{n_3}{n_1} \qquad (4\text{-}2)$$

The angle θ in Fig. 4-1 must be equal to or greater than the largest of these two critical angles if light is to propagate without leaking into the outer layers. To obtain total reflection, the boundaries must be smooth. Otherwise, diffuse reflections will scatter light out of the guiding layer. Similarly, inhomogeneities in the film will scatter light and increase losses. Finally, for efficient transmission the material absorption must be small. Lithium niobate (LiNbO$_3$) and gallium arsenide (GaAs), two popular materials for integrated optics, have losses of around 1 dB/cm and a little more than 2 dB/cm, respectively.[2]

This loss is acceptable over the short lengths involved in an integrated network. The materials used in fibers have much lower loss, as is required for long-distance communications. In the section on critical-angle reflections, we noted that an evanescent field exists beyond the reflecting boundary. Because of this, absorption in the upper and lower layers of the dielectric waveguide must also be low.

The symmetrical structure, where $n_2 = n_3$, is particularly interesting because it closely resembles an optic fiber. The analogous fiber has a core of index n_1 and is surrounded by a cladding of index n_2. The asymmetrical waveguide having $n_3 = 1.0$ is also important. This is the configuration of an integrated optic circuit that is open on top to the air. In this case, n_2 is the refractive index of the substrate. We consider the symmetrical and asymmetrical waveguides in separate sections of this chapter.

The field in the film is a plane wave of the type discussed in Chapter 3, zigzagging back and forth at the angle θ (see Fig. 4-1). Somewhat similarly, we can view the total field as the sum of two uniform plane waves, one traveling upward at an angle θ and one traveling downward at that angle. As presented in Chapter 3, these waves have a propagation factor that can be written as $k = k_0 n_1$, where k_0 is the free-space propagation factor. The propagation factor is drawn in Fig. 4-2 for the two waves. The guided wave's net direc-

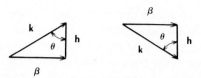

Figure 4-2 Propagation factors for the waves in the slab waveguide. $\beta = k \sin \theta$, $h = k \cos \theta$.

tion of travel is horizontal in this figure. The component of the propagation factor along this direction is

$$\beta = k \sin \theta = k_0 n_1 \sin \theta \qquad (4\text{-}3)$$

We will call this the *longitudinal propagation factor*. Because of the interference between the upward and downward traveling waves, the field is not uniform along the y direction, but varies sinusoidally. This variation is the standing-wave pattern. The field in the film can be written as

$$E = E_1 \cos hy \sin (\omega t - \beta z) \qquad (4\text{-}4a)$$

for modes evenly distributed about the $y = 0$ plane. Fields having an odd distribution also exist, represented by

$$E = E_1 \sin hy \sin (\omega t - \beta z) \qquad (4\text{-}4b)$$

In these equations, E_1 is the peak value of the field and $h = k \cos \theta$ (the vertical component of k). Comparison with Eq.(3-1) for an unguided wave shows the same variation along the direction of travel, except for the replacement of k by β. By making this replacement in Eq. (3-2), we can write the relationship between the waveguide phase velocity v_g and the longitudinal propagation factor. Then,

$$\beta = \frac{\omega}{v_g} \qquad (4\text{-}5)$$

or

$$v_g = \frac{\omega}{\beta}$$

We defined the refractive index as the velocity of light in free space divided by the velocity in an unbounded medium. We now define an *effective refractive index* n_{eff} as the free-space velocity divided by the guided velocity. That is, $n_{\text{eff}} = c/v_g$. Then, using Eq. (4-5), $n_{\text{eff}} = c\beta/\omega$. From Eq. (3-4), the free-space propagation factor is $k_0 = \omega/c$, so that we finally obtain

$$n_{\text{eff}} = \frac{\beta}{k_0} \qquad (4\text{-}6)$$

or, using Eq. (4-3),

$$n_{\text{eff}} = n_1 \sin \theta \qquad (4\text{-}7)$$

The effective refractive index is a key parameter in guided propagation, just like the refractive index is in unguided wave travel. In fact, the effective refractive index changes the wavelength in the same way that the bulk refractive index does. Following Eq. (3-7), the wavelength as measured in the waveguide is

$$\lambda_g = \lambda_0/n_{\text{eff}}$$

Evanescent fields outside the film decay exponentially with an attenuation factor given by Eq. (3-35), with n_3 replacing n_2 for fields in the upper medium and no change for fields in the lower layer. In the upper layer ($y > d/2$),

$$E = E_3 e^{-\alpha(y-d/2)} \sin (\omega t - \beta z) \qquad (4\text{-}8a)$$

In the lower layer ($y < -d/2$),

$$E = E_2 e^{\alpha(y+d/2)} \sin (\omega t - \beta z) \qquad (4\text{-}8b)$$

E_2 and E_3 are the peak values of the fields at the lower and upper boundaries, respectively.

4-2 MODES IN THE SYMMETRIC SLAB WAVEGUIDE

Consider the symmetric waveguide. There will be total reflection for all angles greater than the critical angle and up to 90°. A ray at 90° travels horizontally in Fig. 4-1, straight down the waveguide. Since $\theta = 90°$ for this ray, the effective index is $n_{\text{eff}} = n_1$. We conclude that a ray traveling parallel to the slab has an effective index that depends on the guiding film alone. For a ray at the critical angle, $\sin \theta = n_2/n_1$, so that Eq. (4-7) yields $n_{\text{eff}} = n_2$. The effective index for critical-angle rays depends only on the outer material. Rays at the critical angle travel more steeply, relative to the waveguide axis, than any other trapped rays. We have now determined that the effective refractive index is limited by the indices of the film and its surroundings. All ray angles for propagating waves lie between θ_c and 90°, and the corresponding effective refractive indices are in the range

$$n_2 \leq n_{\text{eff}} \leq n_1 \qquad (4\text{-}9)$$

Mode Condition

It is true that all waves having ray directions between the critical angle and 90° will be trapped within the film by total reflection. It is not true, however, that all these waves will propagate along the structure. In fact, only certain ray directions are allowed. The allowed directions correspond to the modes of the waveguide. We can understand the existence of these modes by analogy with the cavity resonances, covered in Section 3-4. In that section we found that stable interference patterns (the modes of the cavity) occurred only when the phase shift for a complete round trip was equal to an integral number of 2π radians even though all rays were totally reflected. Denoting the round trip phase shift by $\Delta\phi$, the cavity resonance condition can be written as

$$\Delta\phi = m2\pi \qquad (4\text{-}10)$$

where m is an integer. This equation is satisfied by a number of wavelengths for a fixed cavity length. The slab waveguide can also be treated as a cavity, because it has two reflecting boundaries. Instead of waves moving back and forth along the same line, the waves in the slab propagate at some angle. The upward and downward traveling waves still overlap and interfere. The resonance condition, Eq. (4-10), must still be satisfied to obtain a stable interference pattern. In this instance, the phase shift occurs over one complete cycle of the zigzag path, as drawn in Fig. 4-3. This shift is the sum of the phase shift along the path and the phase shift that takes place at each of the two reflecting boundaries. These last two shifts are determined from the reflection coefficient equations. That is, the phase shift due to reflection is the angle of the complex reflection coefficient as computed from either Eq. (3-29) or Eq. (3-30).

We can alter the path length, and thus vary the total phase shift for a fixed wavelength, by changing the ray direction. By do-

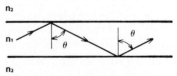

Figure 4-3 One cycle of the zigzag path of a propagating mode. The phase of the wave shifts along its path and at the reflecting boundaries.

ing this we may find that Eq. (4-10) is satisfied for several distinct angles. The waves traveling at these angles are the modes of the waveguide. They are the allowed propagation directions. Waves whose ray angles do not satisfy Eq. (4-10) will diminish quickly owing to destructive interference.

TE and TM Polarization

As in the case of reflection from a plane boundary, we divide the problem into the two possible polarizations, perpendicular and parallel to the plane of incidence. In Fig. 4-1, the yz plane is the plane of incidence. An electric field pointing in the x direction corresponds to the perpendicular, or s, polarization. Waves with this polarization are labeled *transverse electric* (TE) fields because the electric field vector lies entirely in a plane (the xy plane); that is, transverse to the direction of net travel (the z direction). Figure 4-4 illustrates the parallel, or p, polarization. In this case, the electric field is no longer purely transverse. It has a component along the z direction. However, the magnetic field, which points in the x direction for this polarization, is entirely transverse. The p polarization is therefore labeled *transverse magnetic* (TM) in the slab.

TE Mode Chart

For even TE modes (those having even symmetry in the transverse plane), the solution to

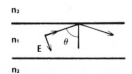

Figure 4-4 The TM wave (p polarization) in the slab waveguide.

Eq. (4-10) is

$$\tan(hd/2) = \frac{1}{n_1 \cos \theta} \sqrt{n_1^2 \sin^2\theta - n_2^2}$$

(4-11)

where $h = k \cos \theta = (2\pi n_1/\lambda) \cos \theta$, and λ is the free-space wavelength. For odd modes, $hd/2$ is replaced by $(hd/2) - (\pi/2)$. If the film thickness is known, then it is difficult to determine the ray angle θ directly from Eq. (4-11). It is simpler to choose various ray angles (between θ_c and 90°) and solve for the corresponding thickness. A plot of the results yields the relationship between thickness and propagation angle. An example will illustrate the method.

For the symmetrical slab, let $n_1 = 3.6$ and $n_2 = 3.55$. These values are characteristic of an AlGaAs double heterojunction laser diode. We will study such a source in Chapter 6. The critical angle for this structure is $\theta_c = \sin^{-1}(n_2/n_1) = 80.4°$. The range of angles for trapped rays is then $80.4° \le \theta \le 90°$, and the refractive index range is $3.55 \le n_{\text{eff}} \le 3.6$. Table 4-1 shows a few of the calculations used in solving Eq. (4-11). The first column in the table is the chosen angle. The effective refractive index, calculated from Eq. (4-7), is in the second column. The right side of Eq. (4-11) is calculated next and appears in the third column. From these values of $\tan(hd/2)$, hd itself is computed. The result, in radians, is listed in the fourth column. Noting that $hd = (2\pi/\lambda)n_1 d \cos \theta$, we can compute d/λ from

$$\frac{d}{\lambda} = \frac{hd}{2\pi n_1 \cos \theta}$$

The denominator of this expression is tabulated in column five. The value of d/λ is then computed by dividing column four by column five. The results are shown in the last column.

TABLE 4-1. TE$_0$ Mode Calculations

θ	n_{eff}	$\tan(hd/2)$	hd	$2\pi n_1 \cos\theta$	d/λ
80.4°	3.550	0	0	3.757	0
82°	3.565	0.651	1.155	3.148	0.367
84°	3.580	1.235	1.780	2.364	0.753
86°	3.591	2.161	2.275	1.578	1.442
88°	3.598	4.653	2.718	0.789	3.445
90°	3.600	∞	3.142	0	∞

If the free-space wavelength is specified, we can then find the thickness. The normalized form, d/λ, is also useful. Results of the calculations in Table 4-1 are plotted in Fig. 4-5 and labeled as the TE$_0$ curve. This type of figure is called a *mode chart*.

We may make several conclusions about the TE$_0$ mode from its mode chart. When the film thickness is very small ($d/\lambda \ll 1$), the ray travels close to the critical angle and the effective index is that of the outer layer, n_2. For the thin film, the wave penetrates deeply into the outer layers, because the rays are near the critical angle. In this case, the evanescent decay is slow, as discussed in Section 3-6. As the thickness increases, the ray travels at a larger angle. That is, the ray travels more nearly parallel to the waveguide axis and the effective refractive index lies between n_1 and n_2. For a very thick film ($d/\lambda \gg 1$), the effective index is that of the film itself. In this case,

the wave in the outer layer decays very rapidly, as discussed in Section 3-6 for evanescent waves traveling at angles far above the critical angle.

Higher-Ordered Modes

Because the tangent function repeats itself, Eq. (4-11) has multiple solutions. For any given value of propagation angle, there is a set of film thicknesses that will allow rays in that direction. In Table 4-1 we took the solution of Eq. (4-11) yielding the smallest value of the normalized thickness, d/λ. Denote this solution as $(d/\lambda)_0$. The other solutions (including both even and odd modes) are then

$$(d/\lambda)_m = (d/\lambda)_0 + \frac{m}{2n_1 \cos\theta} \qquad (4\text{-}12)$$

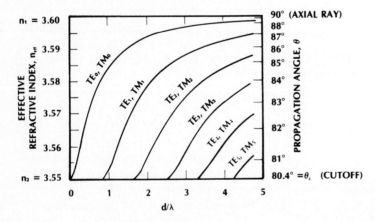

Figure 4-5 Mode chart for the symmetric slab. $n_1 = 3.6$, $n_2 = 3.55$.

where m is a positive integer. Each value of m corresponds to a different allowed mode of the waveguide. The normalized thickness is incremented by the amount

$$\Delta(d/\lambda) = \frac{1}{2n_1 \cos \theta} \qquad (4\text{-}13)$$

between successive modes.

For the symmetrical AlGaAs waveguide, Eq. (4-13) has been computed and added to $(d/\lambda)_0$. The results are tabulated in Table 4-2 for the first six TE modes and plotted on the mode chart of Fig. 4-5. For a fixed thickness and wavelength, the chart shows several solutions. The following example will illustrate this point.

Example 4-1

Find the propagation angles, the effective refractive indices, and the number of TE modes in an AlGaAs waveguide if $d = 1.64$ μm. The free-space wavelength is $\lambda = 0.82$ μm.

Solution:

We first compute that $d/\lambda = 2$. For this value, Fig. 4-5 yields three TE solutions. They are:

TE_0, $n_{\mathrm{eff}} = 3.594$, $\theta = 86.7°$

TE_1, $n_{\mathrm{eff}} = 3.578$, $\theta = 83.7°$

TE_2, $n_{\mathrm{eff}} = 3.557$, $\theta = 81.1°$

Three TE modes can exist simultaneously in this waveguide. The modes travel at different angles and with different effective refractive indices.

In Example 4-1 the TE_3 mode could not propagate because d/λ was not large enough. This mode, and all higher-ordered modes (modes with larger values of m), are cut off. Cutoff occurs when the propagation angle for a given mode just equals the critical angle. Putting this information into Eq. (4-11) yields the condition at cutoff for the mth TE mode

$$(d/\lambda)_{m,c} = \frac{m}{2\sqrt{n_1^2 - n_2^2}} \qquad (4\text{-}14)$$

If d/λ is less than this value, the mth mode will not propagate. We can determine the number of propagating modes allowed by a given film thickness by solving this equation for m. The highest-ordered mode that can propagate has a value of m given by the integer part of

$$m = \frac{2d\sqrt{n_1^2 - n_2^2}}{\lambda} \qquad (4\text{-}15)$$

Because the lowest-ordered mode has $m = 0$, the number of propagating TE modes is the integer value of

$$N = 1 + \frac{2d\sqrt{n_1^2 - n_2^2}}{\lambda} \qquad (4\text{-}16)$$

in the symmetrical slab.

TABLE 4-2. TE$_m$ Mode Calculations

θ	n_{eff}	TE$_0$ $(d/\lambda)_0$	$\Delta(d/\lambda)$	TE$_1$ $(d/\lambda)_1$	TE$_2$ $(d/\lambda)_2$	TE$_3$ $(d/\lambda)_3$	TE$_4$ $(d/\lambda)_4$	TE$_5$ $(d/\lambda)_5$
80.4°	3.550	0	0.836	0.836	1.672	2.508	3.344	4.18
82°	3.565	0.367	0.998	0.365	2.363	3.360	4.358	5.356
84°	3.580	0.753	1.329	2.082	3.410	4.739	6.068	7.397
86°	3.591	1.442	1.991	3.433	5.424	7.415	9.406	11.40
88°	3.598	3.445	3.980	7.425	11.40	15.38	19.36	23.34
90°	3.600	∞	∞	∞	∞	∞	∞	∞

To minimize the number of modes, we can make d/λ small or make n_2 close to n_1. If we wish to propagate only the TE_0 mode, then Eq. (4-14) tells us we must have

$$\frac{d}{\lambda} < \frac{1}{2\sqrt{n_1^2 - n_2^2}} \qquad (4\text{-}17)$$

This cuts off the $m = 1$ mode and all higher-ordered ones.

Example 4-2

Compute the largest thickness that will guarantee single TE mode operation at 0.82 μm in the AlGaAs slab waveguide.

Solution:

By using Eq. (4-17), we find the maximum thickness to be

$$d = \frac{0.82}{2\sqrt{3.6^2 - 3.55^2}} = 0.686 \ \mu\text{m}$$

Notice how thin the film must be if we wish to restrict the waveguide to just one propagating TE mode.

A *multimode waveguide* is one that supports more than one propagating mode. As Fig. 4-5 reveals for such a waveguide, at a fixed thickness the higher-ordered modes propagate at smaller angles than the lower-ordered modes. This means that rays of higher-ordered modes travel more steeply with respect to the waveguide axis than do rays of lower-ordered modes. This situation is illustrated in Fig. 4-6.

TM Mode Chart

Now we will consider the mode chart for TM modes. The solution to Eq. (4-10) for even

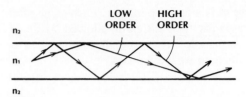

Figure 4-6 Ray paths of high- and low-order modes.

modes having this polarization is

$$\tan (hd/2) = \frac{n_1}{n_2^2 \cos \theta} \sqrt{n_1^2 \sin^2 \theta - n_2^2} \qquad (4\text{-}18)$$

For odd modes, $hd/2$ is replaced by $(hd/2) - (\pi/2)$. Equation (4-18) differs from the TE mode equation. However, if n_1 is close to n_2, then the difference is negligible. This condition is satisfied for the AlGaAs example having $n_1 = 3.6$ and $n_2 = 3.55$, so the TM results are nearly identical to those developed for the TE case. This is why the curves in Fig. 4-5 are labeled as both TE and TM. Each curve represents two modes. They have about the same effective index and propagation angle, but their electric field vectors point in orthogonal directions. Two modes having the same propagation factor are said to be *degenerate*. In the present example, TE and TM modes of the same order are nearly degenerate.

Even when n_1 is not close to n_2, the cutoff values for the TE_m and TM_m mode are the same, so that Eq. (4-14) applies for both cases. It then follows that the number of propagating TM modes equals that of the TE modes as given by the integer value of Eq. (4-16). The total number of allowed modes is twice the number of TE modes found from that equation. It is impossible to obtain single-mode operation by merely making the film small, because both the TE_0 and TM_0 modes propagate for negligibly small thicknesses. A single mode can be obtained in a film obeying Eq. (4-17) by polarizing the incoming light in the direction corresponding to just the TE_0 or just the TM_0 mode. Discontinuities or imperfections in the waveguide may depolarize the light and excite the unwanted mode, so care must be taken when using this technique.

Mode Pattern

The variation of light in the plane transverse to the waveguide axis is the *transverse mode pattern*. According to Eq. (4-4), the electric field in the film varies sinusoidally across the transverse plane. Outside the film there is an exponentially decaying evanescent field, given by Eq. (4-8). Penetration into the outer layer increases as the mode order m increases. This occurs because the ray angle gets closer to the critical angle as m increases, and (as discussed in Section 3-6) wave penetration increases as the ray angle approaches θ_c. For a fixed thickness and wavelength, each mode has a different pattern. A few of these are drawn in Fig. 4-7. From this figure we can identify the integer m as the number of times the electric field crosses through zero in the transverse plane. Also, this figure illustrates the even or odd symmetry of the modes.

In practical waveguides, waves are attenuated by absorption and scattering. Material inhomogeneities and boundary imperfections cause scattering. The steeply angled higher-ordered modes travel farther along their zigzagging paths through the film than do the lower-ordered modes, which travel a more direct route from one end of the waveguide to the other. For this reason the higher-ordered modes suffer greater absorption losses. Scattering causes deviations of the ray paths. Modes close to cutoff (these are the higher-ordered modes) already have their rays close to the critical angle. These rays can easily be

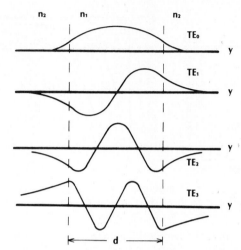

Figure 4-7 Transverse mode patterns in the symmetric slab waveguide.

deflected below the critical angle, at which point the mode energy will radiate into the substrate. Finally, the higher-ordered modes have fields penetrating deeply into the substrate and thus are more susceptible to absorption in that region. We conclude that higher-ordered modes attenuate more quickly than lower-ordered modes.

4-3 MODES IN THE ASYMMETRIC SLAB WAVEGUIDE

The asymmetric slab is the structure commonly used for integrated optic circuits. We will consider a waveguide having $n_1 = 2.29$, $n_2 = 1.5$, and $n_3 = 1.0$. This represents a zinc sulfide (ZnS) film on a glass substrate, with the upper surface of the film open to the air. The value of $n_1 = 2.29$ is appropriate for ZnS at 1.06 μm. Between 0.6 and 1.4 μm, n_1 ranges from 2.37 to 2.28. The critical angle at the ZnS-air interface is $\theta_c = \sin^{-1}(1/2.29) = 25.9°$, while at the ZnS-glass interface it is $\theta_c = \sin^{-1}(1.5/2.29) = 41°$. The ZnS film traps rays between 41° and 90°. Rays between

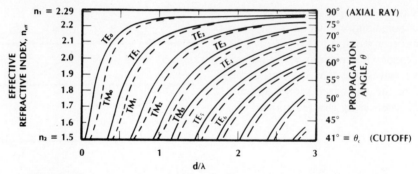

Figure 4-8   Mode chart for the asymmetric slab waveguide. $n_1 = 2.29$, $n_2 = 1.5$, $n_3 = 1.0$.

25.9° and 41° are totally reflected at the ZnS-air boundary but not at the ZnS-glass boundary. These light rays will leak into the substrate, causing high losses. We will only consider totally trapped waves in the remainder of this section. The limits on n_{eff} are found from Eq. (4-7), $n_{eff} = n_1 \sin \theta$. When $\theta = 90°$, $n_{eff} = n_1$. At the critical angle for the ZnS-glass interface, $\sin \theta = n_2/n_1$, so that $n_{eff} = n_2$. The effective index for the asymmetric slab has the range

$$n_2 \leq n_{eff} \leq n_1 \qquad (4-19)$$

The mode chart appears in Fig. 4-8. It was developed from solutions similar to Eqs. (4-11) and (4-18). Because of this similarity, and because of their complexity, the equations leading to Fig. 4-8 have been omitted. The mode chart shows only the first four TE and TM modes. For the asymmetrical slab, cutoff of the lowest-ordered mode (TE₀) does not occur at zero thickness, as it does for the symmetrical case. From the mode chart, if $d/\lambda < 0.05$, then no propagating waves exist. The entire waveguide is beyond cutoff.

Because n_1, n_2, and n_3 are not close, the TE and TM modes are not degenerate. They are well separated. A truly single-mode waveguide exists if the TM₀ mode (but not the TE₀ mode) is cut off. From Fig. 4-8 this occurs if $d/\lambda < 0.12$, the cutoff value for the

TM₀ mode in this example. For wavelengths on the order of 1 μm, a single-mode ZnS slab would have a thickness less than 0.12 μm. Integrated optic circuits are normally single-mode, asymmetric structures. Thin films capable of single-mode propagation are routinely prepared by using techniques such as diffusion, RF (radio-frequency) sputtering, vacuum evaporation, and ion bombardment.

Mode patterns for the asymmetric slab are similar to those of the symmetric waveguide. The mode indicator m is still the number of zero crossings. The asymmetry causes the fields to have unequal amplitudes at the two boundaries and to decay at different rates

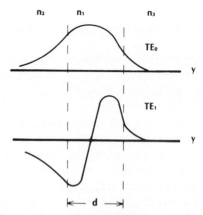

Figure 4-9   Transverse mode patterns in the asymmetric slab waveguide.

in the upper and lower layers. Mode patterns illustrating these features are sketched in Fig. 4-9.

4-4 COUPLING OF THE WAVEGUIDE

Several possibilities exist for coupling light into the dielectric slab waveguide.[3] We will investigate the techniques of edge, prism, and grating coupling.

Edge Coupling

At first glance, direct edge coupling (or butt coupling), as illustrated in Fig. 4-10, appears to be simple and efficient. In the figure, a laser diode or a light-emitting diode is attached to the edge of the film. Several problems become evident upon close scrutiny. For efficient transfer of light from the source to the film, the emitting region of the source must be no larger than the film. Otherwise, the source emits some of its output into the nonguiding layers. This light is lost. As we have seen, slabs suporting only a few propagating modes have film thicknesses on the order of 1 μm. A source having dimensions comparable to 1 μm would have a very low output power because the power-generating capability of a light source is proportional to its size.

A second problem involves the difference between the transverse radiation pattern of the source and the mode patterns of the allowed waveguide modes. Perfect coupling requires that these patterns be identical. That is, they must be matched. Another way of describing this problem involves the rays associated with the different modes. We have found that each allowed mode corresponds to a plane wave, zigzagging through the film at a characteristic angle θ. To excite any particular mode, we need to have a plane wave incident on the slab such that the internal angle is the desired value of θ. This is illustrated in Fig. 4-11, where the incident beam is in a medium whose refractive index is n_0. Quite often, the incident region is air ($n_0 = 1.0$). Let us find the incident angle α_o corresponding to the internal angle θ. By using Snell's law [Eq. (2-3)], $n_0 \sin \alpha_o = n_1 \sin \alpha_1$, or

$$n_0 \sin \alpha_o = n_1 \sin(\pi/2 - \theta) = n_1 \cos \theta \tag{4-20}$$

If α_o is increased sufficiently, then θ will drop below the critical angle and the wave will not propagate. The largest value of α_o occurs when θ is equal to the critical angle, θ_c. At this angle, $\sin \theta = n_2/n_1$ (assuming $n_2 > n_3$), so that $\cos \theta = \sqrt{n_1^2 - n_2^2}/n_1$. Putting this into Eq. (4-20) yields the numerical aperture of the waveguide.

$$\mathrm{NA} = n_0 \sin \alpha_o = \sqrt{n_1^2 - n_2^2} \tag{4-21}$$

A wave incident at an angle beyond that obtained from this equation will not be guided

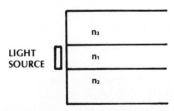

Figure 4-10   Direct edge coupling.

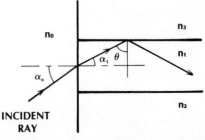

Figure 4-11   Incident and internal ray directions.

by the film. This maximum value of α_o is called the waveguide *acceptance angle*. The numerical aperture, applicable when describing the light-gathering ability of any optic system, was discussed in Section 2-4. LEDs and laser diodes emit over a range of angles. If this range is greater than the waveguide acceptance angle, then some power is lost. Only the light falling within the acceptance angle can be captured. We might also be concerned with incident rays within the acceptance angle that do not produce rays in the film at one of the discrete allowed angles. Power from these rays will also be rejected.

We may note what follows, however, while referring to a typical mode chart, such as Fig. 4-5. For small values of the normalized thickness, only a few modes exist, and they have propagation angles that are widely spaced. The incident rays must match these angles for acceptance. In a waveguide supporting many modes, the discrete allowed angles are very close together. The mode chart illustrates this conclusion for large values of the normalized thickness. If d/λ is large enough, then the angles are so close together that any angle between θ_c and 90° is allowed. In this case, the waveguide captures all the incident light within the acceptance angle. We note that the numerical aperture is a useful measure of the angular light-gathering ability of waveguides that are thick enough to support many modes. This is the case for many optic fibers. For a thin film, the accepted power depends on the match between the incident ray directions and the allowed modes of the waveguide. When only one mode (or just a few) can exist, the match between the incident field pattern and the mode pattern is critical in determining coupling efficiency. In the highly multimode case, the incident energy distributes itself among the various modes.

It is convenient to define the *fractional refractive index change* as

$$\Delta = \frac{n_1 - n_2}{n_1}$$

If the two indices are nearly equal, the numerical aperture in Eq. (4-21) can be written in terms of this parameter as

$$NA = n_1\sqrt{2\Delta}$$

Light not trapped by the film may still be observed in the waveguide. Recall that non-trapped rays merely do not reflect 100% of the light. They do reflect some amount. We visualize these rays as zigzagging down the film with continually reduced amplitude owing to the radiation loss occurring at each reflection. This *radiation mode* has negligible amplitude at the end of a long waveguide, but it can be significant within a short distance from the point of excitation. Some rays might even be trapped owing to critical-angle reflections at the outside boundaries of the upper and lower waveguide materials. Such a mode (called a *cladding mode* when observed in a fiber) is illustrated in Fig. 4-12 for a symmetrical slab waveguide.

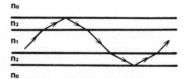

Figure 4-12   Cladding mode in a symmetrical slab waveguide. $n_1 > n_2 > n_0$.

Example 4-3

Compute the numerical aperture and acceptance angle for the symmetrical AlGaAs slab waveguide where $n_1 = 3.6$, $n_2 = n_3 = 3.55$, and $n_0 = 1$.

Solution:

By using Eq. (4-21), NA $= n_0 \sin \alpha_o = \sqrt{3.6^2 - 3.55^2} = 0.598$, so that $\alpha_o = 36.7°$. A thick film would accept all the light incident within a range of $\pm 36.7°$.

Suppose that the waveguide is allowed to radiate into a region n_0 as indicated in Fig. 4-13. Rays radiate at angles equal to those of the accepted incident rays. Thus, for a highly multimode film, Eq. (4-21) predicts the range over which rays will emerge from the slab. In the last example, a thick film will radiate light over a range of $\pm 36.7°$. A thin film will radiate light in a pattern corresponding to the discrete propagating modes. The exact distribution of light radiated by any one mode is determined by diffraction.

From Eq. (4-21) we see that a large acceptance angle corresponds to a large difference between n_1 and n_2. Although a large index difference increases the light-collection efficiency, it also increases the number of modes, as we found from Eq. (4-16).

Another loss to be considered in edge coupling is the transmission loss that occurs whenever a wave strikes a boundary between two dielectric media. The loss is computed from Eq. (3-28) for normal incidence. For the ZnS film, the reflectance is $R = (1 - 2.29)^2/(1 + 2.29)^2 = 0.154$. About 15% of the light is reflected, leaving 85% of

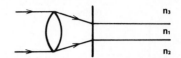

Figure 4-14   Edge coupling using a lens.

the power to enter the film. For an AlGaAs film ($n_1 = 3.6$), the reflectance is 0.319, so that 32% of the light is reflected. Antireflection coatings can reduce these losses.

Despite the objections raised, it may be reasonable to consider butt coupling for low-power applications. The advantages are the simplicity of the design and the compactness of the completed structure.

The problem arising when the source (or the beam) is larger than the film can be solved by using a lens to reduce the beam size, as in Fig. 4-14. For films on the order of 1 μm or less, the alignment is critical. Micromanipulators are often used to orient the light beam, the lens, and the film for optimum coupling efficiency. Other losses occur when the film edge is not perfectly flat and clean and when the beam pattern of the field incident on the waveguide does not match the field pattern of a propagating mode. In Fig. 4-14, the beam incident on the lens appears to be collimated. Gas lasers produce beams that are easily collimated. The transverse distribution of light from gas lasers closely resembles the field of the lowest-ordered TE_0 mode, so that highly efficient coupling into a single-mode waveguide is possible. Light from a laser diode is less well ordered, resulting in reduced coupling into a single-mode film.

Prism Coupling

Prism coupling (Fig. 4-15) is a commonly used technique that relieves the critical alignment problems of edge coupling. It is a practi-

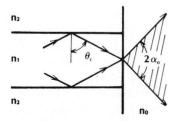

Figure 4-13   Emission from a highly multimode waveguide.

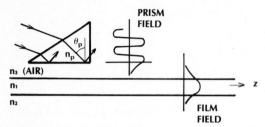

Figure 4-15   Prism coupling. The n_3 region is air.

cal method for introducing light into an integrated optic circuit when the region above the guiding film is air. An incident collimated laser beam enters the prism and undergoes critical-angle reflection at the base. As we know, this produces a standing-wave pattern in the prism caused by the interference between the incident and reflected waves. In addition, a decaying field exists in the air region below the prism base. These fields are sketched in the figure. As we also know, the field patterns for any of the propagating modes in the waveguide extend into the air region above the film. If the air gap is small (on the order of a half wavelength or less), there is an interaction between the decaying fields of the prism and those of the film. That is, there is coupling between the two structures. This coupling causes energy to be fed from the prism into the film. It might appear to be impossible to extract energy past a surface on which critical-angle reflections occur. Remember, however, that the theory of total reflection is based on a boundary between two infinitely extended media. Placing the slab waveguide near the prism base changes the problem, but only slightly. The extraction of energy when critical-angle reflection occurs is called *frustrated total internal reflection*.

For strong coupling, the field being added to the film at any point along the waveguide must be in phase with the wave already present. In other words, the longitudinal propagation factor of the wave in the prism

must equal that of the wave in the film. This is called the *synchronous* (or *phase-matching*) *condition*. By using Eq. (4-3) for the prism yields $\beta_p = k_0 n_p \sin \theta_p$. For the film, $\beta = k_0 n_1 \sin \theta$. The synchronous condition is then

$$n_p \sin \theta_p = n_1 \sin \theta \qquad (4\text{-}22)$$

This condition simply means that the phase shift along the waveguide axis is the same for the propagating mode and the wave feeding it.

The relationship in Eq. (4-22) has important consequences. Any particular mode has a fixed value of θ. To excite this mode, θ_p must be adjusted to satisfy the synchronous condition. This is easily done by varying the angle of the laser beam incident on the prism. The polarization of the laser beam must also match that of the desired mode. Notice that single-mode excitation is particularly convenient with prism coupling. A perfect multimode waveguide, excited by a prism, contains only the single excited mode. If the waveguide has imperfections, or if discontinuities are present, then power may be scattered into the other allowed modes.

We have seen that θ_c is close to $90°$ when the film and substrate have nearly equal refractive indices. For example, for the AlGaAs waveguide discussed in Section 4-2, $\theta_c = 80.4°$. In cases such as this, $\sin \theta \simeq 1$ in Eq. (4-22), so that $n_p = n_1 /\sin \theta_p$. This leads to the condition

$$n_p > n_1 \qquad (4\text{-}23)$$

Therefore, high refractive index materials must be used in the prism coupler. Such materials are not easily obtained. Rutile, having a refractive index above 2, is often used with high-index films, and flint-glass prisms are convenient with lower-index films, such as glass. It should be remembered that the refractive indices of all the materials used in the waveguide and prism vary with wavelength, so that adjustments may have to be made when Eq. (4-22) is to be satisfied at different wavelengths.

The input beam must be positioned correctly relative to the prism for maximum coupling efficiency. Consider excitation by a laser having a Gaussian beam whose spot size is w. Chapter 2 contains a description of such beams. Figure 4-16 shows two possible beam placements. In Fig. 4-16(a), the input beam terminates before the prism does. Consider the region from the right edge of the beam to the end of the prism. Over this length, there is coupling from the film back into the prism. This attenuates the field in the waveguide, as indicated in the figure. There is a corresponding loss in the overall coupling efficiency. More efficient coupling is obtained if the input beam is moved closer to the prism end. Optimum coupling occurs when the input beam actually extends a bit beyond the prism end, as illustrated in Fig. 4-16(b). The power in the input beam diminishes beyond its center so that very little power couples into the film from the right edge of the beam. Meanwhile, the field in the film is substantial at the adjacent waveguide position. The strong waveguide field couples back into the prism. Beyond some point, the coupling into the prism

exceeds the coupling into the film, so that optimum efficiency is obtained by terminating the coupling process slightly before the input beam itself terminates. The maximum coupling is about 81% for the situation described.

Because coupling between the prism and the film is reciprocal, the prism can be used as an output coupler, as illustrated in Fig. 4-17. The synchronous condition still applies, so that modes having different internal angles leave the prism in different directions. This property permits an experimental determination of the modes actually present in the waveguide. The number of output beams yields the number of modes, and the beam angles indicate the particular modes. When projected onto a screen, the output beams form lines called m lines. The notation m corresponds to the subscript on the TE_m and TM_m mode designations. All of the power can be extracted from the waveguide, because the coupling is continuous along the base of the prism. The only requirement is that the base be long enough. The resultant beam leaving the prism will not be Gaussian shaped, but will be somewhat skewed, as illustrated in Fig. 4-17. Reciprocity tells us that if a beam of this shape were incident on the prism, then the coupling would be 100% efficient. It is the difference between the optimum beam shape and the symmetrical Gaussian beam usually available that accounts for maximum coupling of no more than 81%.

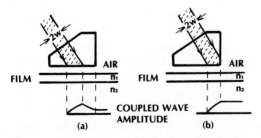

Figure 4-16 Placement of the input beam for a prism coupler. (a) Energy is fed back into the prism from the film. (b) The beam is positioned for maximum efficiency. The amplitude of the wave in the film is shown in both cases.

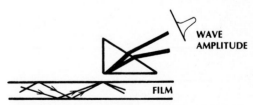

Figure 4-17 Prism output coupling. The two propagating modes couple out of the prism at different angles. The shape of the top beam is sketched.

Grating Coupling

There are several apparent disadvantages of the prism coupler. The need for high-index prisms and the slight wavelength dependence have already been mentioned. The small air gaps are critical and may be difficult to construct and maintain. Finally, this coupler is not itself integrated. A preferred coupler would be more compatible with the integrated optic structure. It would be flat and could be constructed directly on the integrated optic substrate. Such a coupler is the *dielectric grating* shown in Fig. 4-18.

In Fig. 4-18(a), a periodic array of dielectric bars forms the grating, and Fig. 4-18(b) shows a phase grating consisting of a dielectric layer having a periodic variation in refractive index. The bar grating can be formed by exposing a thick photoresist to periodic light and etching away the unexposed regions. The grating profile, shown as rectangular, may actually be sinusoidal, triangular, or another shape. The phase grating can be formed by exposing a layer of dichromated gelatin to the periodic light pattern. This produces a periodic variation in refractive index.

A grating diffracts an incident beam into one or more transmitted waves. If any of these waves has a longitudinal propagation factor equal to that of a propagating mode, then coupling occurs and that mode is excited. The re-

quirements of incident beam placement are the same as those for the prism coupler. Figure 4-18 shows the slight beam overshoot required for highest efficiency. Again, the maximum efficiency is 81% for a symmetrical Gaussian beam.

Modification and variations of the edge, prism, and grating couplers are possible. Other types of coupling, such as coupling directly from a fiber, coupling between two separate integrated circuits, and coupling between adjacent films on the same substrate are also important in applications of integrated optics.

4-5 DISPERSION AND DISTORTION IN THE SLAB WAVEGUIDE

In Section 3-2 we found there was spreading (distortion) of a waveform that traveled through a medium whose refractive index varied with wavelength. Pulse spreading is present in any dielectric structure containing dispersive material. There are two additional sources of distortion in waveguides such as the dielectric slab and the optic fiber: waveguide dispersion and multimode distortion.

Waveguide Dispersion

Figure 4-5 illustrates that the effective refractive index for any particular mode varies with wavelength for a fixed film thickness, even if the film and substrate materials are not dispersive. This is *waveguide dispersion*. The variation in n_{eff} causes pulse spreading just like the variation in n does. When, as is the usual case, the waveguide materials are dispersive, waveguide and material dispersion are present simultaneously.

The amount of pulse spreading caused by the waveguide follows the same equation as

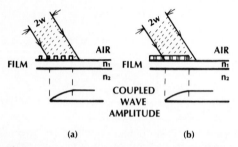

Figure 4-18 Grating couplers. (a) Periodic dielectric array. (b) Dielectric layer having a periodic variation in refractive index.

that developed for material dispersion, with the effective refractive index replacing the material refractive index. By referring to Eq. (3-14), we have

$$\Delta(\tau/L) = -\frac{\lambda}{c} n''_{\text{eff}} \Delta\lambda = -M_g \Delta\lambda \qquad (4\text{-}24)$$

for waveguide dispersion. In this equation, $\Delta\lambda$ is the source linewidth, $n''_{\text{eff}} = d^2 n_{\text{eff}}/d\lambda^2$, and $M_g = \lambda n''_{\text{eff}}/c$. This last term can be obtained from the mode charts by plotting the slopes of n_{eff}, just as n'' in Eq. (3-13) was found from the refractive index curves. We will not obtain numerical results for waveguide dispersion in the slab. Our main purpose in studying this subject is to understand the phenomenon involved, because it occurs and is important in optic fibers.

Multimode Distortion

When numerous modes are propagating in the slab, they all travel with different net velocities with respect to the waveguide axis. An input waveform distorts during propagation because its energy is distributed among several modes, each traveling at a different speed. Parts of the wave arrive at the output before other parts, spreading out the waveform. This is *multimode distortion,* or *modal distortion.* Although often called *multimode dispersion,* it is preferable to use dispersion only when referring to wavelength-dependent phenomena. It is very important to realize that multimode distortion does not depend on the source linewidth. A pulse from a perfectly single-frequency source ($\Delta\lambda = 0$) would still suffer multimode spreading, and the material and waveguide spreading would be zero. Of course, multimode distortion will not occur if the waveguide allows only one mode to propagate. This is the advantage of single-mode waveguides.

The amount of modal spreading is easily developed for the dielectric slab. We merely find the difference in the travel time between a mode propagating along the waveguide axis and one propagating at the steepest angle with respect to the axis. This latter mode will have rays at the critical angle. The axial mode will reach the end of the waveguide first. Considering a symmetric waveguide of length L, the axial travel time will be L/v or

$$t_a = \frac{L}{c} n_1 \qquad (4\text{-}25)$$

The critical-angle ray will arrive last among the many modes, because it travels the farthest, zig-zagging back and forth down the waveguide. Referring to Fig. 4-19, we see that the total distance this ray travels is $L n_1/n_2$. Traveling along the ray path at velocity $v = c/n_1$, its travel time is

$$t_c = \frac{L}{cn_2} n_1^2 \qquad (4\text{-}26)$$

The pulse spread per unit length, $(t_c - t_a)/L$, is then

$$\Delta(\tau/L) = \frac{n_1(n_1 - n_2)}{cn_2} \qquad (4\text{-}27)$$

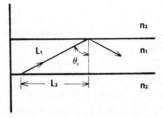

Figure 4-19 A critical-angle ray travels a distance $L_1 = L_2/\sin\theta_c$ to traverse an axial distance L_2. Since $\sin\theta_c = n_2/n_1$, $L_1 = L_2 n_1/n_2$. The total ray path for a waveguide of length L is then $L n_1/n_2$.

This can be written in terms of the fractional refractive index change as

$$\Delta(\tau/L) = n_1\Delta/c$$

and in terms of the numerical aperture as

$$\Delta(\tau/L) = \frac{NA^2}{2cn_1}$$

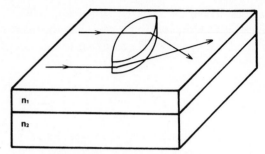

Figure 4-20   Integrated optic lens.

when the two indices are nearly equal.

In a multimode waveguide, all three pulse-spreading mechanisms exist simultaneously: material dispersion, waveguide dispersion, and multimode distortion. We will postpone numerical evaluation of these effects and their interaction until after we have investigated the propagation characteristics of optic fibers directly. We might mention, however, that because distortion increases with path length, and because integrated optic circuits are normally quite short, distortion is not as much of a problem in integrated optics as it is in fibers (which may be many kilometers long). The study of distortion is easier in the integrated optic format than in the cylindrical fiber, however.

4-6 INTEGRATED OPTIC COMPONENTS

Integrated optic networks[4] use both passive and active components. Passive devices include directional couplers, isolators, filters, lenses, and prisms. Directional couplers interconnect adjacent thin-film waveguides. Isolators are one-way transmission lines. They reject reflected light that can upset operation of a light source. Passive devices are generally constructed by slightly changing the waveguide structure. As we can see from the mode charts, varying the thickness of a film will change its effective refractive index. Thickness variations will therefore cause ray deflections, following Snell's law. An integrated optic lens, formed in this way, is drawn in Fig. 4-20.

A passive directional coupler is illustrated in Fig. 4-21. It directs some of the input light to port 2 and some to port 3. Ideally, no light reaches port 4. The percentage of light coupled into the adjacent waveguide can be varied from 0 to 100 by varying the length L of the coupling region. Coupling occurs between the adjacent waveguides owing to the overlapping of their evanescent fields, the same phenomenon as in prism coupling. As in that device, efficient coupling requires that the propagation factor be the same for each waveguide. This is the phase-matched condition. It is satisfied for identical waveguides, because they have the same effective refractive index.

Active devices can be separated into two categories: those that control light and those that convert light. The first group includes devices for beam switching, deflection and scanning, and light modulation. The second group consists of sources (which convert electricity to light), and photodetectors (which convert light to electricity).

Active control devices depend on the availability of materials that are electro-optic or acousto-optic. *Electro-optic* materials change their refractive indices in proportion to

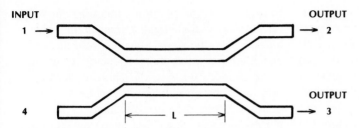

INPUT

1 →

OUTPUT

→ 2

OUTPUT

4

→ 3

L

Figure 4-21   Integrated optic directional coupler.

an applied electric field. *Acousto-optic* components rely on the interaction between an acoustic wave, excited piezoelectrically on the surface of the waveguide, and the light beam. All the active control functions can be produced either electro-optically or acousto-optically.

A electro-optic switch shown in Fig. 4-22 is similar to the directional coupler drawn in the preceding figure. For the switch, the film must have a strong electro-optic effect. Lithium niobate is suitable. The waveguides in Fig. 4-22 are embedded in the substrate. This is a common structure. It can be obtained by diffusing titanium into the $LiNbO_3$ substrate to produce a region of higher refractive index. Electrodes are placed on top of the waveguide, as indicated in the figure by the crosshatching. The coupling length is adjusted so that all the light is transferred to the second waveguide when no voltage is applied. In this state, the

two identical waveguides are phase matched. When a voltage is applied, the resulting electric fields cause the refractive indices of both guides to change. This change is in opposite directions for the two guiding films because the applied electric field points oppositely in them, as indicated in the figure. The refractive indices of the two waveguides are now unequal, destroying the phase-matching condition and reducing the crosscoupling to zero. All the light now continues along the input waveguide.

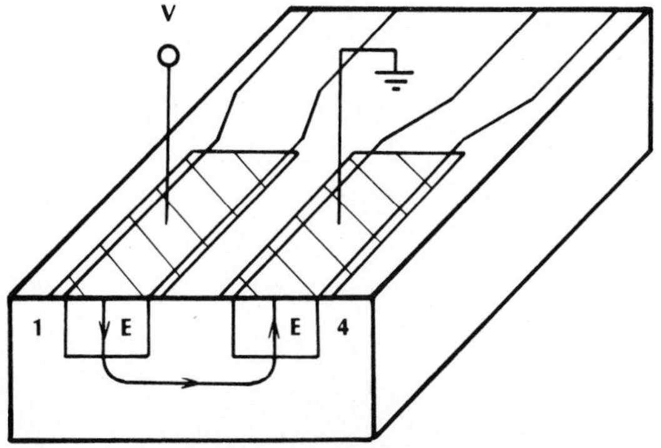

The switch in Fig. 4-22 can also operate as a modulator. The intensity of the light in either channel is controlled by the applied voltage.

Another integrated-optic modulator design is pictured in Fig. 4-23. This device, called a Mach-Zehnder interferometer, consists of parallel titanium-diffused waveguides on a lithium niobate substrate. As indicated, an incoming beam is split evenly along the two paths and then recombined at the output. With no voltage applied to the electrodes and identical lengths along the two parallel paths, the two beams recombine in phase at the out-

Figure 4-22   Electro-optic switch.

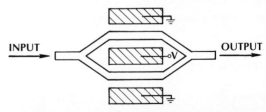

INPUT →

OUTPUT →

V

Figure 4-23   Integrated optic Mach-Zehnder interferometer modulator.

put port (constructive interference), and maximum transmission occurs. However, by applying a suitable voltage to the electrodes, a 180° phase difference between the two beams can be obtained. In this state the beams destructively interfere and the output is minimal. Devices operating in the gigahertz range and requiring around 10 V have been developed.[5]

Modulators and switches external to the source are important for several reasons. Modulation can be produced externally at higher frequencies than can be obtained by direct modulation of light-emitting diodes and laser diodes. Multigigahertz bandwidths are possible with integrated optic modulators. Also, direct modulation of a laser diode can change its output wavelength and linewidth, while external modulation does not. In some applications these changes are undesirable. Examples of such applications include optical multiplexing of closely spaced channels (see Section 9-5), operation at the wavelength of minimum material dispersion, and coherent detection systems (see Section 10-5).

Optoelectronic integrated circuits (OEICs) combining optical and electronic functions on the same substrate can be constructed by using semiconductor substrates and films.[6,7] Semiconductor materials offer the opportunity of combining optic sources, optic detectors, and electronic circuits on a single substrate. An application of this concept is the photonic repeater, whose components are indicated in Fig. 4-24. This system could take the following integrated form: a GaAs substrate, a GaAs photoconductive detector, GaAs MESFET (metal-semiconductor field-effect transistors)

to perform all the electronic functions, and an AlGaAs laser-diode light source. In operation, a low-level optic signal emerges from the fiber and illuminates the photoconductor, changing its resistance. Because the photoconductor is held at a constant voltage, its current will vary in response to the optic modulation. At this point, the optic signal has been converted to electrical circuit form. The electrical signal is amplified and the enlarged current used to modulate the laser diode. The laser output is an amplified version of the optic input signal. The conversion back to optic form is now complete. The laser output is finally coupled to a fiber for further transmission. This type of integrated system has the advantages of ruggedness, ease of connection into a larger network, and economy if produced in large quantities.

OEIC transmitters and receivers using GaAs for 0.8-0.9 μm systems and InP for 1.3-1.6 μm systems have been described in the literature.[8] This technology, sometimes referred to as *photonic integrated circuits* (PICs), is required to produce the very fast electronic interfaces to the high capacity fibers available.

4-7 SUMMARY

This chapter covered two primary topics: integrated optics and propagation in a waveguide that closely resembles an optic fiber. We barely scratched the surface of thin-film integrated optic technology, presenting only a few of the fundamental concepts. Passive and ac-

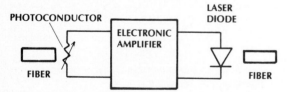

Figure 4-24 Photonic repeater.

tive components useful in optic communications systems were illustrated along with a complete network, the photonic repeater.

We determined what follows for the dielectric slab waveguide:

1. Waves are guided by critical-angle reflections.
2. The waves assume the form of modes. Each mode corresponds to a specific direction of ray travel and has a unique transverse field pattern.
3. The modes are the resonances of the waveguide for ray directions that are inclined with respect to the boundary normal.
4. The effective refractive index of a given mode can be found from a mode chart. The longitudinal propagation factor is obtained directly from n_{eff} using Eq. (4-6).
5. Two orthogonal polarizations exist, denoted as transverse electric and transverse magnetic modes.
6. The number of allowed modes increases with film thickness and with the difference in refractive indices between the guiding film and its surroundings.
7. For a sufficiently thin film, the waveguide may support only a single mode.
8. A pulse is broadened owing to material dispersion, waveguide dispersion, and multimode distortion. The first two phenomena increase with the source linewidth. Multimode distortion is independent of the source linewidth. It increases with refractive index difference, $n_1 - n_2$.
9. Light can be edge coupled only if directed within a range of angles determined by the numerical aperture of the waveguide. The NA increases with refractive index difference.

Most of the results that we obtained by analyzing the symmetrical slab waveguide apply directly to the optic fiber. We use this information in Chapter 5, which covers fibers explicitly. Because they will appear again, we will not repeat in this summary the important equations developed in this chapter. The symmetrical slab and the optic fiber are so similar that the faithful reader already knows a good deal about propagation in a fiber, even before its formal presentation.

PROBLEMS

4-1. For the AlGaAs waveguide in Section 4-2, let wavelength = 0.82 μm, $\theta = 85°$. Plot the peak amplitude of the TE$_0$ mode electric field as a function of the transverse coordinate y. Compute the film thickness and the value of n_{eff}. Use Eq. (3-35) for the decay outside the film.

4-2. Produce a mode chart like that in Fig. 4-5 if $n_1 = 1.48$ and $n_2 = n_3 = 1.46$.

4-3. What is the maximum film thickness if just one TE mode is allowed in the waveguide of Problem 4-2? Wavelength = 0.82 μm.

4-4. How many TE modes can propagate in the symmetrical slab in Section 4-2 (AlGaAs) if $d/\lambda = 5$? Repeat for $d/\lambda = 10$ and 100.

4-5. Determine the film thickness at cutoff for the TE$_0$, TE$_1$, TE$_2$, and TE$_3$ modes. Assume that $n_1 = 1.48$, $n_2 = n_3 = 1.46$, wavelength = 0.82 μm. Plot the transverse mode patterns at cutoff for each of these modes.

4-6. For the slab in Problem 4-2, compute a film thickness such that the only propagating TE mode is the TE$_0$ at 1.3 μm, but at 0.82 μm both the TE$_0$ and TE$_1$ modes propagate. Neglect TM modes.

4-7. Prove Eq. (4-14).

4-8. Show that the NA of the symmetrical slab is equal to $n_1 \sqrt{2\Delta}$ if n_1 is approximately equal to n_2 and Δ is the fractional change in the refractive index.

4-9. For the symmetrical dielectric slab: $n_1 = 1.48$, $n_2 = 1.46$. The slab is surrounded by air. Draw the ray path of a cladding mode if θ (the angle in the film) is 75°. At what values of θ do the cladding modes no longer appear?

4-10. Consider the AlGaAs waveguide described in Section 4-2. The wavelength is 0.82 μm, and $d/\lambda = 10$. This waveguide is excited by a Gaussian-shaped laser beam whose spot diameter is 1 mm. The laser beam is focused onto the edge of the waveguide. Design the coupling arrangement.

4-11. Prove that Eq. (4-21) predicts the maximum exit angle for rays leaving a symmetrical slab waveguide.

4-12. Use Eq. (4-27) to show that the modal pulse spread per unit length can be approximated by the expressions:

$$\Delta(\tau/L) = n_1 \Delta/c$$

$$\Delta(\tau/L) = \frac{NA^2}{2cn_1}$$

4-13. For a symmetrical slab waveguide, $n_1 = 1.48$ (for the film) and $n_2 = 1.46$ (for the substrate and superstrate). The wavelength is 1300 nm.
   (a) Find the thickness of the film if the ray angle is 85° for the TE_0 mode.
   (b) Find the thickness of the film if the ray angle is 85° for the TE_2 mode.

REFERENCES

1. Comprehensive descriptions of the integrated optics field appear in the following books: R. G. Hunsberger. *Integrated Optics: Theory and Technology*. New York: John Wiley, 1984.
   Donald L. Lee. *Electromagnetic Principles of Integrated Optics*. New York: John Wiley, 1986.

2. Rod C. Alferness, "Guided-Wave Devices for Optical Communication," *IEEE J. Quantum Electron*. 17, no. 6 (June 1981): 946–59.

3. W. S. C. Chang, M. W. Muller, and F. J. Rosenbaum, "Integrated Optics." In *Laser Applications*, edited by Monte Ross. New York: Academic Press, 1974, pp. 269–89.

4. Ibid. pp. 289–334.

5. Mach-Zehnder Modulators, Crystal Technology product bulletin. Palo Alto, Calif., May 1986.

6. Nadav Bar-Chaim, Israel Ury, and Amnon Yariv. "Integrated Optoelectronics," *IEEE Spectrum* 19, no. 5 (May 1982): 38–45.

7. Jun Shibata and Takao Kajiwara, "Optics and Electronics are Living Together," *IEEE Spectrum* 26, no. 2 (Feb. 1989): 34–38.

8. Tetsuo Horimatsu and Masaru Sasaki, "OEIC Technology and Its Application to Subscriber Loops," *IEEE J. Lightwave Technol*. 7, no. 11 (Nov. 1989): 1612–1622.

Chapter 5

Optic Fiber Waveguides

We are now ready to address the major item in our communications system, the optic fibers. Although only a few will ever design and fabricate your own fibers, you should have some idea how it is accomplished. Proper choice and proper utilization require a deep understanding of fiber construction and fiber characteristics. With this in mind, we will study the major types of fibers and the properties of waves propagating through them. We will pay particular attention to attenuation, modes, and information capacity. Construction and design of fibers and fiber cables are also discussed.

5-1 STEP-INDEX FIBER

The *step-index* (SI) fiber[1] consists of a central core whose refractive index is n_1, surrounded by a cladding whose refractive index is n_2. Figure 5-1 illustrates the structure, sometimes referred to as the *step-index matched-clad* fiber. As with the dielectric slab, complete guidance requires that the reflection angle θ be equal to or greater than the critical angle θ_c. The critical angle for the SI fiber is given by

$$\sin \theta_c = \frac{n_2}{n_1} \qquad (5\text{-}1)$$

The *fractional refractive index change* Δ is an important fiber parameter. It is given by

$$\Delta = \frac{n_1 - n_2}{n_1} \qquad (5\text{-}2)$$

This parameter is always positive because n_1 must be larger than n_2 for a critical angle to exist. Typically, Δ is on the order of 0.01.

101

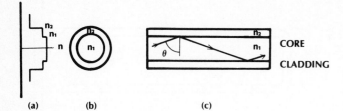

Figure 5-1 Step-index fiber.
(a) Refractive index profile. (b) End
view. (c) Cross-sectional side view.

Efficient transmission requires that the core and cladding be as free of loss as possible. Although the ray diagram implies that the light travels entirely within the core, this is not precisely the case. Actually, some of the light travels in the cladding in the form of an evanescent wave, as discussed in Chapter 4 for the slab waveguide. If the cladding is nonabsorbent, then this light is not lost but travels along the fiber. The evanescent fields decay rapidly, so that no light will reach the edge of the cladding if it is a few tens of microns thick.

The question arises as to the need for the cladding at all. A core of glass surrounded by air satisfies the requirement $n_1 > n_2$, and would indeed guide a light wave. However, severe problems arise when attempting to handle or support this type of structure. Any lossy material attached to the core for support will cause losses in the propagating wave. The freely suspended core could bend or be easily scratched, causing additional losses. The cladding protects the core from contamination and helps preserve its physical integrity.

Step-index fibers have three common forms: a glass core, cladded with a glass having a slightly lower refractive index; a silica glass core, cladded with plastic; and a plastic core, cladded with another plastic. Generally, the refractive index step is smallest for all-glass fibers, a little larger for the plastic-cladded silica (PCS) fibers, and largest for the all-plastic construction. This is due to the limited range of refractive indices available for glasses and the somewhat larger range for

plastics. As with the slab waveguide, modal distortion and numerical aperture increase with the refractive index difference, $n_1 - n_2$. Because of this, the intermodal pulse spread and NA are small for the all-glass fiber, larger for the PCS fiber, and highest for the all-plastic structure. Fibers with little pulse spread have large rate-length products. The NA of these fibers is small, making it difficult to efficiently couple light into them.

The attenuation loss in an all-glass fiber is generally lower than in a PCS or an all-plastic fiber. All-glass losses of a few dB/km and less are available. PCS fibers have losses around 8 dB/km. All-plastic fibers may have losses of several hundred dB/km.

From the information in the preceding paragraphs, we can make a number of conclusions regarding the performance and application of the three types of SI fibers. The following statements apply to fibers that are large enough to support many modes:

1. All-glass fibers have the lowest losses and the smallest intermodal pulse spreading. Because of these properties, they are useful at moderately high information rates or fairly long lengths. 30 MHz × km is an achievable rate-length product. The low NA of the SI glass fiber results in large losses when coupling from a light source. The low transmission loss partially compensates for this problem.

Conventionally, the size of a fiber is denoted by writing its core diameter and then its cladding diameter (both in micrometers) with a slash between them. For example, a 50/125 fiber describes one with a 50-μm core and a 125-μm cladding. Typical dimensions of SI fibers are 50/125, 100/140, and 200/230.

2. Because PCS fibers have higher losses and larger pulse spreads than all-glass fibers, they are suitable for shorter links. Their higher numerical apertures increase the source coupling efficiency, but this advantage is lost in a long fiber owing to increased absorption. PCS fibers are normally suitable choices when the path lengths are less than a few hundred meters. Core diameters of 200 μm are typical for PCS fibers. The large core diameter improves the source coupling efficiency.

3. All-plastic fibers are limited to very short paths by their high propagation losses. Path lengths are usually less than a few tens of meters. Large cores and large numerical apertures make plastic fibers usable because of the resulting high coupling efficiencies. Core diameters as large as 1 mm are typical.

Numerical apertures, acceptance angles, and fractional refractive index changes for fibers representative of all-glass, PCS, and all-plastic constructions are listed in Table 5-1. The numerical apertures and acceptance angles were computed from Eq. (4-21), NA = $\sin \alpha_o = \sqrt{n_1^2 - n_2^2}$, assuming that air surrounded the input end of the fiber. Since the core and cladding refractive indices are nearly equal for most all-glass fibers, the approximate result NA = $n_1\sqrt{2\Delta}$ is valid for them. Only rays emitted within a cone having a full

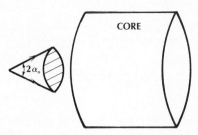

Figure 5-2 Acceptance cone for trapping of light by a step-index fiber.

angle $2\alpha_o$ will be trapped by a SI fiber, as illustrated in Fig. 5-2. Typical LEDs and laser diodes emit light over a wide angular range, often larger than the acceptance angles in Table 5-1. The numerical results in Table 5-1 show the clear advantage of a fiber having a larger NA, and thus a larger acceptance angle, for improved light collection. In Section 8-5 we consider the source-fiber coupling losses quantitatively.

Review of the step-index structure indicates that light can also be trapped by total internal reflection at the outer boundary of the cladding if the material surrounding the cladding has a lower refractive index than the cladding itself. Figure 5-3 illustrates the possible ray paths. In the example shown, the ray angle at the core-cladding interface is less than the critical angle, so some light is transmitted into the cladding. This light strikes the outer surface of the cladding beyond the critical angle for that boundary and totally reflects back toward the fiber axis. The light represented by this ray never leaves the fiber and is thus guided by it. This example illustrates the existence of cladding modes. Cladding modes are characterized by rays traveling along paths

TABLE 5-1. Typical Step-Index Fiber Characteristics

Construction	n_1	n_2	NA	α_0	Δ
All-glass	1.48	1.46	0.24	13.9°	0.0135
PCS	1.46	1.4	0.41	24.2°	0.041
All-plastic	1.49	1.41	0.48	29°	0.054

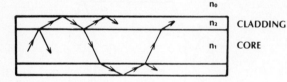

Figure 5-3 Ray paths of cladding modes. At the core-cladding interface there is partial reflection, accounting for the multiple ray paths.

that cross the fiber axis at angles greater than those of the modes guided by the core. They are excited by light introduced into the fiber end at angles beyond the acceptance angle. They also begin at discontinuities, such as splices and connectors, where light may be deflected beyond the core-mode angles.

The light traveling in a cladding mode attenuates more rapidly than light in a core mode because the outer boundary of the cladding is normally in contact with a lossy material. In addition, small bends in the fiber reduce the ray angle below that for total reflection, causing radiation losses. We can often observe power in cladding modes at points close to the light source. This power attenuates so rapidly that the cladding modes are insignificant at the end of a long fiber.

Example 5-1

Suppose that the glass fiber in Table 5-1 is surrounded by air. Compute the critical angle at the cladding-air boundary.

Solution:

Again by using the critical-angle equation, we find $\theta_c = \sin^{-1}(1/1.46) = 43°$.

This should be compared with a core mode, where $\theta_c = \sin^{-1}(1.46/1.48) = 80.6°$. Recalling that θ is the ray angle as measured from the boundary normal, we can see how much more steeply the cladding-mode rays travel, relative to the fiber axis, than the core-mode rays.

Cladding modes are eliminated in some fibers by coating the cladding with a material having a refractive index equal to, or greater than, that of the cladding itself. In such a fiber, referred to as one with a *matched buffer*, a critical angle does not exist at the outer boundary of the cladding.

5-2 GRADED-INDEX FIBER

The *graded-index* (GRIN) fiber has a core material whose refractive index varies with distance from the fiber axis. This structure, illustrated in Fig. 5-4, appears to be quite different from the SI fiber. We will show how the GRIN fiber guides light by trapping rays, not unlike the operation of a SI waveguide. The

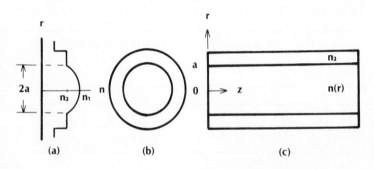

Figure 5-4 Graded-index fiber. (a) Refractive index profile. (b) End view. (c) Cross-sectional view.

index variation is described by[2]

$$n(r) = n_1 \sqrt{1 - 2(r/a)^\alpha \Delta} \qquad r \le a \qquad \text{(5-3a)}$$

$$n(r) = n_1 \sqrt{1 - 2\Delta} = n_2 \qquad r > a \qquad \text{(5-3b)}$$

where

n_1 = refractive index along the fiber axis

n_2 = refractive index outside the core (cladding index)

a = core radius

α = parameter describing the refractive index profile variation

Δ = parameter determining the scale of the profile change

Solving this last equation for the scale factor Δ yields $\Delta = (n_1^2 - n_2^2)/2n_1^2$. For the usual case in which the two indices are nearly the same, this reduces to the approximate result $\Delta = (n_1 - n_2)/n_1$. We now recognize Δ in Eq. (5-3) as the fractional refractive index change, first introduced in Chapter 4 and repeated in Eq. (5-2).

Light rays travel through the fiber in the oscillatory fashion of Fig. 5-5. The changing refractive index causes the rays to be continually redirected toward the fiber axis, and the particular variations in Eqs. (5-3a) and (5-3b)

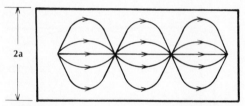

Figure 5-5 Ray paths along a GRIN fiber.

cause them to be periodically refocused. We can illustrate this redirection by modeling the continuous change in refractive index by a series of small step changes, as shown in Fig. 5-6. This model can be made as accurate as desired by increasing the number of steps. Many GRIN fibers resemble this step model because their cores are fabricated in layers. The bending of the rays at each small step follows Snell's law (Eq. (2-3)). As described in Section 2-1, rays are bent away from the normal when traveling from a high to a lower refractive index. With this in mind, the ray trace in Fig. 5-6 becomes reasonable. A ray crossing the fiber axis strikes a series of boundaries, each time traveling into a region of lower refractive index, and thus bending farther toward the horizontal axis. At one of the boundaries away from the axis, the ray angle exceeds the critical angle and is totally reflected back toward the fiber axis. Now the ray goes from low- to higher-index media,

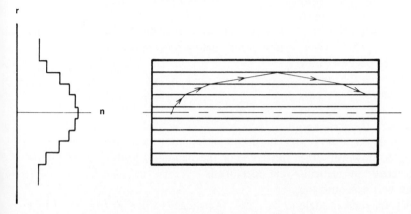

Figure 5-6 Step model of a GRIN fiber.

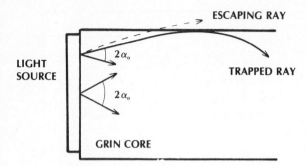

Figure 5-7   The acceptance cone angle $(2\alpha_o)$ decreases as the excitation point moves away from the fiber axis.

thus bending toward the normal until it crosses the fiber axis. At this point, the procedure will repeat. In this manner, the fiber traps a ray, causing it to oscillate back and forth as it propagates down the fiber.

Rays crossing the axis nearly horizontally in Fig. 5-5 are turned back after traveling only a short distance away from the axis. Steeper rays travel farther from the axis. Some rays may start out so steeply that they will not be turned back at all. They are never bent enough to suffer critical-angle reflections. These rays would not be trapped. We now see that only rays within a limited angular range will propagate along a GRIN fiber. The SI and GRIN fibers have this property in common. A GRIN fiber has a numerical aperture and a related acceptance angle. The expression for the NA depends on the parameters α and Δ.

In the preceding, we considered only rays that excite the fiber at its center point. Suppose that a ray enters away from the axis, as do the upper rays shown in Fig. 5-7. These rays are not bent very much because they travel only a short distance through the core in the transverse direction. If one of these rays enters nearly horizontally, then it may be bent enough to be redirected toward the axis and continue through the waveguide. At some relatively small entrance angle, however, the bending is insufficient to create a critical-angle reflection, and the ray will pass into the cladding. We conclude that the entry angle

yielding trapped rays decreases as the excitation point moves away from the fiber axis. In other words, the acceptance angle and numerical aperture decrease with radial distance from the axis. Coupling from a planar light source butted against a GRIN fiber is pictured in Fig. 5-7. The relative sizes of the acceptance cone angles are indicated. Coupling is more efficient near the axis than farther out. This is unlike the behavior of the SI fiber, for which the NA remains the same, regardless of the entry point. For this reason, the coupling efficiency is generally higher for SI fibers than for GRIN fibers, when each has the same core size and the same fractional refractive index change.

When $\alpha = 2$ in Eq. (5-3), the core index becomes

$$n(r) = n_1\sqrt{1 - 2(r/a)^2\Delta}$$

For $\Delta \ll 1$, as is usually the case, this variation is adequately represented by

$$n(r) = n_1[1 - (r/a)^2\Delta] \quad \text{for} \quad r \le a \quad (5\text{-}4a)$$

$$n_2 = n_1(1 - \Delta) \quad\quad \text{for} \quad r > a \quad (5\text{-}4b)$$

This index distribution is called the *parabolic profile*. For the parabolic profile, the numerical aperture is

$$\text{NA} = n_1(2\Delta)^{1/2}\sqrt{1 - (r/a)^2} \quad (5\text{-}5)$$

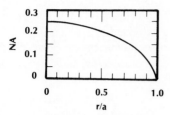

Figure 5-8 Numerical aperture of a parabolic index fiber. $n_1 = 1.48$, $\Delta = 0.0135$.

We have plotted this function in Fig. 5-8 for $n_1 = 1.48$ and $\Delta = 0.0135$. These values yield $n_2 = 1.46$. The axial NA is 0.24, the same as for the step index fiber in Table 5-1. The NA drops off to zero at the edge of the core. Only a ray perfectly parallel to the fiber axis will be guided if it enters the waveguide at this point.

The axial NA for the parabolic fiber ($r = 0$ in the preceding equation) is NA $= n_1 \sqrt{2\Delta}$, identical to that of the all-glass step-index fiber.

For the parabolic profile, the ray position (as measured radially outward from the fiber's axis) is given as a function of its position (z) along the length of the fiber by the expression

$$r(z) = r_0 \cos\left(\sqrt{A}z\right) + \frac{1}{\sqrt{A}} r_0' \sin\left(\sqrt{A}z\right)$$

$$(5\text{-}6a)$$

and the slope of the ray at any point along its trajectory is

$$r'(z) = -\sqrt{A}\, r_0 \sin\left(\sqrt{A}z\right) + r_0' \cos\left(\sqrt{A}z\right)$$

$$(5\text{-}6b)$$

where $A = 2\Delta/a^2$, r_0 is the initial position of the ray, and r_0' is its initial slope. As suggested by several of the problems at the end of this chapter, these equations can be used to plot the ray paths in a GRIN fiber and to derive the NA formula given by Eq. (5-5).

The refractive index profile given by Eq. (5-3) is quite general. We have already seen how it reduces to the parabolic profile. It also includes the SI fiber, as we can show by letting $\alpha = \infty$. Substituting infinity for α in Eq. (5-3) yields $n(r) = n_1$ within the core. The index remains at n_2 in the cladding.

Typical sizes of multimode GRIN fibers are 50/125, 62.5/125, and 85/125. Standardizing on a single cladding size has simplified the design and utilization of connectors and splicing devices. Sometimes the NA of the fiber is appended to the size designation. For typical GRIN fibers the designation then becomes 50/125/0.2, 62.5/125/0.275, and 85/125/0.26.

5-3 ATTENUATION

Signal attenuation[3] is a major factor in the design of any communications system. All receivers require that their input power be above some minimum level, so transmission losses limit the total length of the path. There are several points in an optic system where losses occur. These are at the channel input coupler, splices, and connectors and within the fiber itself. In this section we will study the losses associated with the fiber.

We need concern ourselves only with fiber losses in a range of wavelengths from about 0.5 to 1.6 μm. This is the range within which fiber communications is most practical. Reasons for this include the ability to construct low-loss fibers and efficient sources and detectors in this region and the difficulty of doing so outside this region. Details supporting this conclusion appear in the remainder of this section and in the chapters covering optic sources and photodetectors.

As was mentioned earlier, fibers are made of plastics or glasses. Requirements for the material include low loss and the ability to be formed into long hairlike fibers. Additionally, the material must be capable of slight variations so that two refractive indices, one for the core and one for the cladding, can be obtained. For a graded-index fiber a continuous variation in index must be possible. Step-index fibers can be made from plastic or glass. Graded-index fibers are normally glass. Glass fibers generally have lower absorption than plastic fibers, so they are preferred for long-distance communications.

Glass

The glass of most interest is that formed by fusing molecules of silica (silicon dioxide, SiO_2). The resulting glass is not a compound but a mixture of SiO_2 molecules that have variations in molecular locations throughout the material. This is quite unlike the structure of a crystal, in which the locations of the component atoms form fixed and repetitive patterns. To obtain different refractive indices, other materials are added to the mixture. This doping is done with titanium, thallium, germanium, boron, and other materials. The result is a high-silica-content glass, which can

be formed into a low-loss fiber if high chemical purity is achieved.

The losses occurring in glass fibers can be classified as *absorption, scattering,* and *geometric effects.*

Absorption

Even the purest glass will absorb heavily within specific wavelength regions. This is *intrinsic absorption,* a natural property of the glass itself. Intrinsic absorption is very strong in the short-wavelength ultraviolet portion of the electromagnetic spectrum. The absorption, owing to strong electronic and molecular transition bands, is characterized by peak loss in the ultraviolet and diminishing loss as the visible region is approached. The ultraviolet is far removed from the region where fiber systems are operated, so this loss is unimportant. The tail end of UV absorption probably extends into the visible region, but is generally considered to contribute very little loss at this point. Ultraviolet absorption is indicated in Fig. 5-9.

Intrinsic absorption peaks also occur in

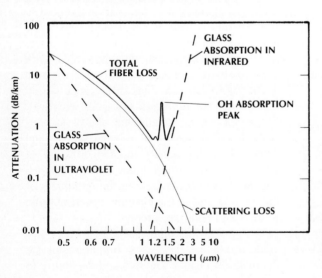

Figure 5-9 Attenuation of a germanium-doped silica glass fiber. [From H. Osanai, T. Shioda, T. Moriyama, S. Araki, M. Horiguchi, T. Izawa, and H. Takara. "Effects of Dopants on Transmission Loss of Low-OH-Content Optical Fibers." *Electronics Letters,* 12, no. 21 (Oct. 14, 1976): 550. Adapted with permission.]

the infrared. The peaks are between 7 and 12 μm for typical glass compositions, far from the region of interest. The infrared loss is associated with vibrations of chemical bonds such as the silicon-oxygen bond. Thermal energy causes the atoms to be moving constantly, so the SiO bond is continually stretching and contracting. This vibration has a resonant frequency in the infrared range. As illustrated in Fig. 5-9, the edges of this absorption mechanism extend downward in wavelength toward the region where fiber systems operate. They contribute a small loss at the upper limit of our range, 1.6 μm. In fact, they prohibit the use of silica fibers much beyond this wavelength.

We conclude that intrinsic losses are mostly insignificant in a wide region where fiber systems can operate, but these losses inhibit the extension of fiber systems toward the ultraviolet as well as toward longer wavelengths.

Impurities are a major source of loss in any practical fiber. Two types of impurities are particularly bothersome: transition metal ions and OH ions. Metal impurities, such as Fe, Cu, V, Co, Ni, Mn, and Cr, absorb strongly in the region of interest and must not exceed levels of a few parts per billion to obtain losses below 20 dB/km. Such purity has been achieved for high-silica-content fibers, so little loss is actually observed.

The loss mechanism in the metals involves incompletely filled inner electron shells. Absorption of light causes electrons to move from a lower-level shell (low-energy state) to a higher-level one (higher-energy state). The added electron energy is obtained from the incident light. The allowed transition energies correspond to photons whose frequencies are in the region of interest for fiber communications.

From a practical point of view, the most important impurity to minimize is the hydroxyl ion (OH). The loss mechanism for the OH ion is the stretching vibration, just as for the absorption of the SiO bond. The oxygen and hydrogen atoms are vibrating owing to thermal motion. The resonant frequency occurs at a wavelength of 2.73 μm. Although the peak absorption lies at 2.73 μm (outside our band of interest), the overtones and combination bands of this resonance lie within the range of interest. The most significant OH losses occur at 1.37, 1.23, and 0.95 μm when OH ions are embedded in a silica fiber. OH absorption peaks can be observed on the fiber-loss curve in Fig. 5-9. To achieve results like those shown, the OH impurity must be kept to less than a few parts per million. Special precautions are taken during the glass manufacture to ensure a low level of OH impurity in the finished product. Dry fibers have particularly low OH levels, and wet fibers just a bit more. Within the low-intrinsic-loss region, OH absorption dictates which wavelengths must be avoided for most efficient propagation.

Atomic defects also contribute to fiber absorption. As an example, titanium (Ti^{4+}), used to dope glass, does not absorb. During fiberization (forming the hairlike fiber from the preformed glass) reduction of some Ti^{4+} atoms to the Ti^{3+} state occurs. In the latter state, titanium absorbs heavily. This reduction process can be minimized by proper manufacturing techniques.

Irradiation of the glass by x-rays, gamma rays, neutrons, and electrons also creates absorbing atomic defects. High-purity, high-silica-content fibers are more resistant to irradiation defect absorption than plastic fibers or less pure glasses. PCS fibers are also somewhat radiation resistant because of their pure silica cores. Radiation losses are higher around 800 nm than at the longer wavelengths. These losses decrease up to about 1300 nm and then raise a bit at 1550 nm.

Losses of radiation-hardened (high silica–content) fibers are in the range of 2.5 dB/km per kilorad of radiation dosage. Commercial nonhardened fibers may have losses ten times this amount. Because radiation losses depend on so many variables (e.g., impurity and dopant levels, water content, wavelength, and the rate at which the radiation dosage is delivered) and because the attenuation is nonlinear (the losses do not increase linearly with the applied dosage), it is difficult to extrapolate useful radiation data from results published under circumstances that differ from those in any particular application.

Contamination of the glass may occur if it is melted in a metal container, such as a platinum crucible. The metal contaminant, when it is an oxide, does not absorb. If the glass is reduced, then atoms of the free metal get into the glass, yielding considerable loss. Fiber manufacturing processes that do not require crucibles solve this problem. Because the reduction to a free metal depends on temperature, keeping the metal below some critical temperature negates the problem. In any high-quality, low-loss fiber, attenuation owing to this effect is negligible.

Rayleigh Scattering

Molecules move randomly through the glass in the molten state during manufacture. The heat applied provides the energy for the motion. As the liquid cools, the motion ceases. Upon reaching the solid state, the random molecular locations are frozen within the glass. This results in localized variations in density and, thus, local variations in refractive index throughout the glass. These variations may be modeled as small scattering objects embedded in an otherwise homogeneous material. The size of these objects is much smaller than optic wavelengths.

A beam of light passing through such a structure will have some of its energy scattered by these objects, as illustrated in Fig. 5-10. This type of loss is known as *Rayleigh scattering,* which applies whenever a wave travels through a medium having scattering objects smaller than a wavelength. Because Rayleigh scattering is proportional to λ^{-4}, it becomes increasingly important as the wavelength diminishes. The scattering loss dependence is indicated in Fig. 5-9.

There is another cause of scattering loss. When a fiber material consists of more than one oxide, concentration fluctuations of the constituent oxides may occur. This is not a problem of imperfect chemical bonding of the various components. In this case the actual glass composition varies from place to place within the glass. Again, we have a localized refractive index variation resulting in a Rayleigh loss following the λ^{-4} dependence.

The Rayleigh scattering loss can be approximated by the expression

$$L = 1.7(0.85/\lambda)^4$$

where λ is in micrometers and the loss L is in dB/km. The corresponding electric field attenuation coefficient, defined in Eq. (3-8), is $\alpha = L/8.685$ in units of km^{-1}. Thus, both the loss in dB/km and the attenuation coefficient are proportional to λ^{-4}. The relationship between L and α is derived by noting that the

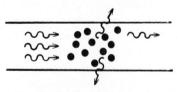

Figure 5-10 Rayleigh scattering, showing attenuation of an incident stream of photons owing to localized variations in refractive index.

beam intensity varies as the square of the electric field amplitude.

It is clear that scattering severely restricts use of fibers at short wavelengths. Below 0.8 μm, the loss owing to this effect alone builds to a prohibitive value for long-distance propagation. On the other hand, as the wavelength increases, the scattering loss diminishes. This effect provides an incentive to work at wavelengths beyond 0.8 μm. In fact, if fiber losses of less than 0.05 dB/km are ever to be attained, the wavelength of operation will have to be greater than 2 μm. Glasses composed of materials other than silica will be required. Glasses based on the light and heavy metal halides (such as flouride glasses) are candidates for this purpose but require further development.

The density and compositional losses just described are intrinsic losses. They cannot be removed by any processing techniques. They can be removed only by actually changing the composition. The scattering losses introduced by these two phenomena are considered to be a minimum below which a fiber cannot be manufactured for a given glass.

Inhomogeneities

Material inhomogeneities unintentionally introduced into the glass during manufacture also cause scattering losses. Imperfect mixing and dissolution of chemicals can cause inhomogeneities within the core. Imperfect processing can produce a rough core-cladding interface. The scattering objects in these instances are larger than the optic wavelength. Unlike Rayleigh scattering, the losses introduced by large objects are independent of wavelength. In addition, these losses can be controlled by proper manufacturing techniques.

Geometric Effects

Bending a fiber causes attenuation. Two types of bends are *macroscopic* and *microscopic*. Macroscopic refers to large-scale bending, such as that which occurs intentionally when wrapping the fiber on a spool or pulling it around a corner. As a practical example, 125-μm diameter fibers can be bent with radii of curvature as small as 25 mm with negligible loss. Typically, breaking will not occur unless the bend radius is much smaller. For example, the fiber will not fracture unless the bend radius is less than 10 mm. This example illustrates the great flexibility of glass fibers, allowing them to be installed where frequent bending is required.

Loss is not the only adverse affect of bending. In addition, bending reduces the fiber's tensile strength. A fiber's strength depends on the microscopic flaws located on its surface. These flaws will grow over time if the fiber is subjected to stress (or moisture), weakening the fiber. Thus, the stress owing to bending may cause the early failure of a fiber. For commercial 125-μm fibers, the minimum bend radius of 25 mm that assures negligible loss also ensures negligible strength loss.

Bending loss can be explained in several ways. In Fig. 5-11, a trapped ray proceeds through a SI fiber, striking the core-cladding interface at an angle $\theta_1 > \theta_c$ (critical angle), so that total reflection occurs. This same ray enters the bend and strikes the interface at an angle θ_2, which is clearly less than θ_1 and which may be less than the critical angle. The angle θ_2 diminishes as the bend radius decreases. At some bend radius, θ_2 becomes

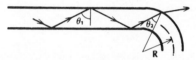

Figure 5-11   Radiation at a bend.

smaller than the critical angle, total reflection does not occur, and a portion of the wave is radiated. Higher-ordered modes (which travel close to the critical angle) are more susceptible to this type of loss than lower-ordered modes.

Radiation at a bend can be explained from another point of view. Consider the wave nature of light rather than the ray nature of light. When a wave moves around a bend, the light at the outside of the bend must move faster than the light on the inside of the curve. The smaller the bend radius, the faster the light on the outside must move to keep up. The necessary speed may exceed the velocity of light, at which point the light radiates away. The situation is like a line of linked skaters rotating around one end of the line. The skater on the far end must move faster than all the others. At some velocity, this skater can no longer keep up and breaks away from the line. The skating line represents the wavefront within the core.

Microscopic bending often occurs when a fiber is sheathed within a protective cable. The stresses set up in the cabling process cause small axial distortions (*microbends*) to appear randomly along the fiber. The micro-

bends couple light between the various guided modes of the fiber and cause some of the light to couple out of the fiber. Because of this effect, a fiber having a certain attenuation when unsheathed often has an increased loss after the cabling process. This effect can be eliminated by using the loose-tube cable construction described in Section 5-8.

Total Attenuation

Combining all the loss phenomena, except cabling, results in the attenuation curve shown by the solid line in Fig. 5-9. This curve shows how the low-loss region for silica-glass fibers is bounded on the short-wavelength side by scattering and on the long-wavelength side by infrared absorption. Figure 5-12 expands our view of the low attenuation regions where fiber transmission is most practical. The minimum loss for silica fibers is about 0.15 dB/km at 1.55 μm. Spectral attenuation curves for several fiber cables of the type available commercially are shown in Figs. 5-13–5-15. These curves are for glass, PCS, and plastic cables, respectively.

The glass fibers are pure silica and doped silica. The fibers in Fig. 5-13 are a

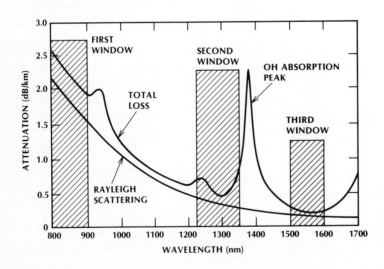

Figure 5-12 Attenuation of a silica glass fiber showing the three major wavelength regions at which fiber systems are most practical.

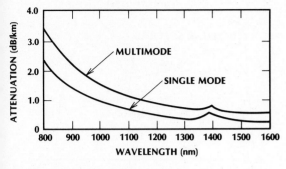

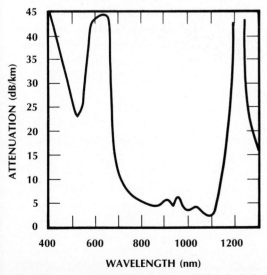

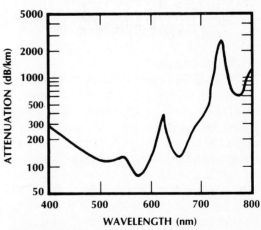

Figure 5-13 Spectral attenuation for all-glass fibers (Corning Glass Works).

multimode GRIN fiber having an 85-μm core diameter and a single-mode fiber whose spot size (the fiber's spot size is defined in the next section) is 5 μm when operated at a wavelength of 1300 nm. The larger attenuation for the multimode fiber is due to the increased loss associated with the propagation of higher-ordered modes. An explanation of this preferential attenuation appears in Section 5-6.

The PCS fiber has a pure silica core, and a thin hard polymer coating serves as the cladding. When the hard polymer (rather than a soft plastic) cladding is used, the resulting fiber is sometimes called a *hard-clad silica* (HCS) fiber.

The plastic fiber has a polymethyl methacrylate core and fluoropolymer cladding. The curves for the PCS and all-plastic fiber are for step-index fibers.

For glass fibers, the small loss between 800 and 900 nm makes this region practical, even for long-distance links. This region is sometimes called the *first transmission window*. In the range 1300–1600 nm, glass losses are lower. This region is divided into two parts by the OH absorption peak that occurs just below 1400 nm. In this range we have the *second transmission window* around 1300 nm

Figure 5-14 Spectral attenuation for hard-clad silica fiber (Ensign-Bickford Optics Co.).

Figure 5-15 Spectral attenuation for an all-plastic fiber cable (Mitsubishi Rayon America, Inc.).

and the *third transmission window* around 1550 nm. PCS fibers generally have more attenuation than glass fibers. Operation of these fibers is possible in the infrared (around 800 nm) and in the visible spectrum over moderate path lengths (up to a few kilometers). Losses for all-plastic fibers are quite high. Only short transmission paths (on the order of tens of meters) are possible with these fibers. LEDs and laser diodes are available that emit at wavelengths around 650–670 nm. These sources are compatible with plastic fibers (with losses as low as 120–160 dB/km at these wavelengths) for short distance data links.

Fiber attenuation is measured in several ways. In the most straightforward method, the power coming out of a long fiber wound on a spool is measured with an optical power meter. The fiber is then cut at a point near the launch point and the power measured again. The loss (in dB/km) is the ratio of the two power measurements (in decibels) divided by the length of fiber (in kilometers) remaining on the spool.

The cutback technique is not difficult to perform if both ends of the fiber to be measured are readily accessible. This may not be the case for an installed fiber, where the ends may be more than 1 km apart. In this situation, an *optical time-domain reflectometer* (OTDR) is often used.

The OTDR requires that only one end of the fiber be available for the measurement.[4] The OTDR transmits an optical pulse down the fiber and measures the reflections. Reflections occur owing to discontinuities (such as splices, connectors, and fiber breaks) and to Rayleigh scattering. Rayleigh scattering provides a continuous return signal from all points along the fiber to the OTDR receiver. The time delay of the reflections is a measure of their location along the fiber. For a section of fiber with no discontinuities, the return amplitude decreases in proportion to the fiber's

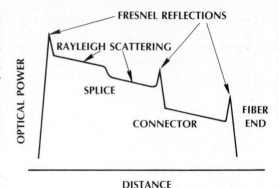

Figure 5-16 Optical time-domain reflectometer display.

attenuation loss (as illustrated on the OTDR display in Fig. 5-16). The fiber loss is proportional to the slope of the return signal as plotted on the OTDR display. Discontinuities cause localized step reductions in the return signal. The steps are indicated on the figure and are a measure of the loss caused by the splices, connectors, and breaks. Note the spike in the plot of the return power at the discontinuities (just before the step reductions). Such spikes are caused by the Fresnel reflections at such discontinuities. According to our development in Section 3-5, they amount to about 4% at any glass-to-air (or air-to-glass) interface at normal incidence. Normal incidence is usually the case because of the low NA of most fibers.

In summary, the OTDR measures fiber, splice, and connector loss. In addition, it measures the location of splices, connectors, and breaks.

5-4 MODES IN STEP-INDEX FIBERS

The mode chart for step-index fibers appears in Fig. 5-17. This chart is similar to the symmetrical slab mode chart in Fig. 4-5. One dif-

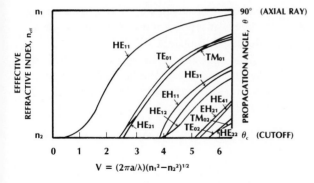

Figure 5-17   Mode chart for step-index fibers. (The HE_{11} mode cuts off at $V = 0$.) (From Donald B. Keck, "Optical Fiber Waveguides," in *Fundamentals of Optical Fiber Communications*, 2nd ed., edited by Michael K. Barnoski, New York: Academic Press, Inc., 1981, p. 18. Reproduced with permission.)

ference is that the fiber chart has been normalized by plotting the effective refractive index as a function of the parameter V. V, called the *normalized frequency*, is given by

$$V = \frac{2\pi a}{\lambda} \sqrt{n_1^2 - n_2^2} \qquad (5\text{-}7)$$

where a is the core radius and λ is the free-space wavelength. By using V, a single chart can be drawn that applies for any combination of values of a, λ, n_1, and n_2. As the propagation characteristics that can be deduced from the SI mode chart is discussed, note the many features common to wave travel in the fiber and in the slab.

The chart shows the existence of many modes. TE and TM modes are the transverse electric and transverse magnetic modes as defined in Section 4-2. HE and EH modes are hybrid and both contain components of electric and magnetic fields pointing along the fiber axis. Each curve in Fig. 5-17 actually represents two modes, one orthogonally polarized with respect to the other in the transverse plane.

Conventional fibers do not preserve the polarization of the launched wave. Since orthogonally polarized waves associated with the same mode have the same effective refractive indices, they travel at the same velocity and easily couple energy between themselves.

This exchange occurs at bends, twists, splices, and any other mechanical deformation of the fiber.

The effective refractive indices for all modes lie between the index of the cladding and that of the core. For a given mode, n_{eff} varies with wavelength, producing waveguide dispersion. At a fixed value of V, several modes may propagate, each having a different effective index. This condition leads to modal distortion. The longitudinal propagation factor can be obtained from n_{eff} by applying Eq. (4-6), $\beta = k_0 n_{eff}$. Ray angles are determined from Eq. (4-7).

Propagation in the fiber is quite similar to propagation in the slab. As with the slab, modes are cut off when their rays travel at the critical angle and rays far from cutoff travel close to $90°$ almost directly down the fiber. In addition, decaying evanescent fields exists outside the core for all of the modes; and the closer a mode comes to cutoff, the deeper its wave penetrates into the cladding. Far from cutoff, a propagating wave travels almost entirely in the core.

For large values of V, many modes will propagate. Large V corresponds to a relatively large core radius. When $V > 10$, the number of modes (including all polarizations) is approximated by[5]

$$N = \frac{V^2}{2} \qquad (5\text{-}8)$$

Example 5-2

Compute the number of modes for a fiber whose core diameter is 50 μm. Assume that $n_1 = 1.48$ and $n_2 = 1.46$, as was done for the all-glass fiber in Section 5-1. Let $\lambda = 0.82$ μm.

Solution:

The value of V, from Eq. (5-7), is

$$V = \frac{2\pi(25)}{0.82} \sqrt{1.48^2 - 1.46^2} = 46.45$$

Then, from Eq. (5-8), we find that there are 1078 modes.

It is clear from this example that even a moderately small fiber can support a large number of modes. Because the normalized frequency is proportional to the difference in refractive indices of the core and cladding, keeping this difference small reduces the number of propagating modes.

The lowest-ordered mode for the SI fiber is the HE_{11} mode. Its transverse field pattern, drawn in Fig. 5-18, is approximately Gaussian shaped. That is, the power distribution in the transverse plane is approximated by the Gaussian intensity pattern given by Eq. (2-15). The Gaussian approximation is good when the V parameter is between 1.8 and 2.4, the re-

gion where (for reasons to be discussed in the next few paragraphs) most single-mode fibers are designed to operate.

The spot size (also frequently called the *mode-field radius*) of the equivalent Gaussian beam is given in terms of the normalized frequency by the expression[6]

$$w/a = 0.65 + 1.619V^{-3/2} + 2.879V^{-6} \quad (5-9)$$

This expression, valid for the range $1.2 < V < 2.4$, is plotted in Fig. 5-19. Note that as the V parameter decreases below 2.4, the spot size increases and eventually becomes much larger than the core radius. In other words, for small values of V the light beam extends significantly beyond the core and into the cladding. In this condition the beam is not as tightly bound to the core and is highly susceptible to bending losses. For this reason, single-mode fibers are normally operated with a value of V in the neighborhood of 2–2.4. Values of V close to 2.4 are avoided to minimize the chances of propagation of more than just one mode.

Single-mode propagation is assured if all modes except the HE_{11} mode are cut off. Referring to Fig. 5-16, this occurs if $V <$

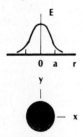

Figure 5-18   Transverse pattern for the lowest-ordered mode in the SI fiber, the HE_{11} mode.

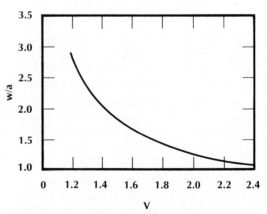

Figure 5-19   Normalized spot size w/a for the lowest-ordered mode in a step-index fiber.

2.405. Combining this result with Eq. (5-7) yields

$$\frac{a}{\lambda} < \frac{2.405}{2\pi\sqrt{n_1^2 - n_2^2}} = \frac{2.405}{2\pi \cdot (\text{NA})} \qquad (5\text{-}10)$$

as the condition for single-mode propagation. This result is very similar to the single-mode condition for the symmetrical slab, Eq. (4-17). If Eq. (5-10) is satisfied, then only the HE_{11} mode may travel through the fiber. Two orthogonally polarized HE_{11} waves can actually exist in the fiber simultaneously, but they have the same n_{eff} and, therefore, travel at the same velocity. This characteristic is more important than the fact that there are actually two modes in most applications.

An exception to this rule occurs when the fiber exhibits significant *birefringence*. Birefringence refers to the phenomenon where the refractive index depends on the direction of wave polarization. In such a case the wave velocity *will* depend on the direction of polarization. Thus the two orthogonally polarized HE_{11} waves will not travel at the same speed. This is a small effect in conventional single-mode fibers. It is enhanced in single-mode *polarization-preserving fibers*, which are designed to maintain the polarization of the launched wave. Polarization is preserved because the two possible waves have significantly different propagation characteristics. This keeps them from exchanging energy as they propagate through the fiber.

Polarization-preserving fibers are constructed by designing asymmetries into the fiber.[7] Examples include fibers with elliptical cores (which cause waves polarized along the major and minor axes of the ellipse to have different effective refractive indices) and fibers that contain nonsymmetrical stress-producing parts. These are illustrated in Fig. 5-20. The shaded region in the bow-tie fiber is highly doped with a material such as boron.

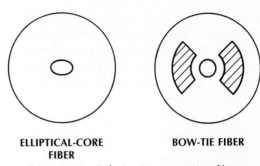

ELLIPTICAL-CORE FIBER   BOW-TIE FIBER

Figure 5-20   Polarization preserving fibers.

Because the thermal expansion of this doped region is so different from that of the pure silica cladding, a nonsymmetrical stress is exerted on the core. This produces a large stress-induced birefringence, which in turn decouples the two orthogonal modes of the single-mode fiber.

Polarization-preserving fibers are required in several applications. These include the fiber optic gyroscope and coherent optical detection systems (these applications are discussed in Section 10-5).

Example 5-3

What is the maximum core radius allowed for a glass fiber having $n_1 = 1.465$ and $n_2 = 1.46$ if the waveguide is to support only one mode at a wavelength of 1250 nm?

Solution:

From Eq. (5-10) we find that $a = 3.96\ \mu m$. The core diameter is then $7.9\ \mu m$.

We conclude that single-mode fibers will be very small. By making n_2 closer to n_1, and by operating at longer wavelengths, the core diameter can be increased. Practical single-mode fibers have core diameters of around 4–12 μm and cladding diameters of 125 μm. Handling waveguides of this size takes care.

Nonetheless, single-mode fibers are preferred for long-path, large-bandwidth applications and for use with integrated optic components. We will compare the bandwidth capabilities of single-mode and multimode fibers in Section 5-6.

5-5 MODES IN GRADED-INDEX FIBERS

We will not produce a mode chart for graded fibers.[8,9] Instead, we will show an explicit expression for the effective refractive index of the allowed modes. This can be done for the parabolic profile described by Eq. (5-4). It is not possible to find an expression for n_{eff} for the general graded-index distribution in Eq. (5-3). The parabolic profile represents a practical GRIN fiber. The results in this section apply specifically to the parabolic fiber. Other GRIN fibers behave somewhat similarly.

The effective refractive index for a mode described by the integers p and q is

$$n_{eff} = \frac{\beta_{pq}}{k_0}$$

$$= n_1 - (p + q + 1)\frac{\sqrt{2\Delta}}{k_0 a} \qquad (5\text{-}11)$$

The lowest mode has $p = q = 0$. The integers p and q are increased separately for each new mode. The factors β and k_0 have the same meaning as before. They are the longitudinal propagation factor and the free-space propagation factor, respectively.

In terms of the normalized frequency, the total number of modes in a multimode GRIN fiber is approximated by $N = V^2/4$ for large values of V.[10] This is half the number of modes in a comparable SI fiber, as determined from Eq. (5-8). For a 50-μm core and $n_1 = 1.48$, $n_2 = 1.46$, the number of modes would be 539 at 0.82 μm.

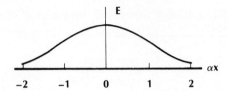

Figure 5-21   Transverse pattern for the lowest-ordered mode in a parabolic profile GRIN fiber.

The lowest-ordered mode has an electric field given by

$$E_{00} = E_0 e^{-\alpha^2 r^2/2} \sin(\omega t - \beta_{00} z) \qquad (5\text{-}12a)$$

where $\alpha = (k_0 n_1/a)^{1/2}(2\Delta)^{1/4}$ and $r^2 = x^2 + y^2$. The transverse pattern, drawn in Fig. 5-21, is circularly symmetric and Gaussian shaped. Figure 5-22 shows the patterns for the $p = 1$, $q = 0$ and the $p = 2$, $q = 0$ modes. These modes are not circularly symmetric, and their patterns are found from

$$E_{10} = E_1 \alpha x e^{-\alpha^2 r^2/2} \sin(\omega t - \beta_{10} z) \qquad (5\text{-}12b)$$

$$E_{20} = E_2 [2(\alpha x)^2 - 1] e^{-\alpha^2 r^2/2} \sin(\omega t - \beta_{20} z) \qquad (5\text{-}12c)$$

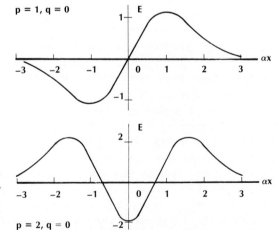

Figure 5-22   Transverse patterns for higher-order modes in a parabolic fiber.

respectively. Comparisons with the patterns of the symmetric slab (in Fig. 4-7) show close similarities. The peak amplitude of each mode depends on the fiber's excitation.

As with all the dielectric waveguides that we have studied, the allowed modes of propagation have effective refractive indices bounded by

$$n_2 \leq n_{eff} \leq n_1 \qquad (5\text{-}13)$$

Cutoff occurs for any mode when its refractive index equals n_2. With this knowledge, we can determine the relationship between core size, wavelength, and the refractive indices at cutoff. If we do this for the (1, 0) mode (i.e., $p = 1$, $q = 0$), we find the condition for single-mode propagation. We proceed by setting $n_{eff} = n_2$, $p = 1$, and $q = 0$ in Eq. (5-11). Recognizing that $k_0 = 2\pi/\lambda$, and solving for a/λ, yields the single-mode condition

$$\frac{a}{\lambda} < \frac{1.4}{\pi \sqrt{n_1(n_1 - n_2)}} \qquad (5\text{-}14)$$

A more precise analysis changes the factor 1.4 to the number 1.2. Once again, we note that making n_1 close to n_2, and operating at longer wavelengths, will permit a larger core size for a single-mode fiber. Comparison of Eqs. (5-10) and (5-14) shows that the maximum value of a/λ for single-mode propagation is 1.6 times larger for parabolic fibers than for SI fibers.

Example 5-4

Consider a GRIN fiber having a parabolic profile with $n_1 = 1.47$ and $n_2 = 1.46$. Compute the fractional refractive index change and the largest core size for single-mode propagation. Calculate the value of n_{eff} for the propagating mode by using the core size just found. The wavelength is 1.3 μm.

Solution:
From Eq. (5-4b) we find $\Delta = (n_1 - n_2)/n_1$, so $\Delta = 0.0068$. From Eq. (5-14) we obtain $a/\lambda = 3.7$ and $a = 4.8$ μm. In Eq.(5-11) we set $p = q = 0$ and $k_0 = 2\pi/\lambda$ to get

$$n_{eff} = n_1 - \frac{\sqrt{2\Delta}}{2\pi (a/\lambda)}$$

for the (0, 0) mode. Evaluating this result yields $n_{eff} = 1.465$.

5-6 PULSE DISTORTION AND INFORMATION RATE IN OPTIC FIBERS

Fiber links are limited in path length by attenuation and pulse distortion. In some applications the signal reaching the receiver is too weak for clear reception, although the received signal shape is not objectionable. When attenuation is the major problem, the system is *power limited*. In Section 5-3 we covered those losses owing to the fiber itself. Later we will need to look at the added losses occurring at the source coupler and at splices and connectors. For some links the power is sufficient but the distorted signal shape precludes correct reconstruction of the transmitted message. Such systems are *bandwidth limited*. In this section we will investigate signal distortion in a fiber, leaning heavily upon the material presented for the slab waveguide in Section 4-5.

Distortion in SI Fibers

Signals are distorted in the SI fiber by material and waveguide dispersion and by multimode pulse spreading. The amount of multimode pulse spreading in a dielectric slab was found to be $\Delta(\tau/L) = n_1(n_1 - n_2)/(cn_2)$ in Eq. (4-27). In terms of the fractional index change Δ and numerical aperture, this can be written as

$$\Delta(\tau/L) = \frac{n_1}{c}\Delta = \frac{NA^2}{2cn_1} \qquad (5\text{-}15)$$

when n_1 and n_2 are nearly equal. By using the values $n_1 = 1.48$, $n_2 = 1.46$, typical of glass fibers, we find that $\Delta(\tau/L) = 67$ ns/km. This is a rather high number. In fact, most SI glass fibers have measured pulse spreads a bit lower, around 10–50 ns/km. The discrepancy arises from several sources: mode mixing and preferential attenuation. *Mode mixing* is the exchange of power between modes. Rays in one mode are deflected (by scattering and at bends and splices) into the paths of other modes. Rays may move from higher- to lower-ordered modes, and vice versa. The result of continued mode mixing is that energy launched in any one mode travels a total zigzag path length that is somewhere between the shortest (axial-mode) path and the longest (critical-angle) path. All rays travel nearly the same total length, reducing multimode pulse spreading considerably. The mode mixing is not perfect, so modal distortion remains the main cause of spreading in SI fibers. Although mode mixing reduces the pulse spread, it is not altogether desirable. Deflections will also direct some rays into paths at less than the critical angle. This light will be lost, increasing the fiber attenuation.

The second source of pulse spread reduction is the greater attenuation suffered by higher-ordered modes. Of all the modes, these travel farthest through the fiber along their zigzag paths and penetrate more deeply into the cladding. They are therefore subject to more absorption. Having smaller ampli-

tudes, they contribute less to the received pulse than do the lower-ordered modes. The derivation leading to Eq. (5-15) assumed that all modes carried the same power. If the higher-ordered modes were neglected owing to their diminished size, then a pulse spread smaller than that predicted by Eq. (5-15) would result. While reducing the spreading, selective absorption increases the total signal attenuation, just as mode mixing does.

There is another reason why Eqs. (4-27) and (5-15) may predict higher modal distortion than actually exists. In a short link (a few tens or hundreds of meters), the light source may only excite several lower-order modes. This is possible with a laser diode whose emission pattern may not fill all the fiber's modes but is also observable with a light-emitting diode. Whereas in a long fiber all modes would eventually be excited owing to imperfections in the path, in the short fiber this will not occur. Thus, the modal spreading is caused by only a few modes whose ray angles (and, thus, velocities) are not too different from each other. We can conservatively use our theoretical results as upper bounds on the modal spreading.

Let us stress that modal distortion does not depend on the source wavelength or on the source bandwidth. This is unlike material and waveguide dispersive spreading that strongly depends on the source wavelength and bandwidth.

The total pulse spreading $\Delta\tau$, owing to both dispersion and modal distortion, is given by

$$(\Delta\tau)^2 = (\Delta\tau)^2_{mod} + (\Delta\tau)^2_{dis} \qquad (5\text{-}16)$$

where $(\Delta\tau)_{mod}$ is the multimode pulse spread and $(\Delta\tau)_{dis}$ is the dispersive spread. This equation is the most general relationship for combining the effects of modal and dispersive pulse spreading. Modal distortion and dispersion do not add algebraically because they are linearly independent processes.[11] Usually, dispersion contributes only a small amount to the total spread in multimode SI fibers. For example, consider 1 km of a typical SI fiber having a total pulse spread of 20 ns/km. In Example 3-2 we found that the pulse spread owing to material dispersion was 2.2 ns/km at 0.82 μm using a LED whose spectral width was 20 nm. From Eq. (5-16) we compute that $(\Delta\tau)_{\text{mod}} = 19.9$ ns, showing the negligible effect of material dispersion in this multimode SI fiber.

As noted in Sections 4-5 and 5-4, waveguide dispersion occurs because the effective refractive index for any one mode varies with wavelength. The amount of pulse spread is given by Eq. (4-24), which we developed for the slab waveguide. The fiber waveguide spread is then

$$\Delta(\tau/L) = -\frac{\lambda}{c} n_{\text{eff}}'' \, \Delta\lambda = -M_g \Delta\lambda \qquad (5\text{-}17)$$

where M_g is the waveguide dispersion and $\Delta\lambda$ is the source linewidth. Typical values of M are shown in Fig. 5-23. The total dispersive spread is computed by adding Eqs. (3-14) and (5-17). Thus, $\Delta(\tau/L)_{\text{dis}} = -(M + M_g)\Delta\lambda$. Comparison of Figs. 3-8 and 5-23 shows that waveguide dispersion is much less than mate-

rial dispersion in the range of wavelengths 800–900 nm. For example, at 0.82 μm, the material dispersion is 110 ps/(nm × km), but the waveguide dispersion is only about 2 ps/(nm × km). Waveguide broadening can be safely neglected except for systems operating in the region 1200–1600 nm.

The pulse spread owing to material and waveguide dispersion is proportional to the source bandwidth. A narrow-linewidth laser diode minimizes this spread. However, modal distortion is usually dominant in a multimode step-index fiber, making the laser diode largely ineffective in reducing the spread. For this reason, less expensive LED sources are normally chosen for systems using multimode, step-index fibers.

Distortion in Single-Mode Fibers

Single-mode fibers have only material and waveguide dispersion. As seen by comparing Figs. 3-8 and 5-23, the major contributor to pulse spreading is material dispersion. This is particularly true in the range 0.8–0.9 μm. The pulse spread per unit length, $M\Delta\lambda$, is plotted in Fig. 5-24 for the single-mode fiber. The material dispersion M was taken from Fig. 3-8 for this plot. Note that the pulse spread becomes smaller for longer wavelengths and narrower source linewidths. The figure shows the advantages of laser diodes. The 3-dB modulation bandwidth-length product, found from Eq. (3-16), is conveniently labeled on the right side of Fig. 5-24.

When the operating wavelength is near 1.3 μm, waveguide dispersion should be considered. At the wavelength at which the material dispersion disappears, waveguide dispersion is significant. Just beyond this wavelength, material dispersion becomes negative while waveguide dispersion stays positive. Cancelation occurs, leaving zero pulse spread

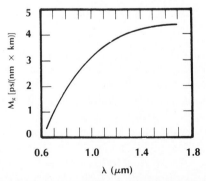

Figure 5-23   Waveguide dispersion in a SI fiber.

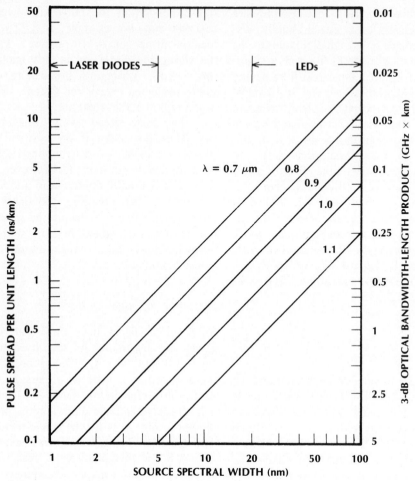

Figure 5-24 Pulse spread for a single-mode fiber. The spread shown is caused by material dispersion.

at a wavelength that is still close to 1.3 μm. At this point, material dispersion causes the shorter wavelengths in the source spectrum to travel faster, while the waveguide dispersion causes these same wavelengths to slow down. In the region where dispersion is very low, fiber attenuation is also low, as noted from Fig. 5-9. Long, high-data-rate systems can be constructed by using single-mode fibers operating from 1.3 to 1.6 μm.

Earlier in this chapter we noted that fibers have their lowest attenuation at 1550 nm. It would be helpful if they also had their lowest dispersion at this wavelength. The addition of waveguide and material dispersion just described gives a clue as to how this might be arranged: Modify the waveguide such that its dispersion just cancels that of the material at the desired wavelength. This has been accomplished by constructing a single-mode fiber with a triangular-shaped refractive index variation[12] (rather than a step-index or graded-index variation). The dispersion curve for a fiber constructed in this way (called a *dispersion-shifted* fiber) is shown in Fig. 5-25. Also shown on the figure is the dispersion characteristic for a *dispersion-flattened* fiber, which also makes use of the cancellation possible between waveguide and material dispersion by appropriately tailoring the refractive-index profile of the fiber. One refractive-index profile that produces dispersion flattening is the so-called *depressed-cladding* fiber, where the core is surrounded by a thin inner cladding whose index is low and an outer cladding whose index is a bit higher.[13] The dispersion-flattened fiber can be used over a wide range of wavelengths in the region 1330–1600 nm because of its uniformly low dispersion.

The core of a single-mode SI fiber must be small, as given by Eq. (5-10). Let us investigate the design of a single-mode fiber having n_1 close to n_2, so that the core can be large. This will simplify manufacture of the fiber and will reduce the tolerances involved in coupling, splicing, and connecting. Assume that $n_1 = 1.465$ and $n_2 = 1.46$. The NA of this fiber is then, from Eq. (4-21), 0.12. From Eq. (5-10) we find $a/\lambda < 3.17$ as the condition for single-mode operation. Suppose that we wish to operate at 0.8 μm. The radius must be less than 0.8(3.17) = 2.54 μm, or a core diameter of about 5 μm. If we change the wavelength to 1.3 μm, this same fiber will still be single mode because Eq. (5-10) will still be satisfied. We can, however, increase the core size at 1.3 μm to 1.3(3.17) = 4.12 μm, a core diameter of about 8.2 μm. This larger fiber will not satisfy the single-mode condition at 0.8 μm. It will carry more than one mode at all wavelengths below 1.3 μm. We have just illustrated the concept of cutoff of a single-mode fiber. The wavelength at which the left and right sides of Eq. (5-10) are equal is the single-mode cutoff wavelength. Wavelengths smaller than the cutoff value will propagate in more than one mode. Solving Eq. (5-10) for the cutoff wavelength of a SI fiber yields

$$\lambda_c = 2.61a(\text{NA}) \qquad (5\text{-}18)$$

While it increases the single-mode core size, making the core and cladding indices close to each other reduces the fiber NA. This makes source coupling more inefficient because of the wide angular spread of the radiation from typical fiber optic emitters.

Distortion in Graded-Index Fibers

GRIN fibers produce much less multimode distortion than SI fibers. We can explain this by considering ray trajectories and velocities in the GRIN fiber. Axial rays travel the shortest route. Rays that cross the fiber axis at large angles travel farther, but they speed up when propagating through regions away from the axis, where the refractive index is lower. (Remember that $v = c/n$.) During the time spent

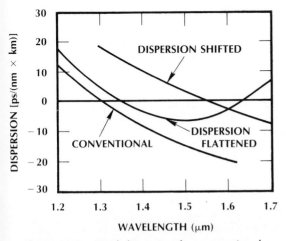

Figure 5-25 Total dispersion for conventional, dispersion-flattened, and dispersion-shifted fibers.

away from the axis, nonaxial rays catch up with axial rays. This process minimizes multimode pulse spreading. Typical multimode GRIN fibers have pulse spreads of just a few nanoseconds per kilometer or less, which is much smaller than the pulse spreads in SI fibers.

An approximate expression for the modal pulse spread in a GRIN fiber is[14]

$$\Delta(\tau/L) = \frac{n_1 \Delta^2}{2c} \qquad (5\text{-}19)$$

Comparison with Eq. (5-15) shows a reduction in pulse spread by the factor $2/\Delta$ when a GRIN fiber replaces a SI fiber. For the all-glass fiber having $n_1 = 1.48$ and $n_2 = 1.46$, we found $\Delta = 0.0135$. The reduction factor is then 148. The step-index spread was previously found to be 67 ns/km, so that the GRIN spread is 0.45 ns/km.

Material and waveguide dispersion can be included in the total GRIN spread by using Eq. (5-16). As with the SI fiber, material dispersion dominates over waveguide dispersion in the short-wavelength region. Referring to Fig. 5-24 (showing the material pulse spread), spreads of a nanosecond per kilometer or less are unobtainable in the range 0.8–0.9 μm by using LEDs. A LED source negates much of the advantage of the GRIN fiber's low modal distortion in the short-wavelength region. Narrow-band laser diodes are more compatible with multimode GRIN fibers. At wavelengths near 1.3 μm, the dispersion is small, making LEDs attractive for use with GRIN fibers.

The α profile in Eq. (5-3) can be optimized for minimum modal distortion. The best value of α depends on the glass composition and on the source wavelength. The parabolic profile ($\alpha = 2$) is close to the optimum.[15]

Length Dependence of the Pulse Spread

Up to now we have been implying that pulse broadening increases linearly with fiber length. Experiments with multimode fibers have shown that this is true for short lengths (usually less than 1 km), but for longer paths broadening does not increase so quickly. Instead, it is proportional to the square root of the length. Figure 5-26 illustrates the difference in these two conditions. The square root dependence arises from mode mixing. Over short paths, the mixing of power among the modes is incomplete. After further travel, an equilibrium modal power distribution is reached. Mixing continues, but the power in any one mode remains the same. In this condition, the $L^{1/2}$ dependence is observed. The length at which equilibrium is reached, called the *equilibrium length* L_e, depends on the particular fiber. In Fig. 5-26, L_e was taken as 1 km. Generally, we can write the modal pulse spread as[16]

$$\Delta\tau = L\Delta(\tau/L) \qquad \text{for} \quad L \leq L_e \quad (5\text{-}20a)$$

and

$$\Delta\tau = \sqrt{LL_e}\,\Delta(\tau/L) \quad \text{for} \quad L \geq L_e \quad (5\text{-}20b)$$

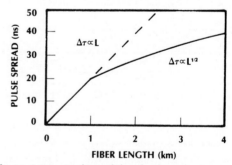

Figure 5-26 Multimodal pulse broadening showing linear length dependence for short paths and $\sqrt{L}$ dependence for longer paths.

where $\Delta(\tau/L)$ is the spread per unit length in the linear region. It was taken as 20 ns/km in Fig. 5-26.

A good fiber has little mode mixing, so equilibrium is established only after travel over a long distance. A fiber with no mode mixing would have infinite L_e and its pulse spread would increase linearly with length. A poor fiber has a lot of mode mixing owing to scattering, microbends, and inhomogeneities. For this fiber, L_e is relatively short. Although the poor fiber has improved bandwidth, its attenuation will be higher than that of better fibers.

Because material and waveguide dispersion are independent of mode coupling, broadening caused by these mechanisms increases linearly with path length. When operating with fiber lengths longer than the equilibrium length, the modal contribution to the spread increases as the square root of the length and the dispersive contribution to the spread increases as the length itself. Thus, we must be careful in the way the total spread and fiber capacity are computed. If both modal distortion and material dispersion contribute significantly to the total pulse spread, then it is always correct to add them together as indicated by Eq. (5-16) *before* applying the bandwidth or rate limits developed in Section 3-2. After computing the total pulse spread, use $f = 0.5/\Delta\tau$ for the optical bandwidth, $f = 0.35/\Delta\tau$ for the electrical bandwidth, $R = 0.35/\Delta\tau$ for the RZ bit rate, and $R = 0.7/\Delta\tau$ for the NRZ bit rate. This procedure must always be followed when operating with fibers longer than the equilibrium length. It will yield results identical with Eqs. (3-16), (3-19), (3-20), and (3-21) based on the spread per unit length only if modal distortion is insignificant or if operating within the equilibrium length of the fiber. Otherwise, the results based on spread per unit length will not be the same and will be in error.

As an example, suppose that modal distortion is so dominant that dispersion can be ignored. Applying Eq. (5-20) to the expression $f_{3\text{-dB}} = (2\Delta\tau)^{-1}$, we find that the optic bandwidth is

$$f_{3\text{-dB}} = \frac{1}{2L\ \Delta(\tau/L)} \qquad (5\text{-}21)$$

for paths shorter than the equilibrium length, and

$$f_{3\text{-dB}} = \frac{1}{2\sqrt{LL_e}\ \Delta(\tau/L)} \qquad (5\text{-}22)$$

for longer paths. A conservative design would ignore the $L^{1/2}$ dependence and simply use Eq. (5-21). This may be necessary if the equilibrium length L_e is unknown. It should be emphasized that Eq. (5-22) only applies if modal distortion is much greater than pulse dispersion.

Example 5-5
Compute and plot the 3-dB bandwidth of a SI multimode fiber whose linear pulse spread per unit length is 20 ns/km and whose equilibrium length is 1 km.

Solution:
For lengths less than 1 km, Eq. (5-21) yields a maximum bandwidth (the 3-dB value) of $25/L$ MHz, with L in kilometers. For lengths beyond 1 km, Eq. (5-22) yields $25/\sqrt{L}$ MHz. These two results are plotted in Fig. 5-27.

Many manufacturers list the frequency-length product directly, rather than the pulse spread per unit length, in their literature. It

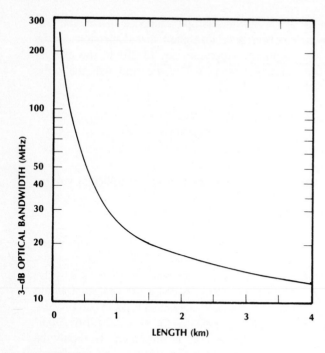

Figure 5-27   Three-dB optic bandwidth for a multimode fiber having $\Delta(\tau/L) = 20$ ns/km and a 1-km equilibrium length.

may not be clear whether the optic or electrical 3-dB bandwidth is specified. The optic bandwidth is larger than the electrical bandwidth, as discussed in Section 3-2.

5-7 CONSTRUCTION OF OPTIC FIBERS

Fibers have been fabricated by a number of techniques. Two methods will be described for directly producing fibers and several methods for producing preforms. Fibers are pulled from preforms in a separate procedure.

Double Crucible

The double-crucible method[17] is illustrated in Fig. 5-28. Molten core-glass is placed in the inner vessel and molten cladding-glass occupies the outer vessel. The two glasses come together at the base of the outer container, forming a glass-cladded core. This molten mixture is pulled into a fiber.

At first glance it would appear that the double-crucible technique could only produce step-index fibers. This is not true. Graded fibers may be produced by allowing the core and cladding-glasses to interdiffuse after they come together. Diffusion causes a gradual change of refractive index between that of the core and cladding-glasses.

With some care, glass can be continually added to the crucibles, making it possible to obtain long continuous lengths of fiber.

Rod In Tube

In the rod-in-tube procedure a rod of core-glass is placed inside of a tube of cladding-glass. Both rods are typically a meter long. The diameter of the core rod may be a few centimeters and the inner diameter of the cladding rod just a bit larger. The end of this combination is heated, softening the glass so that a thin fiber can be pulled from it.

Great care must be taken to ensure that

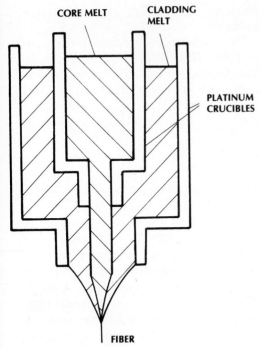

CORE MELT

CLADDING MELT

PLATINUM CRUCIBLES

FIBER

Figure 5-28 Double-crucible fiber fabrication process.

the highly purified glass rods and tubes from another manufacturer and pull the fiber.

Doped Deposited Silica

The most extensively used fabrication processes involve building up a fiber preform by vapor deposition of the glass constituents. This process is called *doped deposited silica* (DDS), *chemical vapor deposition* (CVD), or *vapor phase oxidation* (VPO). Pure silica is used as a base, and small amounts of dopants (such as GeO_2, B_2O_3, and P_2O_5) are added to produce the slight changes in refractive index that are required. The resulting cylindrical preform has the desired refractive index variation, but its cross-sectional area is many times that of the finished fiber. A representative preform has a 1-m length and a 2-cm diameter. This diameter is 160 times that of a fiber having a 125-μm cladding diameter. Continuous fibers of several kilometers can be drawn from preforms of this size.

We will describe three DDS processes: external deposition, axial deposition, and internal deposition.

External Deposition

External deposition[18] by flame hydrolysis, illustrated in Fig. 5-29, is referred to as *external chemical vapor deposition* (external CVD),

contaminants do not enter the empty region between the core and cladding rods. These contaminants would end up as part of the fiber's core-cladding interface, causing undesireable scattering losses.

This technique is probably the easiest method of fabricating fibers for a group that does not produce glass. They simply purchase

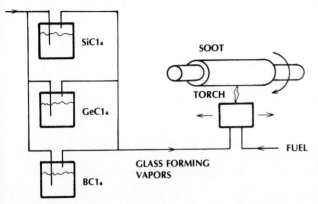

SiCl₄

GeCl₄

BCl₄

SOOT

TORCH

FUEL

GLASS FORMING VAPORS

Figure 5-29 External chemical vapor deposition.

outside vapor phase oxidation (OVPO), *outside vapor deposition* (OVD), and probably other names. The material vapors are oxidized in a flame. The torch moves laterally, depositing the glass particles onto a rotating bait or mandrel. The deposition forms a powder or *soot* on the mandrel. After deposition has been completed, the material is sintered and the bait is removed. The resulting tube is then thermally collapsed (by heating to temperatures high enough to soften it), creating a solid preform.

Axial Deposition

Axial deposition[19] is illustrated in Fig. 5-30. This process, known as *axial vapor deposition* (AVD) or *vapor axial deposition* (VAD), is another form of external deposition. In this case, the deposition occurs on the end of the rotating bait, which is withdrawn as the preform builds up. A very long core preform can be constructed in this manner. The cladding can be deposited on the core by flame hydrolysis. Alternatively, a cladded fiber can be constructed by inserting the core preform inside a lower-refractive-index glass tube and pulling

the fiber from the tube. This is the *rod-in-tube* configuration.

The VAD process produces both SI and GRIN fibers. GRIN fibers result when the deposited particle density varies owing to temperature gradients in the plane perpendicular to the core axis.

Internal Deposition

Internal deposition[20] is illustrated in Fig. 5-31. Its various aliases are *internal chemical vapor deposition* (internal CVD), *modified chemical vapor deposition* (MCVD), and *inside vapor deposition* (IVD). In this process the chemical vapors are deposited on the inside of a glass tube that is rotating in a glass lathe. A traveling oxyhydrogen torch moves along the tube, fusing the deposited material to form a transparent glassy film. Layer upon layer is deposited as the torch repeatedly traverses the length of the tube. Typically, 30–100 layers are deposited. By changing the concentration of dopants, the refractive index can be changed from layer to layer, creating a graded-index profile. Very fine control of the profile can be obtained by this technique.

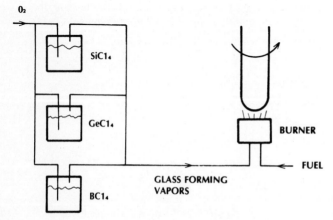

Figure 5-30   Axial vapor deposition.

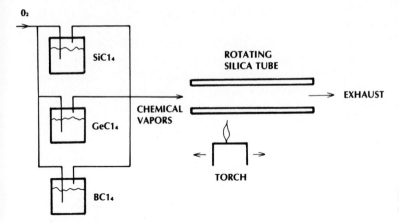

Figure 5-31 Modified chemical vapor deposition.

Deposition is completed before the tube closes. The tube is thermally collapsed into a solid preform before the fiber is drawn.

Increased fabrication rates can be obtained by the *plasma-enhanced* MCVD process (PCVD), shown in Fig. 5-32. The plasma, a region of electrically heated ionized gases, increases the chemical reaction rates within the tube. The deposition proceeds more quickly than with conventional MCVD.

Fiber Drawing

Preforms are drawn into fibers by structures like that shown in Fig. 5-33. The preform is attached to a precision feed that moves it into the furnace at the proper speed. The drawing process is designed to produce fibers with as little variance in diameter as possible. This minimizes fiber attenuation and improves strength. Precise diameter control is also needed to make fibers compatible with precision connectors designed for low connection loss. During pulling, the diameter is continually monitored by an accurate measurement device, such as a laser micrometer.

As shown in Fig. 5-33, a primary coating is applied to the fiber immediately after it has been drawn and measured. The coating is a buffer needed to protect the fiber from moisture and abrasion, which would seriously weaken the fiber. Appropriate coating materials are Kynar, epoxy, silicone RTV, and UV-cured resin. A secondary buffer coating is often applied during the drawing process to improve cushioning, increasing the fiber's

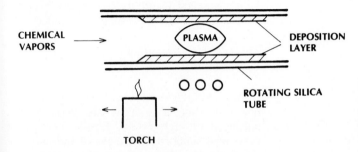

Figure 5-32 Plasma-enhanced modified chemical vapor deposition. The RF heating coil and the torch independently traverse the tube.

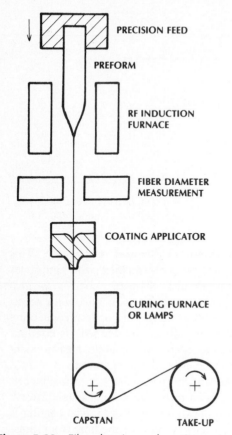

Figure 5-33 Fiber drawing and coating system.

crush resistance. All production fibers are proof tested to meet minimum tensile strength requirements. The test is performed as part of the drawing process after the coating has been applied or as an independent procedure following the pulling procedure.

For economic reasons, a fast pulling rate is desirable. The speed is limited by the need to maintain the precision of the fiber cladding diameter and maintain fiber strength. Drawing speeds of 1 m/s to 10 m/s are usual. Preforms typically have diameters of 1–6 cm in diameter and lengths of 1–2 m. When drawn into 125-μm outer cladding diameters they produce fiber lengths from 15 km to over 100 km.

Plastic-Cladded Silica

Plastic-cladded Silica (PCS) fibers can be made by drawing a pure silica preform in the manner shown in Fig. 5-33. The coating applicator, noted in the figure, contains the plastic cladding material.

5-8 OPTIC FIBER CABLES

The amount of protection a fiber needs varies from one application to another. In a laboratory setting, a fiber protected by a thin buffer coating might be quite serviceable, while a transoceanic fiber would need considerable protection during transportation, installation, and operation. A variety of cable designs have been implemented to meet the requirements of different fiber applications. We will discuss the problems involved in protecting an optic fiber, describe general techniques successfully used in solving these problems, and show a few commercially produced cables as examples.

Cabling should improve the mechanical characteristics of a fiber without causing a deterioration of its optic properties. As mentioned in Section 5-3, cabling can cause microbends in the fiber, increasing its attenuation. Microbends can also occur when the finished cable is stressed by movement of any sort (e.g., when the cable is coiled on a reel). Cables are designed to minimize microbends during construction and limit their occurrence later.

The types of strengthening and protection needed follow:

1. *Tensile strength*. High tensile strength is required when a cable is installed by pulling it through a duct. Tensile members must support the weight of the cable when it is hung in a vertical duct,

when it is suspended between poles, and when it is installed under the ocean. Cables strung between poles might also be stressed by severe ice and wind loading.

2. *Crush resistance.* Cables are often subjected to large lateral forces, which can crush a glass fiber. Some cabled fibers must survive being stepped on or being run over by large vehicles.

3. *Protection from excess bending.* Sharp bends produce two problems: radiation loss at the bend and possible breaks in the fiber. A good cable will be stiff enough to prevent excessive bending but will be flexible enough for easy handling and installation.

4. *Abrasion protection.* Glass fibers will deteriorate severely if they suffer abrasions. Small defects caused by abrasions can propagate through the glass and increase losses significantly.

5. *Vibration isolation.* Vibration will increase fiber losses. Cables are designed to cushion the fiber, damping out excessive motion.

6. *Moisture and chemical protection.* Moisture and chemicals degrade glass fibers after prolonged exposure. Some cables guard the fiber against contact with these contaminants.

In addition to being strong and chemical resistant, good fiber cables are light, small, flexible, flame retardant, rodent resistant, and temperature insensitive.

Several general structural forms that produce adequate cables have evolved. Among the variations are the following ones:

1. Single-fiber cables and multifiber cables
2. Tightly packed fibers (referred to as *tight buffer*) and loosely held fibers (called *loose-tube* buffer).
3. Centralized strenghtening members and externally located strengthening members

4. Dielectric strengthening members and metallic strengthening members
5. Circular geometries and ribbon geometries

We will now discuss these options.

If only one fiber is required, then a single-fiber cable is certainly the best choice. In some instances, future needs might be economically accommodated by installing a multifiber cable. Unused fibers can be used later. The cost of transporting and installing a multifiber cable is not much more than the cost of transporting and installing a single-fiber cable. A multifiber cable makes better use of space than does a single-fiber cable because the fibers share common strengthening members. As the number of fibers in a cable increases, the cost per fiber decreases. Multifiber cables are ideal for trunk transmission links in which many messages travel the same route. Simple two-fiber cables are designed for duplex communications systems. One fiber handles the transmission in one direction. The other fiber carries signals in the opposite direction.

As noted previously, fibers are coated with a buffer immediately after being drawn. The buffered fiber may be completely enclosed in a cushioning material as the next step in the cabling process. This is the *tight buffer* construction. Soft plastic can be used for the coating. The cushioning helps minimize microbending and provides crush resistance and vibration isolation but adds little to the cable's tensile strength. An alternative to the tight buffer cushioning is illustrated in Fig. 5-34, where the fiber lies loosely inside a surround-

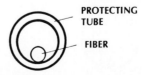

Figure 5-34   Loose-tube construction.

ing plastic tube. This is the *loose-tube* construction. The fiber can adjust itself within the tube when the cable is distorted. Microbending is almost completely eliminated by this technique. Moisture protection can be added by placing a foam or jelly inside the tube. In another form of the loose construction, the fiber lies in a large slot provided in a central strengthening member, as shown in Fig. 5-35. In the figure, four fibers are accommodated. Tape surrounds the slotted core, keeping the fibers in their grooves. The fibers can freely slide within the slots when the cable itself is pulled, twisted, or bent. Cables with loosely held fibers are normally larger than those with tightly held fibers.

Generally, fibers installed outdoors are subjected to greater stresses than are those installed indoors. For example, mechanical deformations of the cable owing to temperature variations (causing nonuniform expansions and contractions of different cable components) are much more likely when the cable lies outdoors than indoors. Since fibers in the loose construction are much less affected by mechanical deformations of the surrounding cable, the loose fiber is preferred for almost all outdoor applications.

When there will be multiple fibers in a cable, it is necessary to color each fiber so that the fibers can be identified separately. A thin layer of coloring is placed on the buffer for this purpose. The coloring layer can be easily removed from a small section of the fiber by wiping with a solvent, such as acetone, or by scraping.

Strengthening members are added to fiber cables to help fibers withstand pulling, shearing, and bending. Steel wires and textile fibers are the most popular materials for this purpose. The strengthening materials should be strong and light. Steel is strong but heavier than the textile fibers. Steel is found in some commercial cables. The textile fiber Kevlar, a very strong polymer, is one of the most frequently used strengthening materials. Its effective strength-to-weight ratio is almost four times that of steel. It is commonly applied in filaments that are twisted and stranded around a buffered and cushioned fiber. It can also be braided around a tube in the loose-tube construction.

A light-duty cable can be completed by surrounding the Kevlar braiding with an outer jacket. The jacket provides cut and abrasion resistance. Materials such as polyurethane, polyethylene, polyvinyl chloride (PVC), and Hytrel have been successfully employed in commercial cables.

A representative light-duty cable is sketched in Fig. 5-36. This cable weighs 12.5 kg/km and can be safely bent to a radius of 5 cm. It contains a single, tightly packed fiber and an external strengthening member. The term *external* means other than at the center of the cable. The cable shown in the figure can withstand a tensile load of 400 N during installation and can be loaded up to 50 N in operation.

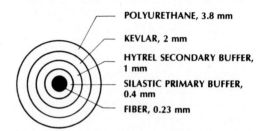

Figure 5-36 Light-duty, tight-buffer fiber cable (Siecor Corporation). The dimensions given are the diameters.

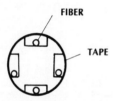

Figure 5-35 Loose fiber, slotted construction.

The tensile strength of a cable is the axial force it can tolerate. In commercial literature, this force may be given in any of three different units: newtons, kilograms, or pounds. Force and mass are related by Newton's second law:

$$F = ma \qquad (5\text{-}23)$$

Since the gravitational acceleration is $a = 9.8$ m/s^2, we see that 1 kg produces a force of 9.8 N. Conversely, a mass of $(9.8)^{-1} = 0.102$ kg produces a force of 1 N. With this equivalence in mind, we find that the 50-N load for the cable in Fig. 5-36 is equal to the stress produced by $50(0.12) = 5.1$ kg. The relationship between the pound and the newton is 1 N = 0.225 lb. A 50-N load converts to $50(0.225) = 11.25$ lb. To summarize, equal forces are produced by 1 N, 0.225 lb, or 0.102 kg.

Our next cable, in Fig. 5-37, contains six fibers and has a centralized steel strengthening member. The steel core provides a breaking strength of nearly 5000 N. The fibers are individually buffered and strengthened. A corrugated aluminum sheath provides resistance to crushing forces and to water seepage. This cable has an outer diameter of 16.5 mm and weighs 185 kg/km. It is possible to include insulated copper conductors (for electri-

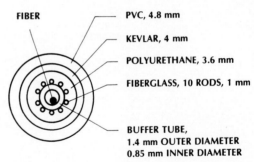

FIBER
PVC, 4.8 mm
KEVLAR, 4 mm
POLYURETHANE, 3.6 mm
FIBERGLASS, 10 RODS, 1 mm

BUFFER TUBE,
1.4 mm OUTER DIAMETER
0.85 mm INNER DIAMETER

Figure 5-38 Loose-tube cable (Siecor Corporation). The dimensions given are the diameters. The coated fiber has a 0.153-mm diameter.

cal transmission) in the space between the fibers and within the Mylar wrapping. The conductors may be used for low-rate signaling or for transmitting power to a distant location, as might be required for a remote repeater. This very strong cable can be placed in service by conventional pulling equipment designed for installation of metal transmission lines.

In Fig. 5-38 we show an example of the loose-tube construction. The strength is provided by 10 fiberglass rods that are embedded in polyurethane. A two-fiber version of this cable is drawn in Fig. 5-39.

The next cable we wish to show demonstrates the ribbon construction.[21] It is drawn in Fig. 5-40. This cable was developed for the telephone system, in which large numbers of channels need to be transmitted along a common path between interchanges. There are up

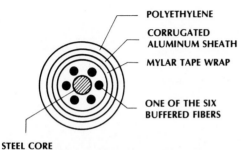

POLYETHYLENE
CORRUGATED ALUMINUM SHEATH
MYLAR TAPE WRAP
ONE OF THE SIX BUFFERED FIBERS
STEEL CORE

Figure 5-37 Multifiber cable having a centralized strengthening member and an armored sheath (Valtec Corporation).

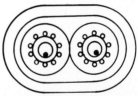

Figure 5-39 Loose-tube, two-fiber version of the cable in Fig. 5-38.

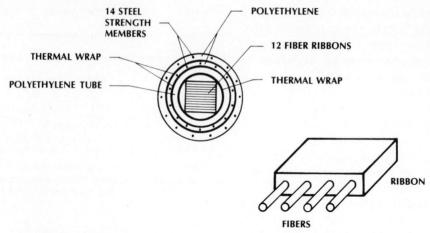

Figure 5-40 144-fiber ribbon cable. The outer diameter is 12 mm. The ribbon is constructed by sandwiching up to 12 buffered fibers between two adhesive-backed polyester tapes.

to 12 fibers in each thin ribbon. In one version the fibers are individually buffered with a polymer coating and then placed in a flat array and held in position by sandwiching between a top and bottom layer of adhesive-backed tape. The fibers are color coded for identification. Up to 12 of the ribbons are stacked as indicated in the figure, producing a rectangular structure containing 144 fibers. A total of 28 external steel-strengthening members is embedded in the surrounding polyethylene sheath. This sturdy cable makes very efficient use of the space it occupies, packing 144 fibers within a diameter of 12 mm.

A particularly interesting application of fibers has occurred in the utilities industry. The utilities have constructed telecommunications networks using existing and newly installed overhead transmission and distribution facilities. For new installations the fiber is embedded in the overhead power ground wire (OPGW) cable as illustrated in Fig. 5-41. For

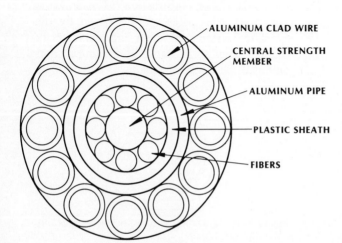

Figure 5-41 Fiber optic overhead power ground wire.

existing installations, a fiber cable can be lashed to the previously installed overhead ground wire. The fiber's imunity to electromagnetic interference makes it suitable for communications in the noisy environment surrounding power transmission lines.

The final cable to be discussed illustrates one developed for an extremely harsh environment, the ocean floor. As indicated in Fig. 5-42, the cable contains six fibers. They are embedded in an elastomer, which cushions them and minimizes microbending losses. The fibers are helically wound around the central

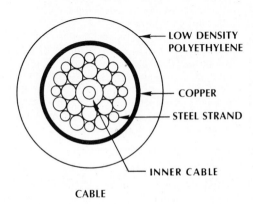

CABLE

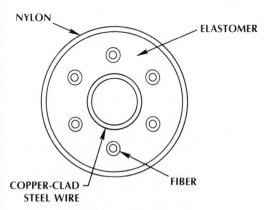

INNER CABLE DETAIL

Figure 5-42   Undersea fiber cable. [From Peter K. Runge and Patrick R. Trischitta, "The SL Undersea Lightwave System," *Journal of Lightwave Technology* LT2, no. 6 (December 1984): 744-53. © 1984 IEEE.]

steel wire. The combination of elastomer cushioning and the helical winding approximates the protective effects of the loose-tube construction. Numerous steel strands provide cable strength. Electrical power is needed to supply regenerators for long under-sea links. This power is carried by the copper conductor shown on the figure. The copper cylinder also serves as a water and hydrogen diffusion barrier. The outer diameter of this cable is only 21 mm.

In the many years since fibers were first proven to be practical, numerous cables were developed for various fiber applications. This development continues as new applications and environments emerge. We have only illustrated a small percentage of the cables available. However, the illustrations do point out most of the general features to be found in all useful cable designs.

5-9 SUMMARY

Knowledge of the material in this chapter allows you to choose the proper fiber for any particular application and account for its behavior in the system. In principle, the fiber and the cable structure can be chosen separately. Any particular fiber can be encased by any of the constructions shown in Section 5-8. In practice, a manufacturer is not likely to offer all combinations of fiber and cable, but custom designs may be practical.

The optical performance of a fiber is characterized by its attenuation, pulse spreading, and numerical aperture. In a power-limited system, fiber attenuation is more critical than pulse spreading. The NA is directly related to the source-coupling efficiency, so it is important in a power-limited system. For long high-rate links, the spreading may be the chief concern, and losses may be secondary in importance. Figure 5-43 summarizes the phe-

(a) MATERIAL DISPERSION

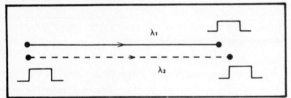

(b) WAVEGUIDE DISPERSION

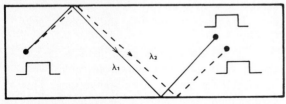

(c) MODAL DISTORTION

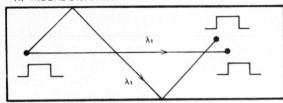

Figure 5-43 Schematic description of the three contributions to pulse spreading. The space occupied by an entering pulse is shown on the left. At some later time this pulse exits the fiber, occupying the space indicated on the right side of the figure. (a) Pulses at different wavelengths have different velocities. (b) Pulses at different wavelengths (but propagating in the same mode) must travel at slightly different angles, resulting in a difference in net axial velocities. (c) A pulse at a single wavelength splits its power into modes that travel at different axial velocities because of the path differences.

nomena of material, waveguide, and multi-mode pulse spreading.

We will now list, and briefly comment upon, the choices that exist in picking a suitable fiber.

1. *Multimode step-index and graded-index fibers*. GRIN fibers can transmit at higher information rates than SI fibers. SI source coupling is normally more efficient, while losses for the two fiber types are the same. GRIN fibers are designed for low pulse distortion, making them appropriate for long-distance, high-rate applications.

2. *Multimode and single-mode propagation*. Some systems will perform adequately with multimode fibers. They are larger and easier to handle than single-mode fibers. The advantage of single-mode fibers is their large information capacity, resulting from the absence of modal pulse spreading. Long, large information capacity systems require these fibers.

3. *Materials*. The choices in this category are glass, plastic-cladded glass, and plastic. Glass has the lowest attenuation, making it the choice for long paths. Although PCS fibers have higher losses, their larger numerical apertures makes coupling more efficient. PCS fibers are used for moderate path lengths. Plastic fibers have large losses. However, their large cores and high numerical apertures make them convenient and efficient for short runs.

4. *Wavelength of operation*. Operation in the short-wavelength range (800–900 nm) has proven quite practical. Losses and pulse spreading are low enough to produce long-distance, high-rate systems. Sources and detectors in this range are readily available. In the longer wavelength range (1300–

TABLE 5-2.   Representative Characteristics of Commercial Fibers

Description	Core Diameter (μm)	NA	Loss (dB/km)	$\Delta(\tau/L)$ (ns/km)	$f_{3\text{-dB}} \times L$ (MHz $\times$ km)	Source	Wavelength (nm)
Multimode							
Glass							
SI	50	0.24	5	15	33	LED	850
GRIN	50	0.24	5	1	500	LD	850
GRIN	50	0.20	1	0.5	1000	LED, LD	1300
PCS							
SI	200	0.41	8	50	10	LED	800
Plastic							
SI	1000	0.48	200	—	—	LED	580
Single Mode							
Glass	5	0.10	4	<0.5	>1000	LD	850
Glass	10	0.10	0.5	0.006	83000	LD	1300
Glass	10	0.10	0.2	0.006	83000	LD	1550

1600 nm), both attenuation and dispersion are reduced. Operation in this region is attractive for very-high-rate, long-distance links.

We have now addressed several of the problems introduced in the summaries of earlier chapters. Specifically, we have investigated the choice of operating wavelength and the specification of a suitable fiber and cable, topics brought up in Section 1-6. As suggested in Section 2-6, we have studied how light travels within a fiber and how the fiber NA is determined.

In Table 5-2 we have compiled representative numerical values of important properties for the various fibers introduced in this chapter. Within each category, a number of designs have been commercially produced, so that somewhat different characteristics may be found. when searching manufacturers' literature for specific fibers. The table is useful as a guide and for numerical examples worked in later chapters. In compiling the table, Eq. (3-16) was used to relate the 3-dB bandwidth and the pulse spread. The data rate can be obtained from Eq. (3-20) or (3-21). We had to

list the operating wavelength because both attenuation and distortion vary with it. The type of source is also listed. LEDs are generally suitable for multimode SI fibers in which modal distortion is dominant. A narrow-band laser diode would not significantly reduce the total spreading in this case. When material dispersion is dominant, as in a GRIN or single-mode fiber, the spreading is minimized with a LD source. In the long-wavelength region, material dispersion becomes small, so LEDs become suitable for some applications.

Note the absence of bandwidth data for the plastic fiber. The distances for which this fiber is practical are so small that the pulse spreading is generally not a problem.

PROBLEMS

5-1. A silica fiber has an outer diameter of 125 μm. What is the total volume of silica for a 1-km length? This fiber is wound on a spool whose unloaded diameter is 20 cm. The spool height is

10 cm. Compute the diameter of the fully loaded spool.

5-2. Repeat the calculations of Problem 5-1 if the fiber is embedded in a cable having an outer diameter of 1 mm.

5-3. A SI fiber has $n_1 = 1.5$, $n_2 = 1.49$, and core diameter 50 μm. Consider the guided ray traveling at the steepest angle with respect to the fiber axis. How many reflections are there per meter for this ray?

5-4. Plot acceptance angle versus NA for the range $0 \leq$ NA ≤ 1. Let $n_o = 1$.

5-5. Starting with Eq. (5-1), prove that

$$\cos \theta_c = \frac{\sqrt{n_1^2 - n_2^2}}{n_1}$$

5-6. For a GRIN fiber, let $n_1 = 1.5$, $\Delta = 0.01$, $\alpha = 2$, and $a = 50$ μm.
   (a) Plot $n(r)$ within the core to scale.
   (b) Repeat on the same graph, changing α to 10.
   (c) Repeat on the same graph with $\Delta = 0.001$ and $\alpha = 2$.

5-7. Model a parabolic GRIN fiber by the equivalent multiple-step approximation. Let $n_1 = 1.5$ and $\Delta = 0.01$. Divide the radius into 10 equal parts. Consider a ray crossing the fiber axis at 5° with respect to that axis (and inside the fiber). Sketch its progress through the fiber until it turns back and recrosses the fiber axis. (The angles can be magnified on your sketch.) At what value of r/a did the ray turn back?

5-8. For Problem 5-7 suppose that the initial ray angle is increased beyond 5°. What is the maximum angle before being no longer totally reflected? Sketch the ray that travels at this maximum angle.

5-9. Consider a fiber whose core index is 1.5 and whose cladding index is 1.485. The core radius is 100 μm. At what bending radius does a ray traveling along the fiber axis strike the cladding at the critical angle in the bend?

5-10. Consider a SI fiber with $n_1 = 1.5$ and $n_2 = 1.485$ at 0.82 μm. If the core radius is 50 μm, how many modes can propagate? Repeat if the wavelength is changed to 1.2 μm.

5-11. Prove that the maximum value of a/λ is 1.6 times larger for a single-mode parabolic index fiber than for a single-mode SI fiber.

5-12. For a parabolic fiber, plot the transverse patterns for the (0, 0), (1, 0), and (2, 0) modes. Let $a = 25$ μm, wavelength = 0.82 μm, $n_1 = 1.48$, and $n_2 = 1.46$. Put each plot on the same graph.

5-13. For a parabolic fiber, compute the equation for the cutoff value of a/λ for the (5, 5) mode. If all modes up to this one are allowed to propagate, how many allowed modes are there? Compare your result with $N = V^2/4$.

5-14. Give a general equation for the cutoff value of the (p, q) mode in the parabolic fiber.

5-15. For a single-mode and a multimode GRIN fiber, $n_1 = 1.48$, $n_2 = 1.46$, wavelength = 0.82 μm, linewidth = 20 nm.
   (a) Compute the 3-dB bandwidth-length product. Neglect waveguide dispersion.
   (b) Repeat if linewidth = 1 nm.
   (c) Repeat if wavelength = 1.5 μm, linewidth = 50 nm.
   (d) Repeat if wavelength = 1.5 μm, linewidth = 1 nm.

5-16. A SI fiber is single-mode at 1.4 μm, but not at shorter wavelengths, $n_1 = 1.465$, $n_2 = 1.46$. Compute the core radius. Find the number of modes at

0.8, 0.85, and 0.9 μm. (*Hint:* Use the mode chart.)

5-17. A multimode fiber has an equilibrium length of 0.5 km. In the linear region, its pulse spread per unit length is 30 ns/km. The spread is due primarily to modal distortion. Plot the 3-dB electrical and 3-dB optical bandwidths versus fiber length for 0 to 5 km. Also plot the maximum RZ and NRZ data rates.

5-18. Plot the pulse spread per unit length and the 3-dB bandwidth-length product versus linewidth for a single-mode SI fiber at 1.5 μm. Include waveguide and material dispersion.

5-19. A silica multimode step-index fiber has core and cladding refractive indices of 1.46 and 1.459, respectively. Compute the RZ rate-length product of this fiber if the source emits at 1550 nm and has a linewidth of 120 nm.

5-20. Make up a table comparing the advantages and disadvantages of multimode SI fiber relative to multimode GRIN fiber.

5-21. Make up a table comparing the advantages and disadvantages of multimode fiber relative to single-mode fiber.

5-22. Find the total number of propagating modes in a SI fiber having a normalized frequency of 4.5 when the wavelength is 800 nm.

5-23. For a SI fiber, the normalized frequency is 2.2 and the core diameter is 10 μm. Plot the transverse plane intensity pattern. At what radial distance does the field drop to 10% of its maximum value?

5-24. The equilibrium length of a multimode fiber is 2 km. The modal spread is 25 ns for a 1 km length. The light source emits at 800 nm and has a spectral width of 50 nm. Compute the opti-cal 3-dB bandwidth of a 5-km length of this fiber

5-25. A fiber has a NA = 0.2588. A light source is coupled to it which emits 75% of its light into a 60° full-cone angle, 50% into a 30° cone, and 25% into a 15° cone. What is the coupling efficiency when this source and fiber are connected?

REFERENCES

1. D. Gloge, "Weakly Guiding Fibers," *Appl. Opt.* 10, no. 10 (Oct. 1971): 2252–58.

2. D. Gloge and E. A. J. Marcatili, "Multimode Theory of Graded-Core Fibers," *Bell Syst. Tech. J.* 52 (Nov. 1973): 1563–78.

3. John E. Midwinter. *Optical Fibers for Transmission.* New York: John Wiley, 1979, pp. 128–61.

4. R. E. DePuy, "OTDRs Meet the Challenge of Single-Mode Technology," *Laser Focus* 22, no. 3 (March 1986): 120–32.

5. Gloge. "Weakly Guiding Fibers." p. 2256.

6. Luc B. Jeunhomme. *Single-Mode Fiber Optics,* 2nd ed. New York: Marcel Dekker, 1990, pp. 17–20.

7. Masayuki Nishimura, "The Two Modes of Single-Mode Fiber," *Photonics Spectra* 20, no. 6 (June 1986): 109–16.

8. Dietrich Marcuse, *Light Transmission Optics.* New York: Van Nostrand Reinhold, 1972, pp. 263–72.

9. A. K. Ghatak and K. Thyagarajan. *Contemporary Optics.* New York: Plenum, 1978, pp. 301–8.

10. Gloge and Marcatili. "Multimode Theory of Graded-Core Fibers," pp. 1565–69.

11. John Gowar, *Optical Communications Systems,* Englewood Cliffs, N.J.: Prentice Hall, 1984, pp. 56–68.

12. M. A. Saifi and S. J. Jang, "Triangular-Profile Single-Mode Fiber," *Opt. Lett.* 7, no. 1 (Jan. 1982): 43-45.

13. Jeunhomme. *Single-Mode Fiber Optics*. pp. 128–141, 154–159.

14. Stewart E. Miller, Enrique A. J. Marcatili, and Tingye Li, "Research Toward Optical-Fiber Transmission Systems," *Proc. IEEE* 61, no. 12 (Dec. 1973): 1703–51.

15. Donald B. Keck, "Optical Fiber Waveguides." In *Fundamentals of Optical Fiber Communications*, 2nd ed., edited by Michael K. Barnoski. New York: Academic Press, 1981, p. 63.

16. Miller et al. "Research toward Optical-Fiber Transmission Systems." p. 17.

17. Midwinter. *Optical Fibers for Transmission*. pp. 166-78.

18. Michael G. Blakenship and Charles W. Deneka, "The Outside Vapor Deposition Method of Fabricating Optical Waveguide Fibers," *IEEE J. Quantum Electron.* 18, no. 10 (Oct. 1982): 1418–23.

19. Koichi Inada, "Recent Progress in Fiber Fabrication Techniques by Vapor-Phase Axial Deposition," *IEEE J. Quantum Electron.* 18, no. 10 (Oct. 1982): 1424–31.

20. Suzanne R. Nagel, J. B. MacChesney, and Kenneth L. Walker, "An Overview of the Modified Chemical Vapor Deposition (MCVD) Process and Performance," *IEEE J. Quantum Electron.* 18, no. 4 (April 1982). pp. 459–76.

21. Frank J. Dezelsky, Robert B. Sprow, and Francis J. Topolski, "Lightguide Packaging." *Western Elec. Eng.* 24, no. 1 (Winter 1980): 80–85.

Chapter 6

Light Sources

In fiber systems, optic beams generated by light sources carry the information. Laser diodes and light-emitting diodes are the most common sources. Their small size is compatible with the small diameters of fibers, and their solid structure and low-power requirements are compatible with modern solid-state electronics. In the majority of systems, information is put onto the beam by modulating the source input current. External modulation is possible but will not be stressed because it is less important. Our study of LEDs and laser diodes includes operating principles, transfer characteristics, and modulation. We plan to obtain a good idea of the differences between the two sources and what situations call for one or the other.

6-1 LIGHT-EMITTING DIODES

A light-emitting diode[1,2] is a *pn* junction semiconductor that emits light when forward bi-
ased. Figure 6-1 shows the junction, the circuit symbol, and the energy bands associated with the diode. Band theory provides a simple explanation of semiconductor emitter (and detector) operation. Two allowed bands of energies are shown in the figure, separated by a forbidden region (a bandgap) whose width has energy W_g. In the upper-energy level, called the *conduction band,* electrons not bound to individual atoms are free to move. In the lower level, the *valence band,* unbound holes are free to move. Holes have positive charge. They exist at locations where an electron has been taken away from a neutral atom, leaving the atom with a net positive charge. A free electron can recombine with a hole, returning the atom to its neutral state. Energy is released when this occurs. An *n* type semiconductor has a number of free electrons, as pictured in Fig. 6-1. A *p* type semiconductor has a number of free holes. When a *p* type and an *n* type material are brought together without any ap-

141

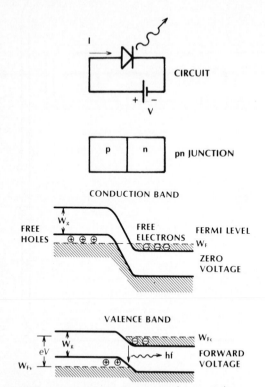

CIRCUIT

pn JUNCTION

CONDUCTION BAND

FREE HOLES

FREE ELECTRONS   FERMI LEVEL

W_F

ZERO VOLTAGE

VALENCE BAND

W_{Fc}

FORWARD VOLTAGE

hf

Figure 6-1 Light-emitting diode. The circled plus and minus signs on the energy band diagram designate free holes and free electrons, respectively.

rier. A forward voltage V separates the Fermi levels of the two materials. The applied voltage decreases the barrier by raising the potential energy of the n side and lowering that of the p side. If the energy supplied (eV) is about the same as the gap energy (W_g), free electrons and free holes will have sufficient energy to move into the junction region as shown on the bottom figure. When a free electron meets a free hole in the junction, the electron can fall to the valence band and recombine with the hole. The energy lost in the transition is converted to optic energy in the form of a photon. In its simplest terms, radiation from a LED is caused by the recombination of holes and electrons that are injected into the junction by a forward bias voltage.

As given by Eq. (1-4), photon energy and frequency are related by $W = hf$. The radiated wavelength is then

$$\lambda = \frac{hc}{W_g} \qquad (6\text{-}1)$$

with the gap energy in joules and the wavelength in meters. Expressing the gap energy in electron volts and the wavelength in micrometers changes Eq. (6-1) to

$$\lambda = \frac{1.24}{W_g} \qquad (6\text{-}2)$$

plied voltage the Fermi levels (W_F) of the p and n materials align producing the energy barrier shown on the figure. The materials for which this figure was drawn were heavily doped, a condition necessary to provide the many electrons and holes needed in the emission process. Free electrons in the n region do not have enough energy to climb over the barrier and move into the p region. Similarly, holes lack sufficient energy to surmount the barrier. The potential energy of holes, being opposite to that of electrons, increases in the downward direction in the diagram. With zero voltage applied to the diode, there is no charge movement because of the energy bar-

Different materials and alloys have different bandgap energies. Common emitter materials, operating wavelengths, and approximate bandgap energies are shown in Table 6-1. Silicon is not listed. Its holes and electrons do not recombine directly, making it an inefficient emitter. The operating wavelength can be chosen for the GaInP, AlGaAs, InGaAs, and InGaAsP devices by varying the proportions of the constituent atoms. This changes the bandgap energy and, according to Eq. (6-2), the emitting wavelength. The red-emitting material, GaInP, is included for oper-

TABLE 6-1.   Light-Emitting Semiconductors

Material	Wavelength Range (μm)	Bandgap Energy (eV)
GaInP	0.64–0.68	1.82–1.94
GaAs	0.9	1.4
AlGaAs	0.8–0.9	1.4–1.55
InGaAs	1.0–1.3	0.95–1.24
InGaAsP	0.9–1.7	0.73–1.35

ation with plastic fibers that have a relative attenuation minimum in this region.[3] The other materials are used with glass fibers.

Figure 6-1 illustrates a *homojunction,* a *pn* junction formed with a single semiconductor. A homojunction LED does not confine its emitted radiation very well. Photons radiate from the edges of the junction and from its large planar surface. This makes coupling to a small fiber very inefficient. Two reasons for this behavior can be identified. First, charge carriers exist over a large area, causing recombination and emission over an extensive region. Second, after the photons are created they diverge over unrestricted paths. These problems are solved by the heterojunction LED, shown in Fig. 6-2. A *heterojunction* is a junction formed by dissimlar semiconductors. The LED pictured in Fig. 6-2 actually contains two heterojunctions and is thus a *double-heterojunction* emitter. The two materials have different bandgap energies and different refractive indices. The changes in bandgap energies create potential barriers for both holes

and electrons. The free charges can only meet and recombine in the narrow, well-defined active layer. Because the active region has a higher refractive index than the materials on either side, an optic waveguide is formed. This is precisely the dielectric slab waveguide studied in Chapter 4. Critical-angle reflections keep some of the photons in the active region, creating a small area of high intensity. The confined emission improves the coupling efficiency, particularly for small fibers.

Power can be coupled to a fiber from the planar surface of an emitting layer or from its edge. The most efficient surface coupler is the Burrus, or *etched-well,* construction, shown in Fig. 6-3. The AlGaAs diode pictured typically emits at 0.82 μm, where glass fibers have low attentuation. Note the insulating SiO_2 layer and the metal coating at the bottom of the diode. The metal contact is circular, extending through a hole in the SiO_2 layer. This construction confines injected charges to a small central portion of the diode. Fibers as small as 50 μm can be attached with relatively efficient coupling because of the restricted emitting area. Most of the emitted radiation will at least strike the fiber core. The power will not be entirely collected by the fiber because of its limited numerical aperture.

An edge-emitting diode is drawn in Fig. 6-4. This device radiates over a smaller cone than does the Burrus diode. The emitting area is rectangular rather than circular. The emit-

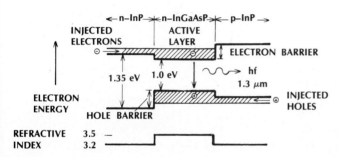

Figure 6-2   Double-heterojunction emitter. The cross-hatched regions represent the energy levels of the free charges. The junction on the right forms an energy barrier that prohibits electrons from crossing into the *p* region, the junction on the left prohibits holes from crossing into the InP *n* region. Recombination occurs only in the active InGaAsP layer. This LED emits at wavelengths around 1.3 μm.

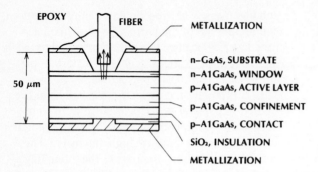

EPOXY FIBER METALLIZATION

50 μm

- METALLIZATION
- n–GaAs, SUBSTRATE
- n–A1GaAs, WINDOW
- p–A1GaAs, ACTIVE LAYER
- p–A1GaAs, CONFINEMENT
- p–A1GaAs, CONTACT
- SiO₂, INSULATION
- METALLIZATION

Figure 6-3 Etched-well, surface-emitting LED.

ting region's thickness may be on the order of a few micrometers and its width on the order of tens of micrometers. For simplicity, the various layers are not designated explicitly in Fig. 6-4. The metal stripe contact restricts charge carriers in the lateral direction, and heterojunctions confine them in the vertical direction. Heterojunctions guide the wave toward the emitting end of the LED, preventing leakage through the planar surface.

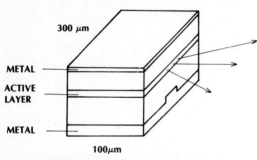

300 μm

METAL

ACTIVE
LAYER

METAL

100μm

Figure 6-4 Edge-emitting diode.

6-2 LIGHT-EMITTING DIODE OPERATING CHARACTERISTICS

The optic power generated by an LED is linearly proportional to the forward driving current. A typical power-current curve is drawn in Fig 6-5. The linear relationship can be understood by the following argument: The current i is the injected charge per second. The number of charges per second is then $N = i/e$, where e is the magnitude of the charge on each electron. If η is the fraction of these charges that will recombine and produce photons, the optic power output will be

$$P = \eta N W_g = \frac{\eta W_g}{e} i \qquad (6-3)$$

proving the linear relationship between optic power and current. In this result, the gap energy is in joules. If it is in electron volts, then the equation simplifies to

$$P = \eta i W_g \qquad (6-4)$$

Variations from perfect linearity are discussed

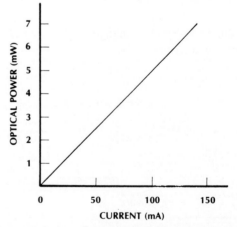

Figure 6-5 Power-current relationship for a LED.

in Section 10-1. The power in Fig. 6-5 is not the power available inside a fiber. The fiber's limited numerical aperture significantly reduces the coupled power. We will determine the coupling efficiency in Section 8-5. A variety of LEDs are available. Typically, they operate around 50–100 mA and require a voltage of 1.2–1.8 V.

Digital modulation is illustrated in Fig. 6-6. The diode is modulated by a current source, which simply turns the LED on or off. Analog modulation (Fig. 6-7) requires a dc bias to keep the total current in the forward direction at all times. Without the dc current, a negative swing in the signal current would reverse bias the diode, shutting it off.

The total diode current is

$$i = I_{dc} + I_{SP} \sin \omega t \qquad (6\text{-}5)$$

and the corresponding optic output power is

$$P = P_{dc} + P_{SP} \sin \omega t \qquad (6\text{-}6)$$

P_{SP} is the peak signal power. We will call it the ac power. Note how the shape of the input

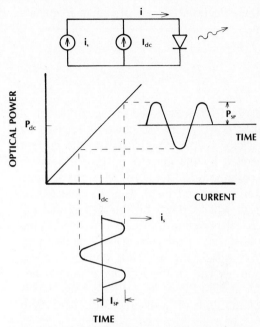

Figure 6-7   Analog modulation of a LED. I_{dc} is the dc bias current and i_s is the signal current. P_{SP} is the peak amplitude of the modulated portion of the output power, and P_{dc} is the average power.

current variation is replicated by the optic power waveform because of the linear power-current relationship. Deviations from linearity distort the signal. When very low distortion is required, the linearity of the proposed source must be evaluated.

In previous chapters we discussed how propagation through fibers limits the information rate. The source may also restrict system capacity. At low modulation frequencies $P_{SP} = a_1 I_{SP}$, where $a_1 = \Delta P/\Delta i$ (the slope of the curve in Fig. 6-7). At higher frequencies, junction and parasitic capacitances short circuit the rapidly varying current, reducing the value of the ac power. However, the major limitation to high-frequency modulation is the *carrier lifetime* τ, the average time for injected charges to recombine. The modulating current must change slowly compared to τ. The carrier-lifetime limited response of an

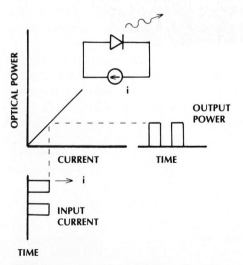

Figure 6-6   Digital modulation of a LED.

LED to electrical signals of radian frequency ω is

$$P_{SP} = \frac{a_1 I_{SP}}{\sqrt{1 + \omega^2 \tau^2}} \qquad (6\text{-}7)$$

Equation (6-7) is sketched in Fig. 6-8. At frequency $\omega = 1/\tau$, the ac power is reduced by the factor 0.707. At a receiver, the current generated by the detector is proportional to the optic power. Therefore, when the ac optic power is reduced by 0.707, the detected ac current will be down by this factor, and the electrical power in the receiver (proportional to the square of the current) will be down by $0.707^2 = 0.5$ (that is, 3 dB down). For this reason, we call $1/\tau$ the 3-dB *modulation bandwidth* of the LED or its 3-dB *electrical bandwidth*. In units of hertz, the 3-dB bandwidth is

$$f_{3\text{-dB}} = \frac{1}{2\pi\tau} \qquad (6\text{-}8)$$

More will be said about the relationship between the bandwidths measured in the optic and electrical domains in Section 12-1. Modulation bandwidths of over 300 MHz have been achieved with surface emitters, but most commercially available LEDs have smaller bandwidths. Typical values range from 1 to 100 MHz.

The *rise time* t_r of a source is the time it takes for the output to change from 10% to 90% of its final value when the input is a step in current. Rise time is illustrated in Fig. 6-9. The input current causes the optic power to rise from zero toward its final steady value. The output shown in Fig. 6-9 is the current waveform generated by the detector used to measure this power. The rise time and the 3-dB electrical bandwidth are related by

$$f_{3\text{-dB}} = \frac{0.35}{t_r} \qquad (6\text{-}9)$$

Typical LED rise times range from a few nanoseconds to 250 ns.

As we know, the optic spectrum of the source directly influences material and waveguide dispersion. Pulse spreading owing to these causes increases linearly with source spectral width. LEDs operating in the region 0.8–0.9 μm generally have widths of 20–50 nm and LEDs emitting in the longer-wavelength region have widths of 50–100 nm. The increased spectral width of a longer-wavelength emitter is compensated by the markedly reduced material dispersion M (shown in Fig. 3-8) in this region.

Coupling efficiency depends heavily on the radiation pattern of an emitter. Surface emitters radiate in what is called a *Lambertian* pattern. In this pattern (illustrated in Fig. 6-10), the power diminishes as $\cos\theta$, where θ

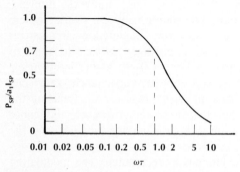

Figure 6-8 Variation of ac optic power with modulation frequency ω.

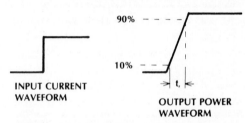

Figure 6-9 Rise time of an optic source.

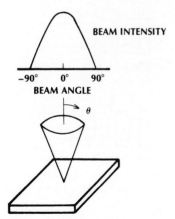

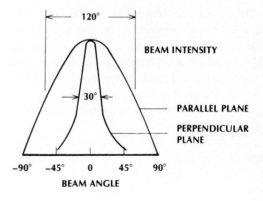

Figure 6-10 Lambertian radiation from a surface-emitting LED. The half-power width is 120°.

is the angle between the viewing direction and the normal to the surface. The emitting surface is uniformly bright, but its projected area diminishes as cos θ when the viewing angle changes, causing the Lambertian power distribution. The power is down to 50% of its peak when $\theta = 60°$. The total halfpower beamwidth is then 120° for a Lambertian emitter. Rays incident on a fiber, but outside its acceptance angle, will not be coupled. Since the acceptance angle for a fiber having NA = 0.24 is only about 14° (total cone angle of 28°), a large amount of the power generated by a surface emitter will be rejected.

Edge emitters concentrate their radiation somewhat more than surface devices, providing improved coupling efficiency. A representative pattern is drawn in Fig. 6-11. The beam is Lambertian in the plane parallel to the junction but diverges more slowly in the plane perpendicular to the junction. In this plane, the modes in the slab waveguide (formed by the refractive index variations in the perpendicular direction) limit the beam divergence. In the parallel plane, there is no beam confinement and the radiation is Lambertian. To maximize the useful output power, a reflector may be placed at the end of the diode opposite the

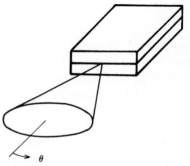

Figure 6-11 Unsymmetric radiation from an edge-emitting LED.

emitting edge. Increased output also occurs if the emitting edge is antireflection coated to reduce reflections at the semiconductor-to-air boundary. Edge emitters having speeds above 500 Mbps have been developed for use with single-mode fibers.[4]

Light-emitting diodes are very reliable and long lasting if operated within the power, voltage, current, and temperature limits specified by the manufacturer. As time goes on, LED output power diminishes. The *lifetime* is the time it takes for the power to reduce to half its initial value. Lifetimes of 10^5 hours (about 11 years) and more are common for good LEDs. Temperatures between $-65°$ and 125°C can be tolerated during operation by some of the diodes, although the output power decreases as the junction temperature rises. A representative decrease is 0.012 dB/°C.[5] Over the 190° range between −65° and 125°C, this represents a 59% change in power. The output power can be maintained at a constant level by increasing the drive current as the temperature increases. Of course, allowing for this type of compensation complicates the transmitter circuitry.

Light emitters come in a variety of packages. In some cases it is up to the purchaser to use skill and ingenuity to efficiently couple the source to the fiber transmission line. In others, the source packaged in a form that makes coupling simple. We will look at a few of the packaging possibilities.

LEDs can be mounted on standard headers, such as the TO-18 (sketched in Fig. 6-12). The header is covered by a metal cap having a clear glass top through which the light can pass. As illustrated in Fig.6-13(a), the radiated beam expands quickly. In addition to the loss of rays beyond the acceptance angle, some of the rays miss the fiber completely. An external lens can be added to the system to reduce the ray angles, but the lens will not reduce the beam diameter (see Fig. 6-13(b)). Part of the light is still lost. The efficiency is improved if the glass cover in Fig. 6-12 is removed (in some designs the metal cap is removable) and the fiber attached directly on, or just above, the emitting diode. Most of the light will now be intercepted by

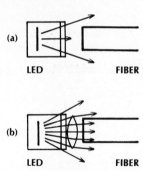

Figure 6-13 Source-to-fiber coupling of a glass-covered LED. (a) Without a lens. (b) With a lens.

the fiber core. Attaching the fiber in this manner is a chore that most users wish to avoid.

Manufacturers also produce diodes in which the glass cover plate in Fig. 6-12 is replaced by a lens. This lens is far from the LED, so the beam diameter leaving the device might still be considerably larger than the fiber. For a large fiber, such as one having a 1000-μm diameter, this construction would be suitable.

Diodes can be purchased with a short length of fiber already attached. This is the *pigtailed* construction. The manufacturer has epoxied the pigtail close to the emitter. The pigtail can be spliced onto the desired transmission fiber. Alternatively, a connector can be attached to the pigtail, allowing quick connection to the rest of the system. A problem arises when the pigtail and the transmission fiber are not identical. If their core diameters or numerical apertures differ, then there will be a loss in power when they are connected. Losses of this type are evaluated in Chapter 8.

Another package is illustrated in Fig. 6-14. In this device a very small lens (a *microlens*) is placed directly on the emitter. This differs from the design in which the lens is away from the LED because in this case, the beam does not enlarge much before collimation. This construction is efficient for fiber

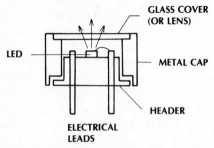

Figure 6-12 LED mounted on a header.

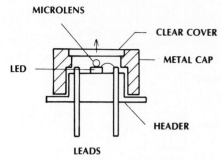

Figure 6-14 Microlensed LED.

core diameters as small as 50 μm and numerical apertures above 0.1.

6-3 LASER PRINCIPLES

We do not have to be experts on lasers to use them in communications systems. On the other hand, a knowledge of laser principles[6] helps to explain laser peculiarities and limitations. The more we know about a device, the less likely we are to make an error using it. Although the semiconductor laser diode is the most suitable laser for fiber communications, two other lasers should be mentioned: the gas laser, operating in the visible spectrum, and the Nd:YAG (neodymium yttrium-aluminum-garnet) laser, emitting at 1.06 μm.

The gas laser, principally the red-emitting helium-neon laser, is used for testing fibers and other fiber optics devices. In a simple test a HeNe laser beam is coupled to a bare fiber to detect a break or crack. If no light emerges from the fiber, then a break has obviously occurred. Small disturbances, such as air bubbles or slight fractures, can be located visually by the localized scattering of light around them. As another example, the numerical aperture of a fiber can be conveniently measured by using the HeNe laser because the NA is independent of the wavelength.

The Nd:YAG laser is a solid-state device. Its 1.06-μm operating wavelength is in a region of lower fiber attenuation and lower material dispersion than the commonly utilized region 0.8 to 0.9 μm. In addition, its spectral width is around 0.1 nm, much narrower than the linewidth of LD. This means that the Nd:YAG laser would greatly increase the bandwidth of a system that was limited by material and waveguide dispersion rather than by modal distortion. This conclusion becomes apparent when looking at Fig. 5-24 for a single-mode fiber at 1.06 μm. The 0.1-nm linewidth produces so little pulse spreading that it does not even appear on the graph. A possible embodiment of the Nd:YAG laser is sketched in Fig. 6-15. The active medium is a thin Nd:YAG rod. It is surrounded by LEDs, which provide the input power. The LEDs emit noncoherent radiation at wavelengths shorter than the coherent 1.06-μm output. Two factors discourage the use of the Nd:YAG laser in fiber systems. First, the complexity and cost of the device is many times that of the LD. Second, the modulation is usually done externally, after the light has been generated. External modulators based on electro-optic and acousto-optic effects do exist, but they are costly and add to the power requirements of the transmitter. It is much simpler to internally modulate LEDs and laser diodes.

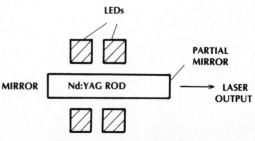

Figure 6-15 Nd : YAG laser.

A few of the characteristics that all lasers possess, and which are important in their utilization, follow:

1. *Pumping threshold.* The power input to a laser must be above a threshold level before the device will emit. This is unlike an LED, which radiates even at very low levels of input current.

2. *Output spectrum.* The laser output power is not at a single frequency but is spread over a range of frequencies. Usually the power does not vary smoothly over this range but is a series of peaks and valleys.

3. *Radiation pattern.* The range of angles over which a laser emits light depends on the size of the emitting area and on the modes of oscillation within the laser.

It is easier to explain these effects for a gas laser than for a laser diode. For this reason the HeNe laser will be analyzed in the rest of this section. We will then apply the results to the LD by analogy.

A HeNe laser is drawn in Fig. 6-16, and a partial energy-level diagram for the helium-neon mixture appears in Fig. 6-17. Many more levels exist, but those shown illustrate the principles of laser action. The levels represent the allowed energy states of electrons in the atom. In simplest terms, each state corresponds to a different orbit and to different spin and angular momentum of an electron.

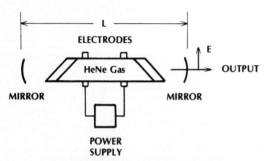

Figure 6-16   Helium-neon laser.

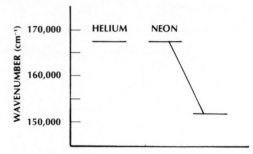

Figure 6-17   Helium-neon allowed energy states.

The allowed energies for atoms in a gas are distinct lines. Solids have bands of allowed energies (such as those shown previously in Fig. 6-1 for semiconductors). It is conventional to give the energy levels in units of inverse wavelength, $1/\lambda$, the photon *wave number*. We can convert the wave number to the corresponding energy in joules by using the relationship $W = hc/\lambda$ developed in Chapter 1; that is, we simply multiply the wave number by hc.

Atoms are normally in their lowest energy state, the *ground state*. In this state their energy is zero. An atom can absorb energy, raising it to an upper level. It is then in an *excited state*. An atom can become excited by absorbing an incoming photon. In this way, atoms in the Nd:YAG laser in Fig. 6-15 rise to high energy levels. In the case of the HeNe laser, the power supply causes an electric discharge current to flow in the gas. The gas atoms become ionized, freeing electrons for movement through the tube. The free electrons gain kinetic energy as they accelerate toward the positive electrode. In collisions with helium atoms, the electrons give up their energy, raising the energy levels of those atoms. This energy transfers to neon atoms when excited helium atoms collide with ground-state neon atoms.

Two of the excited levels for neon are shown in Fig. 6-17. Their energy difference is $15,800 \text{ cm}^{-1}$. This corresponds to a wavelength of $\lambda = 1/15,800 = 6.33 \times 10^{-5}$ cm $= 0.633$ μm. Consider the various possibilities for interaction between photons and the upper and lower of these two excited states.

1. An incoming photon whose wavelength is 0.633 μm can be absorbed by a neon atom that is in the lower excited state. The photon disappears, its energy used to raise the neon atom to the upper level.

2. An atom in the upper level can spontaneously drop to the lower one. The excess energy takes the form of an emitted photon whose wavelength is 0.633 μm. This resembles the process of electron-hole recombination (with the resulting emission of a photon) in the LED. Fluorescent lighting fixtures radiate by *spontaneous emission*.

3. An atom in the upper level may drop to the lower level, emitting a photon having wavelength 0.633 μm, when induced to do so by an incoming photon whose wavelength is also 0.633 μm. This is an example of *stimulated emission*. The stimulated photon will be emitted in phase with the stimulating photon, which continues to propagate.

If there are more neon atoms in the lower excited level than the upper one, then the number of photons entering the gas will decrease because of absorption. On the other hand, if the number of atoms in the upper level exceeds those in the lower one, a condition called *population inversion,* the number of photons will increase as they propagate through the gas because more photons will encounter upper-level atoms (causing generation of additional photons) than will meet lower-level atoms (which would absorb them). We conclude that a medium with population inversion has gain and behaves as an amplifier.

A laser is a high frequency generator, or oscillator. For oscillations to occur, a system needs amplification, feedback, and a tuning mechanism for determining the frequency. For radio-frequency oscillators, an electronic amplifier provides the signal gain, a filter determines the frequency, and feedback results by connecting the amplifier output back to its input. In the case of the laser, the medium provides the amplification. The medium also determines the frequency. It does so through its characteristic energy levels and transitions between levels. Mirrors provide the feedback. Photons bounce off the mirrors and return through the medium for further amplification. One (or both) of the mirrors is partially transmitting to allow a fraction of the generated light to emerge.

Oscillation will not occur until the gain exceeds all the losses in the laser. The losses include absorption (for example, in the medium and at the mirrors), scattering (primarily at the end windows and mirrors for the gas laser), and the extraction of the laser output power at the mirrors. When low voltages are applied to the laser, the gain is less than the loss and the laser output is zero. Spontaneous emissions may be occurring, but the power will be small and the output will not be coherent; that is, the spectral width will be large. When the voltage is increased, more neon atoms are raised to the upper level, increasing the gain. At a certain voltage level the system gain equals the loss and oscillation begins. The laser is at the threshold of oscillation at this stage. Further voltage increases will cause higher power output. The emitted light will now be coherent (the spectral width will be narrow). The concept of a threshold input is important when internally modulating a laser, particularly a laser diode.

The HeNe laser produces red light at 0.633 μm, corresponding to the transition between the two neon levels appearing in Fig. 6-17. The spectral width is small, about 1.98×10^{-3} nm. This represents a band of frequencies 1500 MHz wide. Even though the transition is between two distinct energy levels, the linewidth is not zero because of the thermal motion of the neon atoms in the gas. Each atom acts like a tiny source, generating light when dropping from a higher to a lower energy state. The well-known *Doppler effect* predicts a frequency shift for a source in motion. The random velocities of the atoms produce a range of Doppler-shifted frequencies surrounding the frequency determined by the transition. Stated in a slightly different way, the medium has amplification, not at a single frequency, but over a band of frequencies. Because there are fewer atoms moving at high speeds than lower ones, the amplifier gain drops off away from the center frequency, as sketched at the top of Fig. 6-18.

In Section 3-4 we discussed the resonances, or longitudinal modes, of the cavity formed by two end mirrors. These resonances are pictured below the gain curve in Fig. 6-18. For an output to exist at any frequency, there must be sufficient gain at that frequency and the cavity must be resonant at that frequency. In Fig. 6-18 these conditions are satisfied only at three frequencies, which explains the existence of the three longitudinal modes in the output spectrum. A longer cavity would decrease the separation between

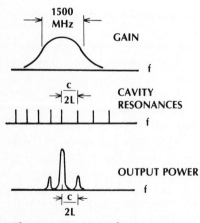

Figure 6-18 HeNe laser output.

modes, permitting more of them within the 1500-MHz-wide gain curve. The output spectrum would then contain more than three longitudinal modes.

Typically the output intensity of a gas laser is Gaussian, as discussed in Section 2-5 and drawn in Fig. 2-26. The divergence angle of a Gaussian beam was given in Eq. (2-17).

Example 6-1

Compute the divergence angle of a HeNe Gaussian beam whose spot size is 25 μm.

Solution:

In Eq. (2-17), $\theta = 2(0.633)/25\pi = 0.016$ rad, or $0.92°$. This is much smaller than the acceptance angle of typical fibers, meaning that all the emitted light rays could be trapped. The only coupling loss would be that caused by reflections at the air-to-fiber interface.

Lasers can radiate in patterns other than Gaussian. The different patterns correspond to the different electromagnetic modes of the laser cavity. These are called the *transverse modes* and are analogous to the modes in dielectric slabs and in fibers, which we studied previously. The Gaussian pattern is the lowest-ordered mode. When higher-ordered modes are allowed, the laser produces a multimode pattern, which is a combination of the individual mode patterns. A multimode beam is larger than a Gaussian beam and diverges more quickly.

6-4 LASER DIODES

Laser diodes[7] and light-emitting diodes have quite similar constructions. The structure of an AlGaAs laser diode is illustrated in Fig. 6-19 and should be compared with the LED in

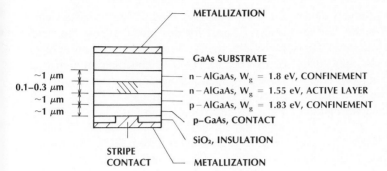

Figure 6-19 Double-heterojunction stripe-contact AlGaAs laser diode. The emitting edge is shown crosshatched in the active layer.

Fig. 6-3. Unlike LEDs, laser diodes are always edge emitters. When forward biased, charges are injected into the active layer where recombination takes place, causing the spontaneous emission of photons. Some of the injected charges are stimulated to emit by other photons. If the current density is sufficiently high, then a large number of injected charges are available for stimulated recombination. The optic gain will be large. The threshold current is reached when the gain is large enough to offset the diode losses. At this point, laser oscillation occurs. The threshold current must be small to prevent overheating of the semiconductor, particularly when operating continuously or at high peak power. A low threshold is achieved by confining the injected charges and the lightwave to the active layer by heterojunctions, as explained in Section 6-1. The heterojunctions provide confinement in the vertical direction in Fig. 6-19. The confinement of charges in the lateral direction is assured by the stripe contact. The charges are injected over the small width of the stripe (about 10–20 μm). They spread only slightly as they move into the recombination layer. The output wavelength, determined by the bandgap energy of 1.55 eV in the active region, is 0.8 μm for the LD in Fig. 6-19.

The lightwave is not entirely confined to the active layer because, as we know from our study of the slab waveguide, an evanescent tail extends beyond the totally reflecting boundaries. This situation is pictured in Fig. 6-20 for the laser diode.

The laser cavity is formed by cleaving the front and back faces of the semiconductor along parallel crystalline planes. The reflectance at the AlGaAs-air interface, as computed from Eq. (3-28), is 32% using a refractive index of 3.6 for the semiconductor. This amount of reflection provides sufficient feedback for oscillation. If desired, the end faces can be dielectric coated to increase their reflectance. Typical cavity lengths are around 300 μm. As with the HeNe laser, multiple cavity resonances produce longitudinal modes in the output spectrum. The longitudinal modes of a cavity were discussed in Section 3-4 and illustrated in Fig. 3-18 for an AlGaAs laser diode. Diodes radiating a spectrum containing numerous longitudinal modes usually have fields made up of several transverse

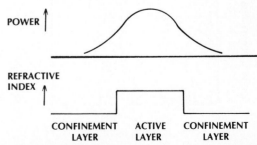

Figure 6-20 Distribution of power near the recombination region.

modes. That is, a multilongitudinal-mode laser may be a multitransverse-mode device. Single-longitudinal-mode lasers provide narrower linewidth, more coherent light than multilongitudinal-mode lasers, making them more suitable for long, high-rate systems because of the reduced material (and waveguide) dispersion.

Single-transverse-mode lasers couple more efficiently into single-mode fibers than do multimode lasers because the source and fiber mode patterns are nearly identical. That is, the output radiation of a single-transverse-mode laser diode has a transverse intensity pattern, which approximates that of the (nearly Gaussian) HE_{11} mode (drawn in Fig. 5-17) of the single-mode fiber. If the spot sizes of the laser and fiber modes are equal, then the radiation and propagation modes will be matched and the coupling efficiency will be high. The spot size of the laser can be adjusted by placing a lens between the laser and the fiber.

6-5 LASER DIODE OPERATING CHARACTERISTICS

The output optic power versus forward input current characteristic is plotted in Fig. 6-21

for a typical laser diode. The threshold current is 75 mA for this diode. Below this level there is a small increase in optic power with drive current. This is noncoherent radiation caused by spontaneous emission in the recombination layer. Spectral measurements would show a sharp decrease in the output linewidth when the current exceeds the threshold value. Threshold currents run from 30 to 250 mA for most diodes. The voltages are on the order of 1.2–2 V at threshold. The forward current increases rapidly with voltage for a diode, as demonstrated in Fig. 6-22, so that only a small increase in voltage beyond the threshold value will bring the current to its operating point. Output powers for continuously running lasers (CW, or *continuous wave*) are typically 1–10 mW. Pulsed lasers operating at low duty cycles can safely produce larger peak powers, but CW lasers that can be turned on and off at high rates are more useful for communications. The operating current is generally about 20–40 mA above the threshold current. Running at currents higher than those suggested by the manufacturer will shorten the lifetime of the diode.

Digital modulation of a laser diode, demonstrated in Fig. 6-23, differs from digital modulation of an LED. A dc bias current I_{dc} is added to place the current at threshold when the signal current i_s is zero. A binary 1 is gen-

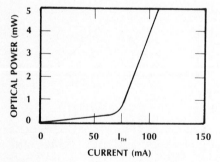

Figure 6-21 Power-current relationship for a laser diode.

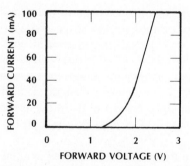

Figure 6-22 Typical voltage-current characteristic for a laser diode.

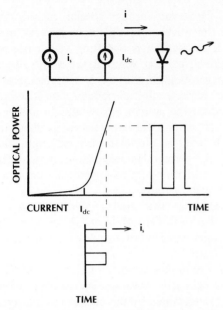

Figure 6-23 Digital modulation of a laser diode.

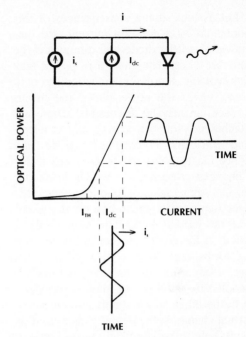

Figure 6-24 Analog modulation of a laser diode.

erated when the signal current contains a positive pulse, as sketched in the figure. When biased near threshold, the diode will turn on quicker and the signal current can be smaller than without the bias.

For analog modulation, Fig. 6-24, the dc bias is moved beyond threshold, so that operation will be along the linear portion of the power-current characteristic curve. The linearity of the laser diode should be carefully checked if the analog signal must be reproduced with low harmonic distortion.

Laser diodes are much more temperature sensitive than LEDs, as Fig. 6-25 illustrates for a representative diode. As the temperature increases, the diode's gain decreases so that more current is required before oscillation can begin. That is, the threshold current becomes greater (increasing about 1.5%/°C). This occurs because of thermal generation of holes in the n layer and electrons in the p layer. These free charges recombine with free electrons and holes outside the active layer, reducing

the number of charges reaching that layer and, consequently, reducing the number of charges available for the production of gain and stimulated emission. In addition, thermally generated holes and electrons in the active layer itself recombine nonradiatively reducing the population inversion. Again, a reduction in gain and an increase in threshold current results.

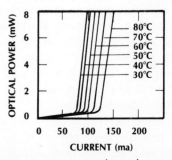

Figure 6-25 Temperature dependence of a laser diode.

Let us look at the consequences of this phenomenon. At a constant current, the output power of a laser diode will diminish if the temperature rises. The change in power may be unacceptable, increasing detection errors in the receiver. If the power drops too much, then reception may be impossible. There are two techniques for overcoming this problem: thermoelectrically cooling the diode and changing the bias current to compensate for the changed threshold. A thermoelectric cooler is a semiconductor junction device that changes temperature depending on the direction of the current flow. The laser diode is mounted on the cooler. A thermistor heat detector forms part of a control circuit that changes the current through the thermoelectric cooler to stabilize the diode's temperature. In the other power stabilization scheme, the actual change in the output is measured by allowing light to radiate from the back end of the laser diode and then by detecting this beam with a photodetector. The dc current is then changed to bring the optic power back to the desired value.

The laser emission wavelength also depends upon the temperature. This arises because of the dependence of the material's refractive index on temperature. As we saw in Section 3-4, the resonant wavelengths and the spacing of adjacent resonant wavelengths are determined by the refractive index of the cavity. As the temperature changes, the refractive index of the guiding layer varies resulting in a shift of the center emission wavelength and a slight change in the spacing of the longitudinal modes of a multimode laser diode. Typical shifts are on the order of a few tenths of a nanometer per degree centigrade (that is, a temperature coefficient of about 0.3 nm/°C). Quite often these wavelength shifts are inconsequential because they are small (a few nanometers) and because photodetectors do not change their response over such a short

range. However, a few special cases exist where the wavelength shift does have an impact. If the system is operating very close to the zero-dispersion wavelength of the fiber, then a wavelength shift of 5–10 nanometers would increase the dispersion considerably and reduce the system bandwidth. In another example, heterodyne systems (described in Section 10-5) require extreme wavelength stability. Temperature-induced wavelength shifts of even a fraction of a nanometer are intolerable for such systems.

AlGaAs laser diodes cover the region 0.8–0.9 μm. InGaAsP laser diodes emit in the longer-wavelength windows between 1 and 1.7 μm.

Laser diodes are much faster than LEDs. This is primarily because the rise time of a LED is determined by the natural spontaneous emission lifetime of the material and the rise time of a laser diode depends upon the stimulated emission lifetime. In a semiconductor, the *spontaneous lifetime* is the average time that free charge carriers (electrons and holes) exist in the active layer before recombining spontaneously. The *stimulated emission lifetime* is the average time that free charge carriers exist in the active layer before being forced to recombine by stimulation. Obviously, for a laser medium to have gain the stimulated lifetime must be shorter than the spontaneous lifetime. Otherwise, spontaneous recombinations would occur before stimulated emission could begin, decreasing the population inversion and prohibiting gain and oscillation. The faster stimulated emission process, which dominates recombination in a laser diode, ensures that a laser diode will respond quicker to changes in the injected current than a light-emitting diode.

The rise times of good laser diodes run between 0.1 and 1 ns. They can be analog modulated at frequencies of several gigahertz. The short rise times are measured with the diodes biased at threshold, as shown in Fig. 6-23. It takes longer to turn the diode on if it is started at zero current. Similarly, the analog modulation rate is determined with the diode biased to some point along the linear portion of the output characteristic, as displayed in Fig. 6-24. Modulation rates in the tens of gigahertz have been demonstrated by using specially designed diodes.

Laser diodes typically possess linewidths of 1–5 nm, considerably smaller than the output spectra of LEDs. The spectral widths are greater than those of gas lasers because the emitting transitions in the semiconductor are between energy bands and the gas transitions are between distinct lines. This phenomena produces linewidth spreading much larger than that caused by the Doppler effect in gases. The spectrum of a representative laser diode operating near 1.3 μm is drawn in Fig. 6-26. The multiple peaks correspond to the longitudinal modes of the device.

When the drive current is just a bit above threshold, laser diodes produce multimode spectra like that shown in Fig. 6-26. As the current increases, the total linewidth decreases and the number of longitudinal modes diminishes. At a sufficiently high current, the spectrum will contain just one mode. Figure 6-27 illustrates the spectrum of a single-longitudinal-mode laser. As expected, its linewidth is much smaller than that of a multi-

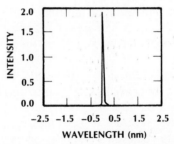

Figure 6-27 Output spectrum of a single-longitudinal-mode laser diode.

mode laser. The linewidth is around 0.2 nm for the spectrum in Fig. 6-27. A single-longitudinal-mode diode would minimize material dispersion in a fiber because of its narrow spectral width.

Laser diodes do not radiate symmetrically. A representative pattern appears in Fig. 6-28. This distribution of light should be compared with the radiation of a surface-emitting LED, in Fig. 6-10, and of an edge-emitting LED, in Fig. 6-11. The light from the laser diode is contained within a much smaller angular region, making coupling to a fiber easier and more efficient. Something else that you may have noticed needs an explanation. The directions of the narrow and broad beams, relative to the emitting edge, are reversed in Figs. 6-11 and 6-28. The light from the LED is noncoherent. The large dimension of the emitting edge is in the plane parallel to the junction and produces the large beam. The narrow dimension of the edge lies in a plane perpendicular to the junction and radiates over a smaller range of angles. Coherent light from the laser follows the laws of diffraction, which we encountered in Section 2-5. We found that the beam divergence was inversely proportional to the dimensions of the radiator. This result only applies for coherent light. It explains the wide beam divergence, corresponding to the narrow dimension of the edge, and the smaller beam divergence, corresponding

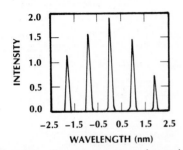

Figure 6-26 Output spectrum of a multimode laser diode.

to the wide dimension of the edge. The diode in Fig. 6-28 has a half-power beamwidth of 10° in the parallel plane and 35° in the perpendicular plane.

The reliability and lifetimes of CW laser diodes have improved greatly since the early 1970s, when the first heterostructure AlGaAs devices were constructed. Lifetimes exceeding 11 years are achieved for diodes operating at room temperature. Diodes degrade faster at elevated temperatures. However, even at 70°C lifetimes of more than 10,000 hours can be expected from good commercial laser diodes.

Like LEDs, laser diodes have been mounted in a variety of packages. These structures must be carefully designed and constructed. Packaging requirements include the following ones:

1. Hermetic seals on all leads. This includes the electrical leads and the fiber (if it penetrates into the diode enclosure)
2. Precise positioning of the laser chip to aid alignment with directly coupled fiber or lens-coupled fibers
3. If desired, provision for a photodetector within the case for monitoring the power emitted from the back face of the laser
4. For operation at high temperatures, the diode can be mounted on a thermoelectric cooler that is located inside the package

A few possible packaging schemes are sketched in Figs. 6-29, 6-30, and 6-31. In Fig. 6-29, the diode rests on a copper heat sink. A lens can be placed outside the window to focus light onto a fiber. Alternatively, the cap can be removed and the fiber can be epoxied close to the laser's emitting edge. In the package of Fig. 6-29, the emitting back face of the diode is blocked off, making it unavailable for monitoring purposes. A fiber pigtail is included in the package drawn in Fig. 6-30. A lens may be located between the diode and the fiber to improve the coupling efficiency. Maximum power is coupled from the pigtail if it is identical to the transmission fiber. The losses incurred when connecting dissimilar fibers are computed in Chapter 8. The customer can splice the pigtail to the transmission fiber or

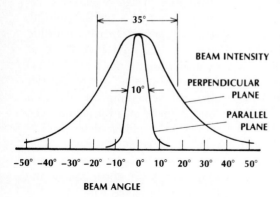

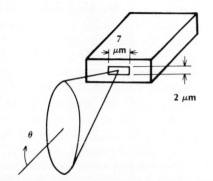

Figure 6-28 Radiation pattern of a laser diode.

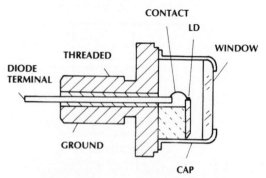

Figure 6-29 Laser diode package.

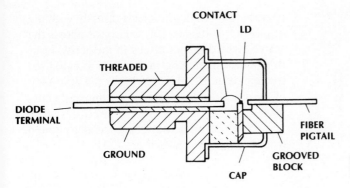

Figure 6-30 Laser diode with an integral fiber pigtail.

attach a connector to the pigtail for ease of connecting and disconnecting the source. The diode manufacturer may even supply an attached connector of the customer's choosing. The wide variety of fiber sizes and connector designs makes it important for the system specialist to specify these components carefully and to understand the losses they will produce.

A laser diode mounted with a power monitor is sketched in Fig. 6-31. The photodetector measures the power radiated from the back face of the emitter. This type of device could be enclosed in a standard electrical structure, such as the multiple-pin dual in-

line package (DIP) illustrated in Fig. 6-32. Pins are provided for connections to the laser, the photodetector, and a thermoelectric cooler and thermistor temperature monitor, if they are included in the package. This assembly plugs into a conventional circuit board.

6-6 DISTRIBUTED-FEEDBACK LASER DIODE

In Fig. 6-27 we showed the output spectrum of a single-longitudinal-mode laser diode. This behavior has been obtained by construct-

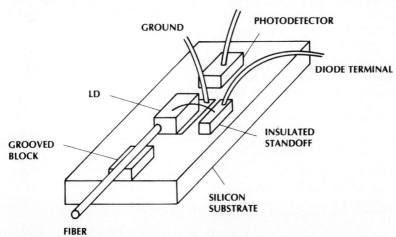

Figure 6-31 Laser diode with integral power monitor and fiber pigtail.

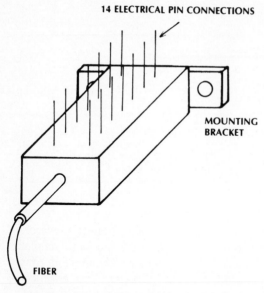

14 ELECTRICAL PIN CONNECTIONS

MOUNTING
BRACKET

FIBER

Figure 6-32   Fourteen-pin dual in-line laser package.

ing the *distributed-feedback* (DFB) laser diode.[8] This laser, drawn in Fig. 6-33, has a corrugated layer etched internally just above the active region. The corrugation forms an optical grating that selectively reflects light according to its wavelength. This grating acts as a distributed filter allowing only one of the cavity's longitudinal modes to propagate back and forth. We might think of the grating and the mirrored cavity each having a set of resonant wavelengths that they will support, but there is only one resonant wavelength that they have in common. This will be the single longitudinal mode of the combined res-

onators. The grating interacts directly with the evanescent mode in the space just above the active layer. It is not placed in the active layer itself because etching in this region could introduce defects that would lower the efficiency of the laser resulting in a higher threshold current.

The operating wavelength is determined from *Bragg's law,* which is

$$\Lambda = m\lambda/2 \qquad (6\text{-}10)$$

where Λ is the grating period (distance between adjacent peaks), λ is the wavelength as measured in the diode, and m is the order of the Bragg diffraction. The wavelength in the material is related to the free-space wavelength by Eq. (3-7). Slightly rearranging that equation yields

$$\lambda = \lambda_0/n$$

To correctly determine the wavelength in the diode, we must use the effective refractive index of the mode propagating in the cavity rather than the index of the bulk material. From our studies in Section 4-2 we know that the effective index lies somewhere between the index of the guiding layer (the active region of the diode) and the cladding layers (just above and below the active layer). Since these indices are not too different from each other, we can easily obtain a pretty good idea of the magnitude of the grating period. The grating period can now be written as

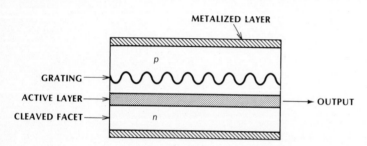

METALIZED LAYER

p

GRATING →

ACTIVE LAYER →

→ OUTPUT

CLEAVED FACET →

n

Figure 6-33   Distributed-feedback laser diode.

$$\Lambda = \frac{m\lambda_0}{2\,n_{\text{eff}}} \qquad (6\text{-}11)$$

An example will illustrate this calculation.

Example 6-2
Compute the grating period for an InGaAsP DFB laser diode emitting at 1.55 μm.

Solution:
From Table 2-1, we see that the refractive index of InGaAsP is about 3.51. We will take the effective index as 3.5 and assume a first-order diffraction ($m = 1$). Then, $\Lambda = 1.55/(2 \times 3.5) = 0.22$ μm. The second-order diffraction, yielding a period of 0.44 μm, is also efficient enough to be used.

DFB lasers have a number of unique properties arising from the grating structure. In addition to their narrow linewidths, which make them attractive for long high-bandwidth transmission paths, they are less temperature dependent than most conventional laser diodes. The grating tends to stabilize the output wavelength, which varies with temperature owing to changes in refractive index. Typical temperature caused wavelength shifts are just under 0.1 nm/°C, which provides 3–5 times better performance than conventional laser diodes. DFB lasers are also more linear in their response than conventional laser diodes. This allows their use in analog systems where a high degree of linearity is required to reduce distortion. Linearity also minimizes intermodulation when several channels are multiplexed for simultaneous transmission. Because of this, DFB lasers have been used successfully for analog modulation of multiplexed television signals.

6-7 OPTICAL AMPLIFIERS

We have indicated that fiber systems are ultimately limited by either bandwidth or attenuation. In either case, for digital systems a regenerator can be inserted in the path to reshape and amplify the pulses. This is done by detecting the optical signal (which converts it to electrical form), determining the presence of ones and zeroes and reconstructing the original optical signal (now at full power and with the pulse-spreading distortion removed) by modulating a light source. Regenerators have successfully been used to extend fiber paths from practical point-to-point limits of a few hundred kilometers to total lengths of thousands of kilometers. For example, transatlantic fiber cables cover over 5000 km and require about a hundred regenerators. While regenerators are invaluable, they are expensive to construct, expensive to install, and expensive to maintain.

If analog modulation is used, then the situation becomes worse for fiber communications. Regeneration is impossible because we do not know what the signal is supposed to look like. (In a digital system we know the data stream consists of only zeroes and ones, allowing us to reconstruct each bit. In an analog system the choices of waveshapes are limitless.) Conversion of an optical analog signal to electrical form for amplification and retransmission is expensive and (probably) noisy.

The preceding discussion leads us to search for an all-optical amplifier. That is, one that amplifies the signal without the intermediate conversion to electrical form. Optical amplifiers will not solve the problem of reconstructing signal waveshapes, but will allow extension of power-limited links. In other words, bandwidth-limited systems will not be helped but power-limited ones will. Because fibers can operate with little band-limiting near the zero-dispersion point of conventional or dispersion-shifted fibers, bandwidth is less of a problem than attenuation. In addition, if the data stream consists of soliton pulses, then no pulse spreading occurs and bandwidth is no longer limited. We discussed soliton propagation earlier in Section 3-2. In the late 1980s several successful amplifiers were developed. Two of them will be examined: the semiconductor amplifier and the erbium-doped fiber amplifier.

From our discussion of laser principles we are led to believe that optical amplifiers can be constructed by using stimulated emission in a material that has gain (e.g., in a pumped semiconductor junction). Essentially, this would be a laser operating without the mirrors (this is a *traveling-wave laser amplifier*) or one with mirrors but operating just below lasing threshold (this is a resonant *Fabry-Perot laser amplifier*).[9] These devices are illustrated in Fig. 6-34. In principle, these structures work, but in practice many problems have kept such devices from becoming

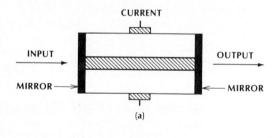

(a)

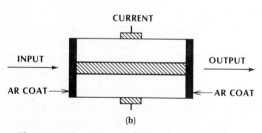

(b)

Figure 6-34 Semiconductor laser amplifiers. (a) Fabry-Perot amplifier. (b) Traveling-wave amplifier.

widely accepted. Achieving enough gain and doing so without adding too much noise has been a problem. Also, the gain of the semiconductor amplifier is polarization dependent.

The erbium-doped fiber amplifier (EDFA) has generated significant interest because of its high gain, large bandwidth, and low noise.[10] This device, shown in Fig. 6-35, consists of an erbium-doped silica fiber that amplifies the incoming light beam. The rare-earth element erbium is the active material. In the silica host it has gain near 1.55 μm when pumped in one of several absorption bands. The most efficient pumping bands are around 980 and 1480 nm. As indicated on the figure, the pump light (from a laser diode) and the signal light are coupled to the doped fiber by a combining wavelength multiplexer. This device is a wavelength-dependent directional coupler of the sort used for wavelength-division multiplexing systems. Its operation will be described in Section 9-6. The pumping light is absorbed by the erbium atoms, raising them to excited states and causing population inversion. The excited erbium atoms are then stimulated to emit by the longer-wavelength photons, amplifing the signal. The signal beam and the pumping beam travel together down the fiber. The signal beam continually increases in strength while depleting the pump power. The second wavelength multiplexer in Fig. 6-35 removes any pump photons not absorbed by the doped fiber so that they do not reach the receiver and interfere with signal detection.

The amplification near 1.55 μm is a perfect match for systems operating at the wavelength of lowest fiber loss. Operating bandwidths of more than 20 nm are achievable, allowing a number of wavelength-division multiplexed channels to be simultaneously amplified. Fiber lengths are typically a few tens of meters. The optimum length depends on the amount of pump power available. Since

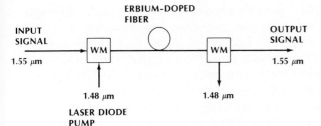

INPUT
SIGNAL

1.55 μm

ERBIUM-DOPED
FIBER

WM

OUTPUT
SIGNAL

1.55 μm

WM

1.48 μm

1.48 μm

LASER DIODE
PUMP

Figure 6-35   Erbium-doped fiber amplifier. The fiber wavelength multiplexers are labeled WM.

the pump power decreases as it travels down through the fiber, eventually it becomes so weak that the gain reduces to zero (and the pumped fiber becomes absorbing rather than amplifying). Net gains of 5–10 dB per milliwatt of pump power have been measured. Because of this, total gains of more than 30 dB have been achieved with pump powers under 10 mW. The EDFA is easily coupled to the fiber transmission line because it is itself a fiber.

All amplifiers (electrical as well as optical) saturate at high levels of output power. Saturation is the decrease in gain that occurs when the amplified power reaches high levels. For the EDFA, the saturation power increases with diode pump power but is expected to reach beyond 50 mW of output power.

The *noise figure F* is a measure of the noise characteristics of an amplifier. It is given by the ratio of the input signal-to-noise ratio to the output signal-to-noise ratio. That is,

$$F = \frac{(S/N)_{in}}{(S/N)_{out}} \qquad (6\text{-}12)$$

It gives an indication of the degradation in a signal owing to amplification. All amplifiers (electrical as well as optical) have noise figures greater than unity, which diminishes the signal quality. Even with this degradation, optical amplification improves performance compared to systems where the receiver amplifies the signal electrically after the detection process. Semiconductor laser amplifiers

generally have noise figures over 4 dB and erbium-doped fibers have achieved noise figures close to the theoretical minimum value of 3 dB.

Example 6-2

An optical amplifier has a noise figure of 3.2 dB. The input signal has a signal-to-noise ratio of 50 dB. Compute the output signal-to-noise ratio.

Solution:

Converting from decibels to ratios, we find the input S/N is 10^5 and the noise figure is 2.089. The output is then $(S/N)_{out}$ = $10^5/2.089$ = 0.4786×10^5. Converting back to decibels, we obtain an output signal-to-noise-ratio of 46.8 dB.

Note from this last example that, if we keep all quantities in decibels, the output signal-to-noise ratio is just the input signal-to-noise ratio minus the amplifier's noise figure. This is a general result. In equation form

$$(S/N)_{out,dB} = (S/N)_{in,dB} - F_{dB} \qquad (6\text{-}13)$$

We will cover noise in a fiber system in more detail in Chapter 11.

6-8 SUMMARY

As an aid in preliminary design, the characteristics of typical semiconductor light sources are listed in Table 6-2. At this point in the discussion we have enough information to select the carrier wavelength, the type of fiber, and the light source. LEDs can be used profitably with both multimode SI or multimode GRIN fibers, but in different regions of the optic spectrum. In SI fibers, modal distortion dominates. Material dispersion, caused by the large spectral width of the LED, is smaller and can usually be neglected. Reducing the material dispersion further by selecting a laser diode serves no purpose. For these reasons, LEDs are normally chosen for multimode SI links. Systems using multimode SI fibers and LED sources will probably remain in the first window (0.8–0.9 μm), where component costs are low. LEDs radiating in the first window are not optimum for GRIN links because material dispersion causes more pulse spreading than the fiber's modal distortion. The advantages of GRIN fiber are mostly lost with this combination of components. However, in the second window (near 1.3 μm) material dispersion becomes minimal, even with a LED source. A GRIN fiber and a LED operating in the long-wavelength region can combine to produce a system transmitting moderately high data rates over fairly long distances.

Because of higher initial costs and increased circuit complexity, laser diodes are only used when necessary. For long, high-capacity systems, they combine effectively with multimode GRIN fibers or single-mode fibers. These systems operate in the first or second window. In the second window, fiber losses are lower, allowing longer transmission paths.

The largest rate-length products are achieved when a single-mode laser diode is matched with a single-mode fiber and operated in the low-loss, long-wavelength windows.

PROBLEMS

6-1. Consider a resistor in series with a capacitor. Let the input be a step of 1 V. Compute and plot the resulting capacitor voltage. Compute the 10–90% rise time of this voltage in terms of R and C.

TABLE 6-2. Typical Characteristics of Diode Light Sources

Property	LED	Laser Diode	Single-Mode Laser Diode
Spectral width (nm)	20–100	1–5	<0.2
Rise time (ns)	2–250	0.1–1	0.05–1
Modulation bandwidth (MHz)	<300	2000	6000
Coupling efficiency[a]	Very low	Moderate	High
Compatible fiber	Multimode SI[b]	Multimode GRIN	Single-mode
	Multimode GRIN[c]	Single-mode	
Temperature sensitivity	Low	High	High
Circuit complexity	Simple	Complex	Complex
Lifetime (hours)	10^5	10^4–10^5	$10^4 - 10^5$
Costs	Low	High	Highest
Primary use	Moderate paths	Long paths	Very long paths
	Moderate data rates	High data rates	Very high rates

[a] Coupling efficiency can be improved with lenses.

[b] First window system.

[c] Second window system.

6-2. For the circuit in Problem 6-1, let the input be $v_{in} = \cos \omega t$. Compute and plot the capacitor voltage versus ω. Show that the circuit's 3-dB bandwidth is $f_{3\text{-}dB} = 0.35/t_r$, where t_r is the 10–90% rise time.

6-3. Photodetected current flows through a resistor R. This current is proportional to the optical power. Prove that a change in optical power (expressed in decibels) is equal to half the change in electrical power (expressed in decibels).

6-4. Assume that the optical power from a LED varies with modulation frequency according to Eq. (6-7). Show that the 3-dB optic and detected-electrical bandwidths are related by $f_{3\text{-}dB}$ (optic) $= 1.73 f_{3\text{-}dB}$ (electrical).

6-5. The optical power versus current relationship for a LED is given by $P = 0.02 \times i$. The maximum allowed power is 10 mW. The LED has a dc bias current and a 1-MHz ac current applied.
   (a) Sketch the power-current curve (this is the diode's transfer characteristic).
   (b) If the peak signal power is 2 mW and the peak total power is 10 mW, compute the total peak current, the dc bias current, the average optical power, and the modulation index (peak signal power)/(average power).
   (c) Repeat if the modulation index is now 100% (and the peak signal power is no longer 2 mW).
   (d) Let the dc current be 50 mA and the peak ac current be 75 mA. Plot the output power versus time for two cycles of the ac signal.

6-6. When a LED has 2 V applied to its terminals, it draws 100 mA and produces 2 mW of optical power. What is the LED's conversion efficiency from electrical to optical power?

6-7. A LED has a 3-dB electrical bandwidth (determined entirely by its carrier lifetime) of 100 MHz. Compute the carrier lifetime. Plot the diode's normalized frequency response (as in Fig. 6-8) for 0–500 MHz.

6-8. Compute the bandgap energy of GaAs in joules.

6-9. A laser diode has a threshold current of 10 mA and a slope of its optical power versus input current of 2 mW/mA. The total diode current in mA is $i = 20 + \sin \omega t$.
   (a) Write the equation of the output power and sketch it.
   (b) Sketch the output power waveform if the current is changed to $i = 10 + \sin \omega t$

6-10. Compare the relative advantages and disadvantages of light-emitting diodes and laser diodes.

6-11. For a Lambertian emitter, compute the total beamwidth to the one-fourth power points on the radiation pattern.

6-12. A GaAs laser diode has a 1.5-nm gain linewidth and a cavity length of 0.5 mm. Sketch the output spectrum including as many details as you can (for example, the emitted wavelengths and the number of modes).

6-13. An erbium-doped fiber amplifier has a noise figure of 6 and a gain of 100. The input signal has a 30-dB signal-to-noise ratio and a signal power of 10 μW. Compute the signal power (in dBm) and signal-to-noise ratio (in dB) at the amplifier's output.

6-14. Erbium amplifiers can operate over bandwidths of about 20 nm (1530–1550 nm). How many 10 GHz channels can fit into this range (by multiplexing) and, thus, be amplified simultaneously?

6-15. Suppose the saturation power of an erbium amplifier is 20 mW and the gain is 5 dB per milliwatt of pump power. The pump power is set at 5 mW. What is the largest amount of input power that can be amplified without driving this amplifier into saturation?

6-16. By using Fig. 6-25, plot the threshold current as a function of the temperature. Determine the change in threshold current per unit of rise in temperature from the resulting plot.

6-17. Suppose a laser diode changes its emission wavelength by 0.5 nm/°C. The unshifted emission wavelength is 1310 nm, and the fiber's zero dispersion wavelength is 1300 nm. The laser's linewidth is 1.5 nm and a temperature shift of 10°C causes an increase in the output wavelength. Compute the decrease in the fiber's 3-dB optical bandwidth arising from this temperature increase. Assume that only material dispersion is important.

6-18. Convert the temperature induced emision shift of 0.5 nm/°C to its equivalent shift in GHz/°C at a wavelength of about 1 μm.

6-19. Compute the grating spacing for an InGaAsP DBF laser diode operating at 1300 nm. Give results for both the first- and second-order diffraction.

REFERENCES

1. H. Kressel. "Electroluminescent Sources for Fiber System." In *Fundamentals of Optical Fiber Communications,* 2nd ed., edited by Michael K. Barnoski. New York: Academic Press, 1981, pp.187–255.

2. H. Kressel, M. Ettenberg, J. P. Wittke, and I. Ladany. "Laser Diodes and LEDs for Fiber Optical Communications." In *Semiconductor Devices for Optical Communication,* edited by H. Kressel. New York: Springer-Verlag, 1980, pp. 9–62.

3. B. V. Dutt, J. H. Racette, S. J. Anderson, and F. W. Scholl. "AlGaInP/GaAs Red Edge-Emitting Diodes for Polymer Optical Fiber Applications." *Appl. Phys. Lett.* 15, no. 21 (Nov. 1988): 2091–2092.

4. Donald M. Fye. "Low-Current 1.3-μm Edge-Emitting LED for Single-Mode Fiber Subscriber Loop Applications," *J. Lightwave Technol.* LT4, no. 10 (Oct. 1986): 1546–51

5. Motorola Semiconductor Technical Data, MFOE3100/D. Phoenix, Ariz., 1986.

6. Introductory texts covering lasers include the following ones: Donald C. O'Shea, W. Russell Callen, and William T. Rhodes. *Introduction to Lasers and Their Applications.* Reading, Mass.: Addison-Wesley, 1977. Joseph T. Verdeyen. *Laser Electronics.* Englewood Cliffs, N.J.: Prentice Hall, 1981.
Ammon Yariv. *Optical Electronics,* 4th ed. New York: Holt, Rinehart and Winston, 1991.

7. Laser diodes are covered in detail in references 1, 2, and 5.

8. N. K. Dutta. "Optical Sources for Lightwave System Applications." In *Optical-Fiber Transmission,* edited by E. E. Basch. Indianapolis, Ind.: Howard W. Sams, 1987, pp. 282–85.

9. K. Nakagawa and S. Shimada. "Optical Amplifiers in Future Optical Communications Systems," *IEEE LCS* 1, no. 4 (Nov. 1990): pp. 57–62.

10. E. Desurvire. "Erbium-Doped Fiber Amplifiers for New Generations of Optical Communication Systems," *Optics Photonics News* 2, no. 1 (Jan. 1991): pp. 6–11.

Chapter 7

Light Detectors

Light can be detected by the eye. The eye is not suitable for modern fiber communications because its response is too slow, its sensitivity to low-level signals is inadequate, and it is not easily connected to electronic receivers for amplification, decoding, or other signal processing. Furthermore, the spectral response of the eye is limited to wavelengths between 0.4 and 0.7 μm, where fibers have high loss. Nonetheless, the eye is very useful when fibers are tested with visible light. Breaks and discontinuities can be observed by viewing the scattered light. Systems, such as couplers and connectors, can be visually aligned with the visible source before the infrared emitter is attached. The remainder of this chapter is confined to an investigation of devices that directly convert optic radiation to electrical signals (either current or voltage) and that respond quickly to changes in the optic power level.

7-1 PRINCIPLES OF PHOTODETECTION

We will look at two distinct photodetection[1,2] mechanisms. The first is the *external photoelectric effect,* in which electrons are freed from the surface of a metal by the energy absorbed from an incident stream of photons. The vacuum photodiode and the photomultiplier tube are based on this effect. A second group of detectors are semiconductor junction devices in which free charge carriers (electrons and holes) are generated by absorption of incoming photons. This mechanism is sometimes called the *internal photoelectric effect.* Three common devices using this phenomenon are the *pn* junction photodiode, the PIN photodiode, and the avalanche photodiode.

167

Important detector properties are responsivity, spectral response, and rise time. The *responsivity* ρ is the ratio of the output current of the detector to its optic input power. In equation form, it is

$$\rho = \frac{i}{P} \qquad (7\text{-}1)$$

The units of responsivity are amperes per watt. In some detector configurations the electrical output is a voltage. In this case, the responsivity is given in units of volts per watt of incident power. The *spectral response* refers to the curve of detector responsivity as a function of wavelength. Because of the rapid change in responsivity with wavelength, different detectors must be used in the two windows of the optic spectrum where fiber losses are low. Within either of the windows, the responsivity at the specific wavelength emitted by the source must be used when designing the receiver.

The rise time t_r is the time for the detector output current to change from 10 to 90% of its final value when the optic input power variation is a step. This is consistent with the definition of the rise time of an optic source given in Chapter 6. Detector rise time is illustrated in Fig. 7-1. The 3-dB modulation bandwidth of the detector is

$$f_{3\text{-dB}} = \frac{0.35}{t_r} \qquad (7\text{-}2)$$

At this frequency, the electrical signal power in the receiver is half of that obtained at very low modulation frequencies, assuming the same amount of optic signal power incident on the detector in both cases.

Other photodetector characteristics will be introduced at appropriate points in the remainder of this chapter.

7-2 PHOTOMULTIPLIER

The vacuum photodiode and photomultiplier tube are not placed in operational fiber communications systems, although they can be useful in testing of fiber components. The high sensitivity of the photomultiplier makes it particularly helpful when measuring low levels of optic power. Photoemissive detectors are somewhat easier to explain than semiconductor devices, and the two have many properties in common. For this reason we will begin our discussion of photodetector operation with the photoemitters.

A vacuum photodiode is sketched in Fig. 7-2. A bias voltage is applied, making the anode positive and the cathode negative. With no light the current passing through the load resistor is zero and the output voltage is zero. When the cathode is irradiated with light, incoming photons are absorbed, giving up their energies to electrons in the metal.

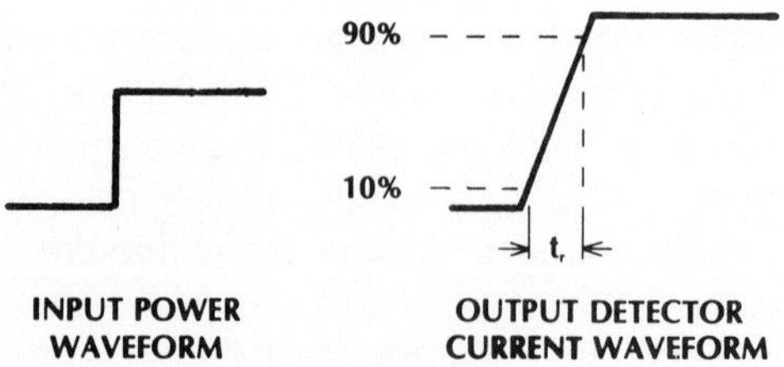

Figure 7-1   Photodetector rise time.

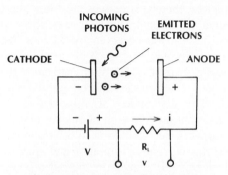

Figure 7-2   Vacuum photodiode.

Some of these electrons gain enough energy to escape from the cathode. These free electrons move toward the anode, attracted by its positive charge. During this movement, positive charge is drawn through the external circuit (i.e., through the load resistor) to the anode, attracted by the approaching negatively charged electrons. In other words, current flows through the circuit. When the electrons strike the anode, they combine with the positive charges and the circuit current stops.

To liberate a single electron from the cathode requires a minimum amount of energy, called the *work function*. An incoming photon must possess at least this much energy to cause electron emission. Denoting the work function by ϕ, the condition for release of an electron is thus

$$hf \geq \phi \qquad (7\text{-}3)$$

The lowest optic frequency that can be detected is $f = \phi/h$. This corresponds to a wavelength of $\lambda = hc/\phi$. If the work function is given in electron volts, then the *cutoff wavelength* (in μm) becomes

$$\lambda = \frac{1.24}{\phi} \qquad (7\text{-}4)$$

Wavelengths longer than this cannot be detected owing to insufficient photon energy. Shorter-wavelength photons, being more energetic, can be detected.

Example 7-1

Cesium, a common photoemissive material, has a work function of 1.9 eV. Compute its cutoff wavelength.

Solution:

From Eq. (7-4), $\lambda = 1.24/1.9 = 0.65$ μm. Only wavelengths shorter than this can be detected by a cesium cathode.

Example 7-1 shows that a cesium detector is not sensitive at all to wavelengths of 0.8 μm and beyond, where fiber systems operate. Vacuum photoemitters other than cesium can detect at wavelengths as large as 1.1 μm, but their responsivities are quite low at the longer wavelengths. In any case, the vacuum photodiode is too large and requires too much voltage (several hundred volts or more) to be practical for fiber communications. We will continue our discussion of this device, however, because of the light it sheds on the operation of the photomultiplier tube and semiconductor detectors.

Not every photon whose energy is greater than the work function will liberate an electron. This characteristic is described by the *quantum efficiency* η of the emitter. It is defined by

$$\eta = \frac{\text{number of emitted electrons}}{\text{number of incident photons}} \qquad (7\text{-}5)$$

We can easily calculate the responsivity, in Eq. (7-1), of a photodetector. Because the optic power is the energy per second being delivered to the detector and hf is the energy per photon, then P/hf is the number of photons per second striking the cathode. With quantum efficiency η, the number of emitted electrons per second is then $\eta P/hf$. Since each electron carries charge of magnitude e, the charge per second (that is, the current) emerging from the cathode is

$$i = \frac{\eta e P}{hf} = \frac{\eta e \lambda P}{hc} \qquad (7\text{-}6)$$

This is the current that flows through the load resistor in the external circuit. The detector behaves as if it were a current source for the receiving circuit. The responsivity is now

$$\rho = \frac{i}{P} = \frac{\eta e}{hf} = \frac{\eta e \lambda}{hc} \qquad (7\text{-}7)$$

The output voltage is

$$v = \frac{\eta e P R_L}{hf} = \rho P R_L \qquad (7\text{-}8)$$

Equations (7-6), (7-7), and (7-8) are valid for photoemissive and semiconductor junction detectors. Equation (7-6) shows that the detected current is directly proportional to the optic power, a property we have been assuming throughout this book.

Example 7-2

Compute the responsivity of a detector having a quantum efficiency of 1% at 0.8 μm.

Solution:

From Eq. (7-7),

$$\rho = \frac{0.01(1.6 \times 10^{-19})(0.8 \times 10^{-6})}{(6.63 \times 10^{-34})(3 \times 10^8)}$$

$$= 0.0064 \ A/W$$

or 6.4 mA/W.

Example 7-3

Use the results of Example 7-2 to compute the voltage across a 50-Ω load resistor when the optic power absorbed by the detector is 1 μW.

Solution:

The current produced by the detector will be $i = (6.4 \times 10^{-3})(10^{-6}) = 6.4$ nA. The output voltage is then $v = (6.4 \times 10^{-9})(50) = 320$ nV, a very small value.

The photomultiplier tube (PMT) has much greater responsivity than does the photodiode because of an internal gain mechanism. A PMT is represented schematically in Fig. 7-3. Electrons emitted from the cathode are accelerated toward an electrode called a *dynode*. The first dynode attracts electrons because it is placed at a higher voltage than the cathode, typically 100 V or more. The electrons hitting the dynode have high kinetic energies. They give up this energy, causing the release of electrons from the dynode. This process is called *secondary emission*. An incident electron can liberate more than one secondary electron, thus amplifying the detected current. The current is amplified at each of the successive dynodes. Each dynode must be at a higher voltage than the preceding one in order to attract (and thus accelerate) the electrons.

The gain at each dynode is the number of secondary electrons released per incident electron. Gains of 2–6 are common. Let us follow the progress of a single photoemitted electron through the multiplier tube. If the gain at each dynode is δ, then the number of electrons emerging from the first dynode is just δ. The number of electrons in the tube after the second dynode is δ^2, after the third δ^3, and so on. When there are N dynodes, the total gain is then

$$M = \Delta^N \qquad (7\text{-}9)$$

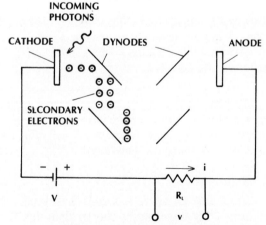

Figure 7-3 Photomultiplier.

and the current through the external circuit is

$$i = \frac{M\eta eP}{hf} \qquad (7\text{-}10)$$

Example 7-4

Compute the current amplification in a photomultiplier tube if the gain at each dynode is 5 and there are nine dynodes.

Solution:

From Eq. (7-9) $M = 5^9 = 1.95 \times 10^6$, a current gain of almost 2 million.

Example 7-5

A PMT with the gain just calculated is used to detect an optic power of 1 μW at 0.8 μm. The cathode is 1% efficient and the load is 50 Ω. Compute the responsivity, current, and output voltage.

Solution:

The numerical values in this example, except for the gain, are the same as those we used in the preceding examples involving the vacuum photodiode. The responsivity is now (1.95×10^6) $(6.4 \times 10^{-3}) = 12.5$ kA/W. The current is $(12.5 \times 10^3)(10^{-6}) = 12.5$ mA, and the voltage is $(12.5 \times 10^{-3})(50) = 0.625$ V or 625 mV. This is a remarkable increase over the 320 nV obtained with the photodiode.

Amplification within a detector, such as occurs in the PMT, is *internal* gain. This is in contrast to *external* gain, which can be obtained from electronic amplifiers following the detector. Internal gain has an important advantage. It increases the signal level without significantly lowering the ratio of signal power-to-noise power. External amplifiers always add noise to the system, diminishing the

signal-to-noise ratio. Because of their high gain, photomultipliers are useful in detecting low levels of radiation and in overcoming noise originating from thermal sources. We consider the effects of noise in more detail in Chapter 11.

Photomultipliers are very fast. Some have rise times of a few tenths of a nanosecond. Their disadvantages include high cost, large size, high weight, and the need for a power supply providing hundreds of volts for bias.

7-3 SEMICONDUCTOR PHOTODIODE

Semiconductor junction photodiodes are small, light, sensitive, fast, and can operate with just a few bias volts. They are almost ideal for fiber systems. We will investigate three forms of these devices: the *pn*, PIN, and avalanche photodiodes.

The simple *pn* photodiode, drawn in Fig. 7-4, illustrates the basic detection mechanism of a junction detector. When *reverse* biased, the potential energy barrier between the *p* and *n* regions increases. Free electrons (which normally reside in the *n* region) and free holes (normally in the *p* region) cannot climb the barrier, so no current flows. The *junction* refers to the region where the barrier exists. Because there are no free charges in the junction, it is called the *depletion region*. Having no free charges, its resistance is high, resulting in almost all the voltage drop across the diode appearing across the junction itself. Therefore, the electrical forces are high in the depletion region and are negligible outside.

Figure 7-4(c) shows an incident photon being absorbed in the junction after passing through the *p* layer. The absorbed energy raises a bound electron across the bandgap

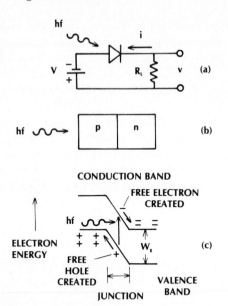

Figure 7-4 Semiconductor junction photodiode.
(a) Reverse-biased diode. (b) *pn* junction
(c) Energy-level diagram.

from the valence to the conduction band. The electron is now free to move. A free hole is left in the valence band at the position vacated by the electron. Free charge carriers are created by photon absorption in this manner. The electron will travel down the barrier and the hole (whose potential energy is opposite that of an electron) will travel up the barrier. These moving charges cause current flow through the external circuit in the same way that photoemitted electrons cause current in a vacuum photodiode. When the free holes and electrons recombine or when they reach the edge of the junction, where the electrical forces are small, the charges cease to move, which stops the current.

What happens when a photon is absorbed in the *p* or *n* regions, outside the junction? An electron-hole pair is created, but these free charges will not move quickly because of the weak electrical forces outside of the junction. Most of the free charges will dif-

fuse slowly through the diode and recombine before reaching the junction. These charges produce negligible current, thus reducing the detector's responsivity. Clearly, this phenomenon makes the *pn* detector inefficient. To increase the response, a preamplifier may be integrated onto the same chip as the diode. The resulting device is an *integrated detector preamplifier* (IDP).

Charge carriers created close to the depletion layer can diffuse toward it and, subsequently, be swept across the junction by the large electrical forces there. An external current is produced, but it is delayed with respect to variations in the incident optic power. Suppose that we wish to measure the rise time of a *pn* photodiode by applying an optic power step input. Some of the photons from the leading edge of the step will be absorbed in the junction, causing almost immediate current flow. However, those photons from the leading edge that are absorbed close to the junction will cause current flow a little later. Therefore, our experiment will show a gradual increase in the current, the maximum being reached well after the input step has been applied. The rise time will be long. Typical *pn* diodes have rise times on the order of microseconds, making them unsuitable for high-rate fiber systems. The PIN diode structure solves the problem of low responsivity and slow response. We discuss this device in the next section.

It is interesting to compare the semiconductor junction used as a light emitter and as a light detector. For emission, the diode is forward biased and charges injected into the junction recombine to produce photons. For detection, the process is reversed. The diode is reverse biased and incoming photons generate electron-hole pairs, producing electrical current. Although not commonly done, a single *pn* device could be designed to be used as both an emitter and a detector.

7-4 PIN PHOTODIODE

PIN photodiodes are the most common detectors in fiber systems. The PIN diode has a wide *intrinsic* semiconductor layer between the *p* and *n* regions, as illustrated in Fig. 7-5. The intrinsic layer has no free charges, so its resistance is high, most of the diode voltage appears across it, and the electrical forces are strong within it. Because the intrinsic layer is so wide, there is a high probability that incoming photons will be absorbed in it rather than in the thin *p* or *n* regions. This improves the efficiency and the speed relative to the *pn* photodiode.

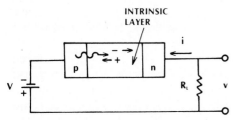

Figure 7-5   PIN photodiode.

Cutoff Wavelength

To create an electron-hole pair, an incoming photon must have enough energy to raise an electron across the bandgap. This requirement, $hf \geq W_g$, leads to a cutoff wavelength

$$\lambda = \frac{1.24}{W_g} \qquad (7\text{-}11)$$

where λ is in μm and W_g is the bandgap energy in electron-volts. This is just like Eq. (7-4) for photoemitters.

Example 7-6

Compute the cutoff wavelength for silicon and germanium PIN diodes.

Their bandgap energies are 1.1 eV and 0.67 eV, respectively.

Solution:
Equation (7-11) yields a cutoff wavelength of 1.1 μm for silicon and 1.85 μm for germanium.

Materials

Silicon is the most practical fiber optic detector in the first window, but Example 7-6 showed that it cannot be used in the long-wavelength second window around 1.3 μm. Germanium and InGaAs diodes introduce more noise than silicon, but they are responsive in the second window. Table 7-1 summarizes the useful ranges of the most common PIN diode materials. The spectral responses of silicon and InGaAs appear in Fig. 7-6. The drop in responsivity at the shorter wavelengths is caused by an increase in the absorption of photons in the *p* and *n* regions.

Silicon and InGaAs have peak quantum efficiencies of about 0.8. By using this value in Eq. (7-7) for silicon at 0.8 μm yields a responsivity of 0.5 A/W. Notice how much greater this is than the 6.4 mA/W that we calculated for a typical vacuum photodiode in Section 7-2. For InGaAs at 1.7 μm, 80% efficiency yields a responsivity of 1.1 A/W. According to the spectral response curve in Fig. 7-6(b), the responsivity is reduced to about 70% of this value, or 0.77 A/W, at 1.3 μm. Germanium's peak response is near 1.55 μm, where its quantum efficiency is around 55%. Equation (7-7) yields a peak responsivity, in this case, of nearly 0.7 A/W.

Current-Voltage Characteristic

The current-voltage characteristic curves for a silicon diode having responsivity 0.5 A/W

TABLE 7-1.   Semiconductor PIN Photodiodes

Material	Wavelength Range (μm)	Wavelength of Peak Response (μm)	Peak Responsivity (A/W)
Silicon	0.3–1.1	0.8	0.5
Germanium	0.5–1.8	1.55	0.7
InGaAs	1.0–1.7	1.7	1.1

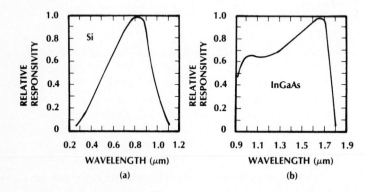

Figure 7-6   Spectral response curves.

are drawn in Fig. 7-7. When reverse biased, the diode is said to operate in the *photoconductive mode*. In this mode, the output current is proportional to the optic power. When no reverse bias is provided, the figure shows that incident optic power results in a forward voltage. This is the *photovoltaic mode,* the basis for solar cells that produce electrical voltages when subjected to optic radiation. Fiber communications detectors work in the photoconductive mode.

Even when there is no optic power present, a small reverse current flows through a reverse-biased diode. This is called the *dark current,* labeled I_D in Fig. 7-7. Dark current is caused by the thermal generation of free charge carriers in the diode. It flows in all diodes, where it is conventionally called the *reverse leakage current*. Its maximum value, occurring at large negative voltages, is the *reverse saturation current*. The dark current, being of thermal origin, will increase rapidly with temperature, sometimes doubling for every 10°C increase near room temperature

(25°C). Dark currents range from a fraction of a nanoampere to more than several hundred nanoamperes. Generally, silicon detectors have the lowest dark currents, InGaAs diodes have somewhat larger dark currents, and germanium diodes have the largest dark currents. This is one of the main reasons that silicon photodiodes are preferred over germanium in wavelength regions where their responsivities are comparable.

As might be expected, a small optic signal might be undetectable because the small photocurrent that it generates could be masked by the dark current.

Example 7-7

Estimate the minimum detectable power for a PIN diode whose responsivity is 0.5 A/W and whose dark current is 1 nA.

Solution:

Assume that we can distinguish the presence of optic power producing a signal

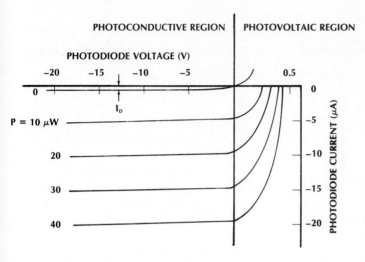

Figure 7-7 Current-voltage characteristic curves for a silicon photodetector.

current equal to the dark current. From Eq. (7-7), then, $P = I_D/\rho = 2$ nW.

In Chapter 11 we quantitatively analyze the limiting effects of dark current on signal-to-noise ratios and digital error rates.

The simplest PIN receiving circuit is drawn in Fig. 7-8 along with an idealized diode characteristic curve. The *loop theorem* (also known as *Kirchhoff's voltage law*) states that the sum of the voltages around a closed circuit must be zero. Applying this theorem to the circuit in Fig. 7-8 yields

$$V_B + v_d + i_d R_L = 0 \qquad (7\text{-}12)$$

Notice that the diode voltage and current have been labeled as being positive in the forward direction. They will both have negative values in this application. Because Eq. (7-12) must be satisfied simultaneously with the characteristic curve, it is also plotted in Fig. 7-8. As an example, we are using a 20-V battery (or dc power supply) and a 1-MΩ load resistor. The resulting straight line, called the *load line*, has a slope equal to $-1/R_L$. It crosses the voltage axis at $-V_B$ (in this example, -20 V), and it crosses the current axis at $-V_B/R_L$ (in this example, $-20\ \mu$A). A transfer characteristic, showing the output voltage v as a function of

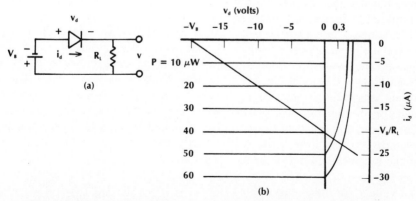

Figure 7-8   (a) Simple PIN circuit. (b) Graphical analysis of the circuit.

TABLE 7-2. Calculating the Transfer Characteristic of a PIN Photodetector

Optic Power (μW)	Diode Voltage (V)	Output Voltage (V)
0	−20	0
10	−15	5
20	−10	10
30	−5	15
40	0	20
50	0.3	20.3
60	0.4	20.4

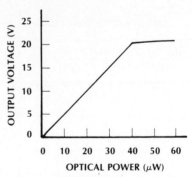

Figure 7-9 PIN photodetection circuit transfer function. $R_L = 1$ MΩ, responsivity = 0.5 A/W.

the input optic power, can easily be developed from Fig. 7-8(b). Table 7-2 summarizes some of the calculations.

Let us illustrate how these numbers were obtained. If the optic power is 10 μW, the load line crosses the PIN characteristic at a diode voltage of −15 V. Because the battery is 20 V, this leaves 5 V across the load R_L (−5 V if we continue to treat the upper terminal in Fig. 7-8(a) as the positive electrode). The other numbers in the table were found in a similar manner. The transfer characteristic appears in Fig. 7-9. When the optic power becomes large enough (over 40 μW in this example), the diode begins to operate in the photovoltaic mode and the transfer characteristic becomes nonlinear as indicated on the curve in Fig. 7-9. Although the usual problem is insufficient power, the designer of short links needs to be careful not to saturate the receiver unintentionally. Saturation refers to the state of operation where the input optical power is so large that the electrical output current and voltage no longer follow the input linearly. Specifically, when saturated the detector response to changes in the input optical power becomes smaller. Saturation not only distorts the signal waveform, it also lowers the response time of the receiver limiting its bandwidth.

The output voltage for the circuit in Fig. 7-8(a) could have been computed from Eq.

(7-8), $v = \rho P R_L$. The graph was used to illustrate saturation at high power levels and the large dynamic range of photodetectors.

We can operate the diode at higher powers and increase the receiver's dynamic range by decreasing the value of the load resistance. For example, changing R_L to 10 kΩ in Fig. 7-8 (a decrease by a factor of 100) would increase the magnitude of the maximum current to $V_B/R_L = 20/10^4 = 2$ mA. Since $i = \rho P$, the maximum current (V_B/R_L) corresponds to a maximum input power of

$$P_{max} = \frac{V_B}{\rho R_L} \qquad (7\text{-}13)$$

With $\rho = 0.5$ A/W, we find a maximum power, before saturation, of 4 mW. Now the dynamic range has been extended by a factor of 100. What is the price paid? The voltage response has been lowered by a factor of 100, as we can see from Eq. (7-8), in which the ratio of output voltage to input optic power is directly proportional to the load resistance. That is,

$$\frac{v}{P} = \rho R_L \qquad (7\text{-}14)$$

The linear relationship between voltage and optic power, as expressed by this equation, or between current and optic power, as expressed by Eq. (7-6), holds over a range of more than 6 decades of optic power for most PIN photodiodes, when not limited by a large load resistor.

Speed of Response

The speed of response is limited by the *transit time*, the time it takes for free charges to traverse the depletion layer. In a PIN diode, the length of the depletion region is just the width of the intrinsic layer. The velocity of the free charge carriers is linearly proportional to the magnitude of the reverse voltage, so higher voltages reduce the transit time. As an example, a depletion width of 50 μm and a typical carrier velocity of 5×10^4 m/s yields a transit time of $50 \times 10^{-6}/5 \times 10^4 = 1$ ns. This is approximately the photodiode rise time. Capacitance also limits the response. We can see this by examining the diode equivalent circuit in Fig. 7-10. C_d is mainly the junction capacitance, formed by the semiconducting p and n layers (serving as electrodes) separated by the insulating intrinsic region. C_d also includes the capacitance of the packaging structure. Analysis of this circuit reveals a 0–63% rise time of $R_L C_d$ (this factor is called the circuit's *time constant*) and a 10–90% rise time of

$$t_r = 2.19 R_L C_d \qquad (7\text{-}15)$$

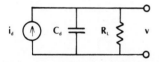

Figure 7-10   Equivalent circuit of a PIN photodiode. C_d is the diode capacitance, i_d is the photocurrent.

The corresponding 3-dB bandwidth can be calculated directly from the circuit or it can be found from Eq. (7-2). The result is

$$f_{3\text{-}dB} = \frac{1}{2\pi R_L C_d} \qquad (7\text{-}16)$$

Photodiodes designed for high-speed applications have capacitances of a few picofarads or less. To obtain low capacitance, the diode's surface area must be small. For efficient coupling, however, the area cannot be reduced below that of the attached optic fiber's core.

The speed of response may be limited by transit time or by the circuit rise time, whichever is larger. Rise times limited by transit time range from about 0.5 to 10 ns for fast PIN diodes. Rise times less than 100 ps have been achieved.

Example 7-8
A PIN photodiode has a capacitance of 5 pF and a transit-time-limited rise time of 2 ns. Compute its 3-dB bandwidth and the largest load resistor that can be used without significantly increasing the rise time.

Solution:
From Eq. (7-2)

$$f_{3\text{-}dB} = \frac{0.35}{2 \times 10^{-9}} = 175 \text{ MHz}$$

To be insignificant, the *RC* rise time from Eq. (7-15) should be less than a fourth of the transit time. The condition $2.19\ R_L C_d \leq 0.5$ ns yields $R_L \leq 46\ \Omega$, a fairly small value.

The criteria for choosing the value of the load resistor are summarized in Table 7-3. The last item in the table, relating to noise, is discussed in Section 11-1.

TABLE 7-3. Criteria for Choosing the Load Resistor[a]

Defining Equation	Conclusion
$v = \rho P R_L$	Choose R_L large for high-output voltages
$P_{max} = V_B/(\rho R_L)$	Choose R_L small for large dynamic range
$f_{3\text{-dB}} = (2\pi R_L C_d)^{-1}$	Choose R_L small for large bandwidth
$i_{NT}^2 = 4kT\Delta f/R_L$	Choose R_L large to reduce the thermal-noise current

[a] v, output voltage; ρ, responsivity; P, optic power; R_L, load resistance; V_B, bias voltage; C_d, diode capacitance; i_{NT}^2, mean square value of the thermal-noise current; k, Boltzmann constant; T, absolute temperature; Δf, receiver bandwidth.

Current-to-Voltage Converter

We may note from Fig. 7-8 that the diode voltage diminishes when the optic power increases. This is because more current is flowing, increasing the voltage across the load resistor and leaving less of the battery voltage for the diode. Nonlinearity begins when the diode voltage drops to zero. We can solve the linearity problem, without using a small load resistor, by using the current-to-voltage converter drawn in Fig. 7-11. The diode is connected to an operational amplifier with a feedback resistor R_F. The properties of this circuit are the following:

1. There is almost no voltage drop across the input terminals of the high-gain operational amplifier. The loop theorem yields $v_d = -V_B$ when applied to the loop consisting of the battery, the diode, and the amplifier's input terminals. That is, the entire battery voltage appears across the diode. This is equivalent to operating along the vertical load line in Fig. 7-12.

2. There is almost no current flowing into the input terminals of the amplifier. The entire diode current flows through the feedback resistor R_F. The voltage across this resistor is $R_F i_d$. Because the negative terminal of the amplifier is nearly at ground potential, the loop theorem shows that the output voltage is also $R_F i_d$. The feedback resistor can be large (hundreds of kilohms if desired) to produce large output voltages without affecting the linearity of the response. The speed of response of this circuit will be limited by the rise time of the feedback resistor combined with the shunt capacitance of the feedback network.

Packaging

Photodetector packages are similar to those used for LEDs and laser diodes but the re-

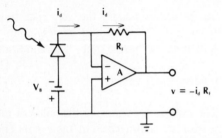

Figure 7-11 Current-to-voltage converter. *A* is an operational amplifier.

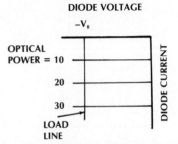

Figure 7-12 Vertical load line seen by the diode in the current-to-voltage converter.

quirements are less critical. The detector's active area is often larger than the core of the incoming fiber, so some lateral misalignment is tolerable. Also, detectors are not restricted by a low numerical aperture, They accept light over a wide angular range. Angular misalignment and mismatch between the NA of the fiber and detector are not severe problems.

Photodiodes are packaged in numerous ways. A few of them follow:

1. The photodiode is mounted on a standard transistor header, much like the LED in Fig. 6-12. A clear glass cover, or a lens, is attached to the metal cap. A lens, if present, will focus light onto the detector's active area. The lens can collect light from a fiber that is larger than the detector, improving the detection efficiency. In some designs, the cap is removable to provide access to the diode.
2. The photodiode package may include a fiber pigtail either with or without a connector on the output.
3. Photodiodes are placed inside dual inline packages (similar to Fig. 6-32) for mounting on printed circuit boards.

An integrated preamplifier is included inside some PIN photodetector packages. This is the IDP structure, mentioned in Section 7-3. The internal amplifier converts the diode photocurrent to an output voltage. The entire device behaves as a low impedance voltage source as viewed by the remainder of the receiving circuit. Receiving circuits are treated further in Chapter 11.

7-5 AVALANCHE PHOTODIODE

The avalanche photodiode (APD) is a semiconductor junction detector that has internal gian, which increases its responsivity over *pn*

or PIN devices. Having gain, the APD is similar to the photomultiplier tube. The avalanche gain, however, is much lower than that available from a PMT, being limited to values of several hundred or less. Nevertheless, the gains that are available make APDs much more sensitive than PIN diodes. As mentioned in Section 7-2, internal gain yields much better signal-to-noise ratios than can be obtained with external amplification. This will be proven in Chapter 11.

Avalanche current multiplication comes about in the following way: A photon is absorbed in the depletion region, creating a free electron and a free hole. The large electrical forces in the depletion region cause these charges to accelerate, gaining kinetic energy. When fast charges collide with neutral atoms, they create additional electron-hole pairs by using part of their kinetic energy to raise electrons across the energy bandgap. One accelerating charge can generate several new secondary charges. The secondary charges, themselves, can accelerate and create even more electron-hole pairs. This, then, is the process of avalanche multiplication.

The accelerating forces must be strong to impart high kinetic energies. This is achieved with large reverse biases, several hundred volts in some instances. The gain increases with reverse bias v_d according to the approximation[3]

$$M = \frac{1}{1 - (v_d/V_{BR})^n} \qquad (7\text{-}17)$$

where V_{BR} is the diode's reverse breakdown voltage and n is an empirically determined parameter that is more than unity. Breakdown voltages of 20 to 500 V occur.

The current generated by an APD with gain M is

$$i = \frac{M\eta eP}{hf} = \frac{M\eta e\lambda P}{hc} \qquad (7\text{-}18)$$

where η is the quantum efficiency when the gain is unity. This result is the same as Eq. (7-10), developed for a photomultiplier. The responsivity is

$$\rho = \frac{M\eta e}{hf} = \frac{M\eta e\lambda}{hc} \qquad (7\text{-}19)$$

Typical avalanche responsivities range from 20 to 80 A/W.

Avalanche photodiodes are usually variations of PIN diodes. The materials used, and thus the spectral ranges, are the same. One form of APD, a *reach-through* diode is sketched in Fig. 7-13. The p^+ and n^+ layers are highly doped, low-resistance regions having a very small voltage drop. The π region is lightly doped, nearly intrinsic. Most of the photons are absorbed in this layer, creating electron-hole pairs. As indicated in the figure, photoelectrons move to the p region, which has been depleted of free charge by the large reverse voltage. In essence, the depletion region at the p-n^+ junction has "reached through" to the π layer. The voltage drop is mostly across the p-n^+ junction, where the resulting large electrical forces cause avalanche multiplication. In this device, multiplication is initiated by electrons. Holes generated in the π layer drift to the p^+ electrode but do not take part in the multiplication process. Structures that limit the initiation of multiplication

to just one type of charge carrier have superior noise characteristics.[4]

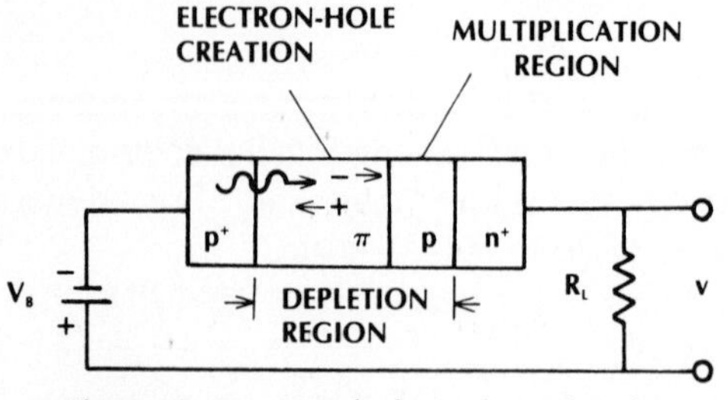

As with the nonmultiplier PIN diode, the response speed of the APD is limited by the charge-carrier transit time and the *RC* time constant. Transit-time-limited avalanche photodiodes are available with rise times as low as a few tenths of a nanosecond. Rise times less than 100 ps have been achieved with both silicon and germanium.

Avalanche photodiodes have excellent linearity over optic power levels ranging from a fraction of a nanowatt to several microwatts. If more than a microwatt is available at the receiver, an APD is usually not needed. At this power level PIN diodes provide enough responsivity and sufficiently large signal-to-noise ratios for most applications.

The gain of an avalanche photodiode is temperature dependent, generally decreasing as the temperature rises. This occurs because the mean free path between collisions is smaller at higher temperatures. Many of the charge carriers do not get a chance to reach the high velocities required to produce secondary carriers. Temperature stabilization, or compensation, may be required in an APD receiver operating over an extended temperature range.

7-6 SUMMARY

The major relationship developed in this chapter was the one between the incident optic power and the electrical current it generates in a photodetector. This relationship can be summarized as $i = \rho P$, where the responsivity ρ is around 0.5–0.7 A/W for PIN diodes and increases by factors up to several hundred for avalanche detectors.

The detector in a fiber communications system will either be an avalanche or a PIN

Figure 7-13 Reach-through avalanche photodiode.

photodiode. The PIN device is cheaper, less sensitive to temperature, and requires lower reverse bias voltage than the APD. The speeds of the two devices are comparable, so the PIN diode is preferable in most systems. The APD gain is needed when the system is loss limited, as occurs for long-distance links. Suppose an APD receiver can clearly detect a signal whose power level is 9 dB below that which can be detected by a PIN diode. If the fiber loss is 3 dB/km, then the APD link can be 3 km longer than the PIN link. Similarly, if repeaters are needed to boost the optic power levels, they can be spaced 3 km farther apart if the APD detector is used in this example.

Although a wide variety of detectors and detector characteristics exist, it is useful to consider the typical values of important photodiode parameters, as shown in Table 7-4. The responsivity given in the table is representative of the value at a wavelength where the detector might be used; that is, near 0.8 μm for silicon and near 1.3 or 1.5 μm for germanium and InGaAs. The responsivity falls off as the wavelength moves toward the edges of the ranges shown, as illustrated in Fig. 7-6.

Some of the information we have gathered on sources, fibers, and detectors is summarized in Fig. 7-14. An initial choice of matched components can be made from this figure. This figure illustrates the many deci-

sions the system designer must take. These include operating wavelength (visible, first, second, or third window); light source (LED or laser diode); fiber material (glass, PCS, or plastic); fiber type (step-index, graded-index, or single mode); and photodetector (PIN or APD). The material in the preceding few chapters will aid in choosing the optimum components. We may need further information, however, for some applications. For example, to determine if an APD is required, we need to know how much power is available at the receiver. This, in turn, requires that we know all the system losses, not just the fiber attenuation. These other losses are caused by source coupling, splices, connectors, and any power division for signal distribution. In addition, we need to determine the effects of noise. These matters will be investigated in succeeding chapters.

PROBLEMS

7-1. How much current is produced by a photodetector whose responsivity is 0.5 A/W if the incident optical power level is −43 dBm?

7-2. Compute the rise time of a photodetector if its 3-dB bandwidth is 500 MHz.

7-3. Calculate the cutoff wavelength and

TABLE 7-4.   Typical Characteristics of Junction Photodetectors

Material	Structure	Rise Time (ns)	Wavelength (nm)	Responsivity (A/W)	Dark Current (nA)	Gain
Silicon	PIN	0.5	300–1100	0.5	1	1
Germanium	PIN	0.1	500–1800	0.7	200	1
InGaAs	PIN	0.3	900–1700	0.6	10	1
Silicon	APD	0.5	400–1000	75	15	150
Germanium	APD	1	1000–1600	35	700	50
InGaAs	APD	0.25	1000–1700	12	100	20

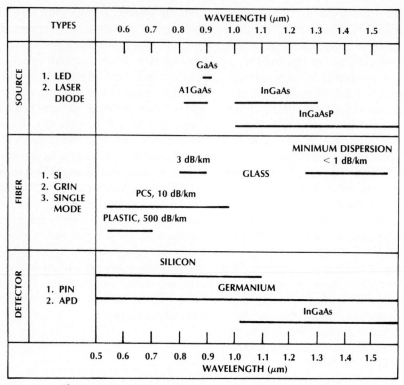

Figure 7-14 Major components of a fiber optic system.

frequency of a vacuum photodetector whose photocathode has a work function equal to 2×10^{-19} J.

7-4. Rewrite Eq. (7-4) for the cutoff wavelength of a photoemissive photodetector so that it will be correct if the work function is given in joules and the wavelength in meters.

7-5. Compute and plot the responsivity of an ideal photodetector (i.e., one whose quantum efficiency is unity) for wavelengths between 0.5 and 1.7 μm. Why does responsivity increase with wavelength?

7-6. Find the output current of a photodetector whose quantum efficiency is 0.9. The wavelength is 1.3 μm and the inci-

dent power level is -37 dBm. Also, compute the resulting output voltage if the load resistance is 50 Ω, 1000 Ω and 1 MΩ.

7-7. Repeat Problem 7-6 if a photomultiplier having four dynodes (each having a gain of 4) is used. The quantum efficiency of the cathode is 0.9.

7-8. Rewrite Eq. (7-11) for the cutoff wavelength of a PIN photodiode so that it will be correct if the bandgap energy is in joules.

7-9. Compute the bandgap energy in joules for (a) silicon and (b) germanium.

7-10. Assume that the dark current of a PIN photodiode is 0.06 nA at 25°C and doubles for every 10° increase. Compute and plot the dark current for temperatures from 25° to 95°C.

7-11. For the germanium PIN photodiode described in Table 7-4, how much optical power is required to produce a photocurrent equal to the detector's dark current?

7-12. Consider a PIN detector circuit just like that drawn in Fig. 7-8. The battery voltage is 10 V, the load resistance is 2 MΩ, the detector's responsivity is 0.25 A/W, and the dark current is 0.5 nA.

(a) Draw the diode's voltage-current characteristic in the photoconductive region for incident power levels ranging from 5 to 50 μW and differing by 5 μW.

(b) Draw the load line on the curve.

(c) Draw a curve of output voltage versus input optical power.

(d) At what value of the optical power does the detector saturate?

7-13. Suppose that a PIN photodiode has a depletion width of 30 μm, a carrier velocity of 5×10^4 m/s, and a junction capacitance of 6 pF.

(a) Compute the transit-time-limited bandwidth.

(b) What is the largest load resistance that will not affect the bandwidth just calculated?

(c) What is the bandwidth if the load resistance is 10,000 Ω?

7-14. Compute the responsivity of a silicon APD operating at 0.82 μm and having a 0.8 quantum efficiency if its gain is 100. How much optical power is needed by this detector to produce 20 nA?

7-15. Compute the responsivity of an InGaAs APD operating at 1.55 μm and having a quantum efficiency of 0.7 if its gain is

10. How much optical power is needed by this detector to produce 20 nA?

7-16. Compare the advantages and disadvantages of PIN photodiodes relative to avalanche photodiodes.

7-17. Compare the relative advantages of fiber systems operating at 800, 1300, and 1550 nm.

7-18. The PIN diode receiver shown in Fig. 7-8(a) has a 1 kΩ resistor. The voltage across the resistor is 0.2 mV when the incident optical power is 800 nW. What is the detector's responsivity?

7-19. A photodetector has the following output current when a step function of input power is applied: $i = 10(1 - e^{-t/\tau})$, where $\tau = 10^{-6}$ seconds. Compute the detector's 3-dB bandwidth.

REFERENCES

1. Photodetectors are covered in numerous books:

   Tien Pei Lee and Tingye Li. "Photodetectors." In *Optical Fiber Telecommunications,* edited by Stewart E. Miller and Alan G. Chynoweth. New York: Academic Press, Inc., 1979, pp. 593–626.

   D. P. Schinke, R. G. Smith, and A. R. Hartman. "Photodetectors." In *Semiconductor Devices for Optical Communication,* edited by H. Kressel. New York: Springer-Verlag, 1980, pp. 63–87.

   Joseph T. Verdeyen. *Laser Electronics.* Englewood Cliffs, N.J.: Prentice Hall, 1981.

   Peter K. Cheo. *Fiber Optics.* Englewood Cliffs, N.J.: Prentice Hall, 1985.

2. Many review articles covering photodetectors for optic communications have appeared in the literature. A few particularly useful ones follow:

Michael Ettenberg and Gregory H. Olsen. "Diode Lasers for the 1.2 to 1.7 Micrometer Region," *Laser Focus* 18, no. 3 (March 1982): 61–66.

Stephen R. Forrest. "Photodiodes for Long-Wavelength Communication Systems," *Laser Focus* 18, no. 12 (Dec. 1982): 81–90.

R. G. Smith. "Photodetectors for Fiber Transmission Systems," *Proc. IEEE* 68, no. 10 (Oct. 1980): 1247–53.

Hans Melchior, Mahlon B. Fisher, and Frank R. Arams. "Photodetectors for Optical Communication Systems," *Proc. IEEE* 58, no. 10 (Oct. 1970): 1466–86.

Stephen R. Forrest. "Optical Detectors: Three Contenders," *IEEE Spectrum* 23, no. 5 (May 1986): 76–83.

3. Schinke et al., "Photodetectors." p. 70.

4. Ibid. p. 68.

Chapter 8

Couplers and Connectors

Connections are normally quite simple in metallic systems. Wires can be spliced very easily by soldering. The splice can even be undone, merely by melting the solder. The losses in a solder joint are so small that they are not usually considered in the system design. Removable connectors for wires are also simple, easy to attach, reliable, economical, and virtually lossless. The favorable attributes of wire connectors are not shared by their fiber counterparts. We shall see what the problems are in splicing and connecting fibers and how these problems can be overcome with sufficient care.

Fiber-to-fiber connections are needed for a variety of reasons. Several fibers must be spliced together for links of more than a few kilometers because only limited continuous lengths of fiber are normally available from manufacturers. Moderate lengths of fiber are easier to pull through ducts than very long cables are, and moderate lengths simplify direct burial or aerial installations.

Coupling of light from a source into a fiber can be very inefficient. We evaluate source-coupling losses and describe techniques for reducing them.

At the receiver, light is coupled from the fiber onto the detector surface. This surface can be chosen to be larger than the fiber core, resulting in very efficient coupling. A small loss owing to reflections at the fiber-to-air and air-to-detector interfaces, does occur. It can be removed by filling the air gap with index-matching material or by antireflection coating the detector surface. In any case, detector coupling is not difficult and need not be discussed further.

8-1 CONNECTOR PRINCIPLES

Losses in fiber-to-fiber connections arise in a number of ways. Core misalignments and imperfections, illustrated in Fig. 8-1, are major factors. A perfect joint would require lateral (or axial) alignment, angular alignment (parallel fiber axes), contacting ends (no gap), and smooth, parallel ends. Coupling efficiency may be reduced when fibers that have different numerical apertures or core diameters are connected. More loss is present when cores having elliptical (rather than circular) cross sections are attached with their major axes unaligned. If the core is not centered in the cladding, and if the outside of the cladding is used as the reference for aligning the joint, then more loss occurs. With care, these problems can be minimized, producing splices with losses on the order of 0.1 dB and reusable connectors with losses less than 1 dB.

Theoretical analyses of the losses caused by the variety of mechanisms just discussed is complicated by the fact that the coupling efficiency depends on the distribution of power across the fiber end face. This pattern is not usually known. It depends on the excitation method and the length of fiber from the excitation point to the joint. For example, in a multimode fiber, mode coupling causes the modal distribution to change along the fiber until the equilibrium length (described in Section 5-6) is reached. Therefore, connector losses depend on the distance between the excitation point and the connector itself. For paths longer than the equilibrium length, the loss would settle to a fixed value. Surprisingly, the coupling efficiency even depends on the length of fiber following the junctions. Higher-ordered modes and cladding modes, which may be excited by imperfections in the joint, are transmitted efficiently only short distances past the connection. The power measured near the connection includes these modes, making it appear that the loss is low. Measurements far from the joint would exclude much of the power in these modes, indicating a higher connection loss.

With these factors in mind, we will proceed with a discussion of the losses under assumed, idealized conditions. Although these conditions may never be exactly met, the results will give us some understanding of the sensitivity of the connections to the various loss mechanisms. This information can be used as a guide by splice and connector designers and by system analysts who must often estimate the total loss in a system.

Lateral Misalignment

A simple model assumes that the power is uniformly distributed over the fiber core. This approximation is most suitable for a multimode step-index fiber. With this assumption, the lateral misalignment loss is simply due to the nonoverlap of the transmitting and receiving fiber cores, as drawn in Fig. 8-2. The coupling efficiency η is the ratio of the overlapping area (shown crosshatched) to the core area. This can be calculated to be[1]

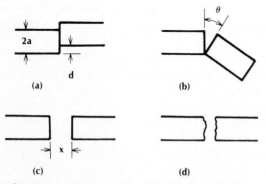

Figure 8-1   Sources of loss in a fiber-to-fiber connection. (a) Lateral misalignment. (b) Angular misalignment. (c) Gap between ends. (d) Nonflat ends.

Figure 8-2 Overlap of transmitting and receiving fibers. The cores are offset by the distance d.

$$\eta = \frac{2}{\pi}\left\{\cos^{-1}\frac{d}{2a} - \frac{d}{2a}\sqrt{\left[1 - \left(\frac{d}{2a}\right)^2\right]}\right\}$$
(8-1)

where the inverse cosine is calculated in radians. The corresponding loss in decibels is

$$L = -10 \log \eta \qquad (8\text{-}2)$$

We will use L for the loss in decibels throughout this book. The axial misalignment loss is plotted in Fig. 8-3. For small displacements $(d/2a < 0.2)$, Eq. (8-1) can be approximated by $\eta = 1 - (2d/\pi a)$.

Example 8-1

What is the allowable axial displacement if the coupling loss is to be less than 1 dB? The core diameter is 50 μm. Repeat for losses of 0.5 and 0.1 dB.

Solution:

Either Fig. 8-3 or Eqs. (8-1) and (8-2) can be used to find what follows:

L (dB)	$d/2a$	d (μm)
1.0	0.16	8
0.5	0.09	4.5
0.1	0.02	1

These numbers show how much care is necessary in properly aligning the axes of the fibers being joined.

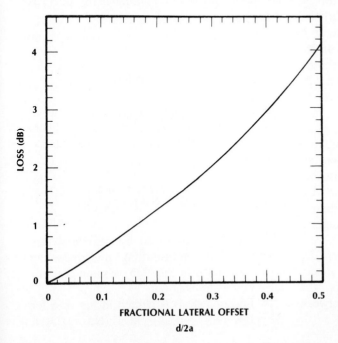

Figure 8-3 Lateral misalignment loss for a multimode SI fiber.

LOSS (dB)

FRACTIONAL LATERAL OFFSET

d/2a

In Chapters 4 and 5 we determined that higher-ordered modes were more heavily attenuated than lower-ordered modes. They also contained more power near the core-cladding interface. We may conclude that the power density at the end of a long fiber will be lower at the edges of the core than at points nearer its center. For small axial displacements, only the edges of the transmitting core miss the receiving fiber. Since the edge contains less power than assumed in developing Eq. (8-1), the actual loss will be less than that predicted by the theory. With the preceding in mind we can use the uniform overlap theory as a conservative estimate of the actual losses when connecting multimode SI fibers.

Multimode graded-index fibers present more of a theoretical problem than do SI fibers because the GRIN numerical aperture varies across the face of the core, as illustrated by Eq. (5-5) for a parabolic profile. When the two fibers meet with no offset, the numerical aperture of transmitter and receiver match at every point within the core. With an offset, however, there is a NA mismatch at nearly every point. At those points where the receiver NA is larger than the transmitter NA, all the power is transferred. At points where the receiver NA is less than the transmitter NA, some of the power is lost. The fractional efficiency at these points is equal to the ratio of the squares of the local numerical apertures. To calculate the coupling efficiency we would have to average the local efficiencies, after weighting them according to the distribution of power across the end face. As we discussed earlier in this section, this distribution is not generally known, which discourages a comprehensive analysis.

The offset loss of single-mode fibers depends on the form of the propagating mode. In both SI and parabolic-index fibers, the beams are nearly Gaussian. The loss between identical fibers is[2]

$$L = -10 \log\left\{\exp\left[-\left(\frac{d}{w}\right)^2\right]\right\} \qquad (8\text{-}3)$$

where w is the spot size, defined in Section 2-5 and computed from Eq. (5-9). For SI fibers operating close to a normalized frequency V of 2.405 (the single-mode cutoff condition developed in Section 5-4), Eq. (5-9) yields a spot size equal to 1.1 times the core radius. The single-mode axial displacement loss is plotted in Fig. 8-4. Because the spot size is only a few microns, we realize that efficient coupling of single-mode fibers requires a very high degree of mechanical precision. For a loss of 1 dB, Eq. (8-3) or Fig. 8-4 yields $d/w = 0.48$. If the spot size is 4 μm, then the allowable misalignment is only 1.9 μm.

Angular Misalignment

The coupling efficiency owing to small angular misalignments of multimode SI fibers is given by[3]

$$\eta = 1 - \frac{n_0\theta}{\pi\,\text{NA}}$$

where n_0 is the refractive index of the material filling the groove formed by the two fibers and θ is the misalignment angle in radians. The loss is

$$L = -10 \log\left(1 - \frac{n_0\theta}{\pi NA}\right) \qquad (8\text{-}4)$$

The efficiency was found by computing the overlap of the transmitting and receiving cones, as drawn in Fig. 8-5, assuming a uniform power distribution.

Equation (8-4) is plotted in Fig. 8-6 for the glass (NA = 0.24), PCS (NA = 0.41),

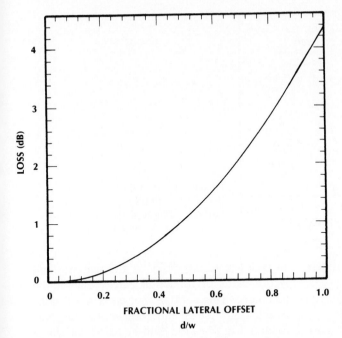

Figure 8-4 Lateral misalignment loss for single-mode fibers whose modal spot size is w.

and plastic (NA = 0.48) fibers listed in Table 5-1. The solid lines were drawn assuming no material in the groove ($n_0 = 1$). The dashed line shows that the angular loss increases when a fluid having a refractive index of 1.5 is present. The purpose of this fluid is explained later in this section. Notice how the angular loss decreases for larger numerical apertures. This is simply explained by reasoning that fibers with high NA spread their transmitting (and receiving) radiation over a wide angular range. Therefore, a small angular error will affect only a small percentage of the total power.

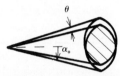

Figure 8-5 Overlap of the transmitting and receiving cones. NA = $n_0 \sin \alpha_0$, θ is the tilt angle. The shaded area indicates the overlap.

Angular misalignment losses for GRIN fibers are not covered because of the theoretical difficulties mentioned earlier in this section.

For single-mode fibers the angular misalignment loss is[4]

$$L = -10 \log\left\{ \exp\left[-\left(\frac{\pi n_2 w \theta}{\lambda}\right)^2 \right] \right\} \quad (8\text{-}5)$$

where θ is in radians, w is the Gaussian spot size, and n_2 is the refractive index of the cladding. The exponent in Eq. (8-5) comes from the ratio of the misalignment angle to the Gaussian beam divergence half angle, which we noted to be $\lambda/\pi w$ in Eq. (2-17). The loss is plotted in Fig. 8-7 for two different SI fibers, both having a normalized frequency of 2.4 and a cladding index of 1.46. As with the multimode case, the loss increases more quickly for the fiber with the smallest numerical aperture. The following example will illus-

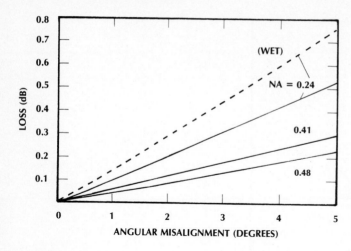

Figure 8-6   Angular misalignment loss for multimode SI fibers. The dashed curve applies when a fluid ($n_0 = 1.5$) fills the gap.

trate some of the calculations made in producing Fig. 8-7.

Example 8-2

A SI fiber has $n_1 = 1.465$, $n_2 = 1.46$, and normalized frequency 2.4. Compute its core radius, numerical aperture, and spot size at 0.8 μm.

Solution:

From Eq. (4-21), NA $= (n_1^2 - n_2^2)^{1/2} = 0.12$. From Eq. (5-7), with $V = 2.4$,

$$a = \frac{\lambda V}{2\pi \sqrt{n_1^2 - n_2^2}} = 2.53 \ \mu\text{m}$$

When $V = 2.4$, $w/a = 1.1$, so the spot size is $1.1(2.53) = 2.78 \ \mu$m.

End Separation

When there is a gap between the fibers being joined, two distinct loss phenomena occur. First, there are two boundaries between the fiber medium and air. In Section 3-5 we computed a reflectance of 4% (0.177 dB) at an air-glass interface, so the two reflecting surfaces together contribute about 0.35 dB loss. One way to eliminate this loss is to fill the gap with

an index-matching fluid, a transparent fluid whose index of refraction equals that of the fiber core. This is often (but not always) done in practical splices and connectors. As indicated in Fig. 8-6, the fluid increases the sensitivity of the connection to angular misalignments.

The second loss mechanism is sketched in Fig. 8-8. When a gap is present, some of the transmitted rays are not intercepted by the receiving fiber. As the gap increases, larger amounts of the transmitted power miss the receiving core because of the beam divergence. Fibers with larger numerical apertures will

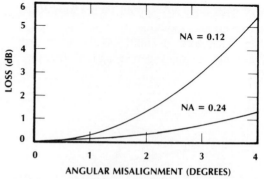

Figure 8-7   Angular misalignment loss for single-mode SI fibers. $V = 2.4$, $w/a = 1.1$, $n_2 = 1.46$, $\lambda = 0.8 \ \mu$m.

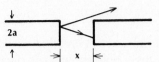

Figure 8-8 A gap allows some of the transmitted rays to escape.

have greater separation losses simply because their beams diverge quicker. Based on the uniform power distribution, the loss for small separations is given by[5]

$$L = -10 \log\left(1 - \frac{x \cdot \text{NA}}{4an_0}\right) \quad (8\text{-}6)$$

where n_0 is the refractive index of the matching fluid. The result is plotted in Fig. 8-9 for the glass, PCS, and plastic fibers in Table 5-1, with no matching fluid ($n_0 = 1$). Examination of Eq. (8-6) shows that an index matching fluid will decrease the gap loss. This is in addition to the reduction in reflection loss because of the fluid. This behavior can be explained by referring to Section 2-1, where we found that rays traveling from a high-index medium (the fiber core) to a lower-index one (air) will be bent away from the normal (as shown in Fig. 8-10). The radiated beam diverges faster in the air region than in the fiber.

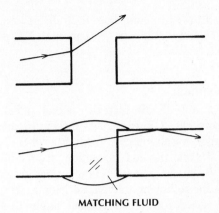

MATCHING FLUID

Figure 8-10 An index-matching fluid decreases fiber-separation loss by reducing the beam divergence.

The fluid keeps this from happening (as shown in the figure). With less beam divergence, more of the transmitted rays are intercepted by the receiving fiber.

Comparison of Figs. 8-3, 8-6, and 8-9 indicates the relative sensitivity of multimode SI fiber connections to the various misalignments. Axial misalignment is by far the most critical error.

Example 8-3

Compute the allowed misalignment for a multimode SI fiber if each type of error

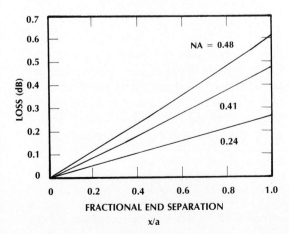

FRACTIONAL END SEPARATION

x/a

Figure 8-9 End-separation loss for multimode SI fibers.

is allowed to contribute 0.25 dB of loss. The core radius is 50 μm and NA = 0.24.

Solution:

For a 0.25 dB loss the lateral offset, from Fig. 8-3, is $d/2a = 0.045$; the angular misalignment, from Fig. 8-6, is 2.4°; and the end separation, from Fig. 8-9, is $x/a = 0.94$. These values result in the following tolerances: lateral offset $d = 4.5$ μm, misalignment angle $\theta = 2.4°$, end separation $x = 47$ μm.

The gap loss for single-mode fibers is[6]

$$L = -10 \log \frac{4(4Z^2 + 1)}{(4Z^2 + 2)^2 + 4Z^2} \qquad (8\text{-}7)$$

where $Z = x\lambda/2\pi n_2 w^2$. This result is plotted in Fig. 8-11 for a fiber having NA = 0.12. For this example, a gap of 10 times the core radius produces a loss less than 0.4 dB. We conclude that the gap is not too critical. As with multimode fibers, axial misalignment is potentially the most serious problem.

Smooth and Parallel Ends

Scattering from a rough end face would cause appreciable loss. Nonparallel ends, formed by end surfaces that are not at right angles to the fiber axis (as illustrated in Fig. 8-12), also add to the loss of a connection. Matching fluid tends to solve these problems by filling in the ragged deviations from flatness and by effectively removing the tilt. For very low connection losses, however, the fiber ends must be smooth and parallel. Specific techniques for achieving this result are described in the next section.

Connecting Different Fibers

Connections between fibers having different numerical apertures or different core diameters are common. For example, a pigtailed source may be available with a fiber that is different from the one being used for the information channel. Unintended diameter differences between fibers of the same construction are also possible.

The loss when transmitting from a fiber of core radius a_1 to one having core radius a_2 is[7]

$$L = -10 \log\left(\frac{a_2}{a_1}\right)^2 \qquad (8\text{-}8)$$

if $a_1 > a_2$. There is no loss if the receiving fiber is larger than the transmitting one. The loss is unidirectional in this respect. This result is simply a calculation of the fraction of the transmitting core area that is intercepted by the receiving fiber, as illustrated in Fig. 8-13. This computation is appropriate for

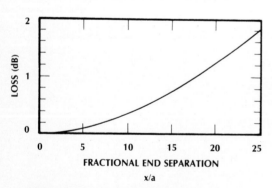

Figure 8-11 End-separation loss for a single-mode SI fiber. $V = 2.4$, $w/a = 1.1$, $n_1 = 1.465$, $n_2 = 1.46$, NA = 0.12, $\lambda = 0.8$ μm.

Figure 8-12 Tilting endfaces contribute to loss.

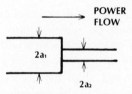

Figure 8-13   Mismatched cores may contribute to loss.

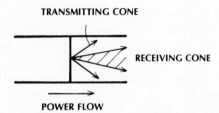

Figure 8-15   Mismatched numerical apertures may contribute to loss.

step-index or graded-index fibers as long as the index profile of the second fiber is a replica of the first (reduced by the ratio of the core radii, a_2/a_1). In this analysis, it is assumed that all the allowed modes are equally excited. Equation (8-8) is plotted in Fig. 8-14.

When transmitting from a higher to a lower numerical aperture fiber, some of the emitted rays will fall outside the acceptance angle of the receiving fiber. This concept is illustrated in Fig. 8-15. The loss is[8]

$$L = -10 \log\left(\frac{NA_2}{NA_1}\right)^2 \qquad (8\text{-}9)$$

if $NA_1 > NA_2$. There is no loss if the NA of the receiver is greater than that of the transmitter. Equation (8-9), plotted in Fig. 8-14, again assumes uniform modal distribution. With this assumption it is valid for multimode

SI and GRIN fibers. For GRIN fibers, the numerical aperture on the axis is used and the profile parameter α, defined in Eq. (5-3), must be the same for both fibers. When power flows from an SI fiber to a parabolic GRIN fiber, both having the same axial NA and the same core radius, the loss is 3 dB. The loss occurs because the NA of the receiving GRIN fiber decreases toward zero at the edge of its core, and the SI fiber radiates with the same NA from all points on its end. Power will flow from the GRIN fiber to the SI fiber without loss.

It is worthwhile to repeat the warning stated near the beginning of this section. The preceding analyses only approximate the behavior of actual fiber-to-fiber joints. The trends we have noted should be accurate, but the numerical results should be used with caution.

8-2 FIBER END PREPARATION

The two distinct methods of fiber end preparation are *scribe-and-break* and *lap-and-polish*.[9] Scribe-and-break is practical when fibers are to be spliced, while lap-and-polish is required when the fiber end is permanently attached to the body of a connector. The procedures we will describe first are applicable to all-glass fibers.

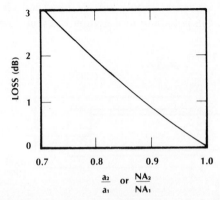

Figure 8-14   Loss caused by unequal core radii or unequal numerical apertures.

In both methods the fiber must first be bared. Any plastic jacketing material, Kevlar stranding, and buffering can be removed by using wire strippers, razor blades, or some other sharp tool. In removing the buffer, great care must be taken to prevent scratching the surface of the cladding. In place of mechanical stripping, a chemical such as methylene chloride can be used to dissolve the coating. This chemical is dangerous to handle and should be applied with proper precautions. The bared glass fiber should next be chemically cleaned (e.g., for example, it could be wiped with isopropyl alcohol).

In the scribe-and-break method, the outer edge of the cladding is nicked by a hard tool, such as a diamond-edged blade (or a sapphire or tungsten-carbide blade). The blade can be pulled across a stationary fiber, or the fiber can be pulled across a fixed blade. In either case, the fiber should be under moderate tension during cutting. After nicking, the tension is increased, by pulling until the fiber breaks. A force of just over 1.47 N (0.15 kg or 0.33 lb of force) is typical. When performed properly, this technique produces a flat, mirror-finish surface. Trained personnel can complete this procedure by hand in just a few minutes, obtaining smooth end faces that are perpendicular to the fiber axis. Commercial equipment is also available that mechanically performs the scribing and breaking. Regardless of the method used, the end must be carefully inspected to verify that a smooth, clean fracture has been obtained.

The scribe-and-break technique is the quickest and cheapest method of preparing fibers for connection. When the fiber is to be part of a demountable connector, however, the lap-and-polish procedure may be followed. A number of different connectors exist, each with an attachment and polishing procedure peculiar to the particular design. A generalized preparation procedure can be de-scribed that applies to most connectors. The bared fiber must be inserted into a ferrule (usually metal or plastic), which will hold the delicate fiber in place, protect it, and mechanically position and align the fiber to prevent the losses discussed in the preceding section. The ferrule is basically a cylindrical tube with a small hole in one end for the fiber and a larger hole in the other end for the cable jacket. Figure 8-16 illustrates the basic concept. In a precision design, a watch jewel placed inside one end of the ferrule accurately positions the fiber. At this point in the end preparation, the fiber protrudes from the ferrule. The bare fiber and its jacketing are epoxied to the ferrule, forming a permanent bond. An epoxy bead is left around the protruding fiber. A removable lapping tool, designed to hold the ferrule securely during polishing, is then attached to the ferrule. It guides the fiber as it is moved across abrasive paper, keeping the fiber perpendicular to the flat grinding surface. The fiber is ground by successively finer abrasive until a polished surface remains. Water is used to lubricate and cool the fiber and to flush residual particles away. The lapping tool, ferrule, and fiber should be rinsed before progressing from one abrasive to another. The final finishing is done with a polishing paste having suspended particles of 0.3–1 μm diameter. After a smooth surface has been obtained, the polishing tool is removed from the

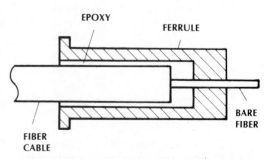

Figure 8-16   Attaching a fiber to a ferrule.

assembly. The flat end of the fiber is now flush with the end of the ferrule and perpendicular to the fiber axis, completing the lap-and-polish procedure.

Plastic-cladded silica and all-plastic fibers are usually prepared by polishing because plastic does not fracture smoothly like glass. The outer coverings (jacket, strength member, and so on) are first removed, exposing a short length of cladded fiber. The fiber and jacket are secured within a ferrule, or by an alternative mounting fixture, or by a removable clamp or vise. The secured fiber is then ground to the desired degree of smoothness in the manner described earlier.

Smooth ends are required for several common fiber measurements. They are needed when experimentally determining the numerical aperture by measuring the emission from a fiber, as illustrated in Fig. 8-17. Measurements yield patterns like those in the figure. The theory we have been using predicts a sharp cutoff in the field pattern at an angle corresponding to the internal critical angle. This theory neglects *skew* rays (rays that do not pass through the fiber axis but that are still guided by the fiber). The experimentally determined acceptance angle is sometimes defined to be the angle at which the radiated power drops to 10% of its peak value. For the measurement in Fig. 8-17, the acceptance angle is 14°, yielding a measured numerical aperture NA = sin 14° = 0.24.

The measurement of fiber attenuation also requires a smooth end face. Loss is found by comparing the power emerging from a long fiber to the power emitted by the same fiber when shortened. The loss per unit length is simply the measured loss, in decibels, divided by the length of fiber removed from the cable between the two power measurements.

8-3 SPLICES

Splices[10,11] are generally permanent fiber joints. (Connectors can be mated and unmated repeatedly and rather easily.) Basic splicing techniques include fusing the two fibers or bonding them together in an alignment structure. The bond may be provided by an adhesive, by mechanical pressure, or by a combination of the two.

Fusion Splicing

Fusion splices are produced by welding two glass fibers, as sketched in Fig. 8-18. Commercial fusion machines use an electric arc to soften the fiber ends. The ends are prepared by the scribe-and-break method. Alignment is obtained by adjusting micromanipulators attached to the fibers. The alignment is visually inspected with a microscope or some other magnifying arrangement.

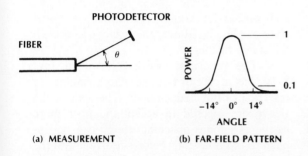

(a) MEASUREMENT   (b) FAR-FIELD PATTERN

Figure 8-17 Determining numerical aperture by measuring the fiber far-field radiation pattern.

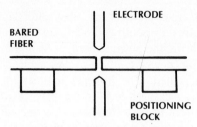

Figure 8-18   Electric arc fusion.

Alignment can also be checked by monitoring the power transmitted past the joint before the fibers are fused. If the transmitter and receiver are far from the splice joint (say, a few hundred meters or more), then this can be a difficult and time-consuming measurement. The solution to this problem is the *light injection and detection* (LID) system. In the LID arrangement, light is inserted into one of the fibers close to the splice joint (about 10–20 cm) and extracted for detection from the other fiber (again close to the joint). The injection and detection are accomplished by winding the fiber in a tight bend around a cylinder having a small radius. The bend is so tight (typically having a radius of just a few millimeters) that energy can be coupled in by placing a light source at the bend in the input fiber and can be coupled out by placing a photodetector at the bend in the output fiber. In many single-mode fibers the buffer coating is transparent so that it need not be removed for the LID coupling. On the other hand, some coatings are covered with a dye (for identification purposes) that may be opaque. If this is the case, the coloring must be removed. A solvent such as acetone will usually work for removal of the dye.

During fusion, surface tension tends to align the fiber axes, minimizing lateral offset. Splices produced by commercial fusion equipment have losses less than 0.25 dB. With care, losses less than 0.1 dB can be obtained. The splice area is protected by covering with such materials as RTV, epoxy, and heat-shrinkable tubing. Fusion works well with all-glass fibers, both multimode and single mode.

Adhesive Splicing

A number of alignment configurations have been suggested for splices by using adhesive bonding. Some of them are sketched in Fig. 8-19. Each of these structures mechanically aligns the fibers and provides strength to the joint. The fibers are held in place by epoxy. Because the epoxy must be cured, these splices cannot be used immediately. Curing times can be reduced by application of heat or, for some epoxies, exposure to ultraviolet radiation.

The V-block is probably the simplest mechanical splice. The bared fibers to be joined are placed in the groove. Angular alignment is particularly well controlled. The two fibers can slide in the groove until they touch. They are then epoxied permanently into position, so end separation errors are minimal. If the epoxy is index matched to the fiber, even small gaps can be tolerated with little loss. Lateral misalignment would be negligible in the groove if both fibers had the same core and cladding diameters and if the cores were centered within the cladding. Offset cores can be detected by rotating the output fiber while monitoring the transmitted power. Identical, well-constructed fibers would produce the same output power for all orientations. None of the splices in Fig. 8-19 can compensate for noncentered cores. A cover plate can be placed over the V-block to protect the splice further.

The precision sleeve, shown in Fig. 8-19, has a central hole just large enough for insertion of the cladded fiber. The ends of the sleeve are tapered to accept the fiber more easily. An index-matching epoxy can be applied to the fiber ends before insertion into the

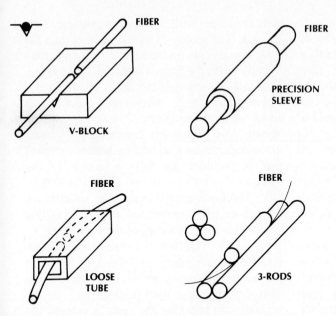

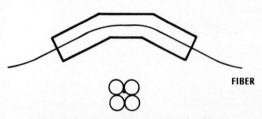

Figure 8-19   Mechanical splices.

sleeve. Alternatively, a hole drilled into the side of the sleeve can be used for observing the contacting fibers and for injecting epoxy or an index-matching fluid. Sleeves may be metal or plastic. In one splicing technique, the sleeve material is a compliant plastic.[12] When the fibers are inserted into the slightly undersized hole, the resilient material forces both fibers to align along a common central axis. Even fibers with unequal cladding diameters will have their axes laterally aligned.

The loose-tube splice, illustrated in Fig. 8-19, is interesting. Two fibers are inserted into the freely suspended tube. Bending the fibers causes the tube to rotate, aligning the fibers in one of the corners. Epoxy secures the aligned fibers.

Three precision glass or metal rods can be positioned as shown in Fig. 8-19 to align fibers. Rod diameters are chosen so that the hole formed at the junction is just large enough to accept the fiber. Index-matching epoxy is applied to the fibers and they are inserted into the hole until they touch. A heat-shrinkable sleeve is placed over the assembly.

When heat is applied, it secures the rods and squeezes them against the fiber.

A splice related to those just described is sketched in Fig. 8-20. Four glass rods are fused together, forming four V-grooves. The spacing between the rods is larger than the fiber. The ends of the bundle are bent so that an entering fiber will be forced into one of the grooves, very much like in the loose-tube arrangement. The glass guide can be prefilled with an index-matched, ultraviolet (UV)-curable epoxy. Prepared fibers are pushed into the flared openings until they touch. The epoxy is exposed to ultraviolet radiation to secure the bond.

Figure 8-20   Bent, fused-rod splice.

A splicing technique that does not use a precision-machined structure to align the fibers directly is the *rotary mechanical splice*, sketched in Fig. 8-21. In this splice, three rods in a bronze alignment clip secure the ferrules. The holes in the ferrules are not centered, so that the two fibers can be aligned by rotating the ferrules while monitoring the transmitted power. Since the ferrules are transparent, they can be fixed in place with an UV-curable epoxy after alignment. The rotary splice is suitable for connecting single-mode fibers because of the active alignment. Losses of less than 0.1 dB can be expected.

The rotary splice has also been manufactured with a single ferrule with an axially centered hole for the fiber. This single piece is constructed in such a way that it can be separated midway along its length by the installer. The two pieces become the ferrules for the two fibers being spliced. The ferrules have tabs attached to them, so that they can be recombined with the same alignment as before separation. The fibers are attached (epoxied permanently) to the ferrules and polished. The tabs on both ferrules allow the installer to mate the two parts in the metal clip, aligning them accurately.

Another splice is essentially a precision sleeve made with elastomeric materials. The elastomer is an elastic material usually made into a cylinder with a V-groove opening along its axis. The groove is a little smaller than the fiber but accepts and centers it by expanding slightly when the fiber is inserted. Gel lubrication eases insertion of the fiber. The fibers are inserted from both ends of the cylinder and touch near its midpoint. The splice can be epoxied for permanent connection. As with all splices, an external splice holder or splice tray is required for full protection of the splice.

Precision glass sleeve and elastomeric splices are also constructed with fixtures that can bind onto the coating of the two fibers being spliced. This has the advantage that no epoxies are required to hold the fibers in place. The splicing operation involves baring a short length of the fiber (half an inch or so), scribing and breaking the ends to prepare the surfaces, inserting the two fibers into the precision sleeve until they touch, and tightening the mechanism that clamps onto the fiber coating. This type of splice is fast to assemble and relatively inexpensive.

Fusion and rotary splicing are examples of connecting techniques that require active alignment for single-mode fibers. Generally, because of the small mechanical tolerances encountered when connecting single-mode fibers, active alignment is required to obtain the highest efficiencies. Single-mode connection schemes that do not use active alignment have somewhat higher losses but have the advantage of convenience.

It should be apparent, from the sampling just presented, that many splicing techniques exist. Designers can choose from a variety of methods already developed or can use their ingenuity to develop new and improved versions.

Good mechanical splices produce losses from less than 0.1 dB to just under 1 dB when identical fibers are connected. To obtain the lowest losses, it is imperative that the fiber

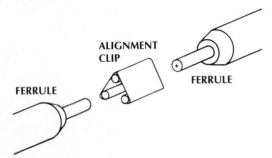

Figure 8-21 Rotary mechanical splice. (From AT&T manufacturer's literature. American Telephone and Telegraph Co., New York, 1985.)

ends be kept clean. It is clear, by comparing the efficiency of actual splices with the misalignment losses enumerated in the preceding section, that mechanical splices provide a high degree of precision positioning.

After fibers have been attached, any bare fiber remaining should be recoated (e.g., by covering with epoxy, lacquer, or RTV). The coating will protect the fiber from abrasion, which could lead to fracture.

We discussed ribbon fibers in Section 5-8. These fibers can be connected by the arrangement shown in Fig. 8-22.[13] The bared fibers are laid in precise grooves, formed by etching silicon chips. These positive chips are grooved on both sides. Another positive chip covers the fibers. The structure is permanently bonded, and the ends are polished. A second fiber ribbon is made the same way. The splice itself is obtained by placing the positive chips onto negative chips as shown in the figure. The negative chips extend across both positive chips, aligning them. The entire splice is held together by metal clips. Although sometimes called a splice, this structure can be repeatedly connected and disconnected with little trouble. An advantage of this technique is that the fibers can be connected in a factory and later spliced in the field. Positive chips can be stacked to produce an array connector for attaching each ribbon in a multiribbon cable.

8-4 CONNECTORS

Rematable attachments have tested the ingenuity of connector designers and the pocketbooks of fiber users. The stringent mechanical tolerances required for efficient coupling make quality connectors difficult to design and expensive to build.

Requirements for a good connector include the following:

1. *Low loss*. The connector assembly must ensure that misalignments are minimized automatically when connectors are mated. Unlike the situation in some splicing arrangements, the joint is not available for viewing within a connector and positioning corrections cannot be made. A system containing several connectors must have efficient ones. For example, if five connectors are used and each has a 2-dB loss, the total loss will be 10 dB, reducing the power available to the receiver by a factor of 10.

2. *Repeatability*. The coupling efficiency should not change much with repeated matings.

3. *Predictability*. The same efficiency should be obtained if the same combinations of connectors and fibers are used. That is, the loss should be relatively insensitive to the skill of the assembler.

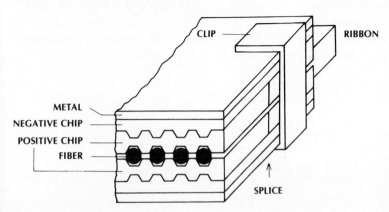

Figure 8-22 Ribbon splice, cutaway view. The position where the fibers meet is marked by the arrow.

4. *Long life*. Repeated matings should not degrade the efficiency or strength of the connection. The loss of a mated connector should not change with time.

5. *High strength*. The connection should not degrade owing to forces on the connector body or tension on the fiber cables.

6. *Compatibility with the environment*. The connection may have to withstand large temperature variations, moisture, chemical attack, dirt, high pressures, and vibrations.

7. *Ease of assembly*. Preparing the fiber and attaching it to a ferrule should not be difficult or time consuming.

8. *Ease of use*. Mating and unmating the connection should be simple.

9. *Economy*. Precision connectors are expensive. Cheaper connectors, normally plastic, may not perform as well.

Most connectors are designed to produce a butt joint, placing the fiber ends as close together as possible. Butt designs include the straight-sleeve, tapered-sleeve, and overlap connectors. A lensed connector is an alternative to the butt configuration. The connector assemblies to be described in the remainder of this section are meant to illustrate the general approaches that have been successful in joining fibers. The descriptions do not give complete details of specific commercial connectors but include features found in many of them.

Butt connectors generally consist of a ferrule for each fiber and a precision sleeve into which the ferrules fit. Figure 8-23 illustrates the straight-sleeve concept. Some straight ferrules are designed like SMA coaxial connectors. Axial and angular alignment are obtained from the smooth fit of the ferrules into the tubular sleeve. Close tolerances are obviously required. The end separation is determined by the length of the ferrule beyond a gap alignment lip and by the length of the sleeve. Threaded caps fit against a guide ring and screw onto the sleeve, securing the connection. The cable in Fig. 8-23 can be epoxied to the tube, crimped to it, or both, to provide strength. An alternative design permits the Kevlar braid to be crimped to the ferrule, as sketched in Fig. 8-24, for added strength. Tension on the cable is transferred to the strong Kevlar member, not to the fragile fiber, providing strain relief.

Tapered-sleeve (*biconical*) connectors, illustrated in Fig. 8-25, can have molded plastic parts. A tapered sleeve accepts and guides the tapered ferrules. Very little abrasive wear occurs with repeated mounting and demounting of a tapered connector. Cables are affixed to the ferrules by adhesives or by crimping, similar to the cable attachment in the straight-sleeve connector. The fiber end separation may be completely determined by the mechanical structure, as it is in Fig. 8-25 where a guide ring prohibits the fibers from drawing closer. If the guide ring were not restrained by the sleeve (e.g., if the length of the alignment sleeve in Fig. 8-25 were too short), then the gap would depend on the amount of tightening of the securing screw caps.

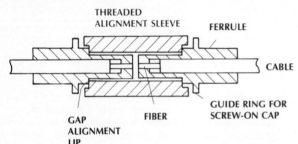

THREADED ALIGNMENT SLEEVE   FERRULE

CABLE

GUIDE RING FOR SCREW-ON CAP

GAP ALIGNMENT LIP   FIBER

Figure 8-23   Straight-sleeve connector.

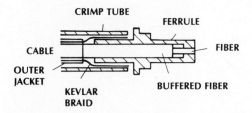

Figure 8-24 The Kevlar-braid strength member can be crimped to the ferrule.

Most connectors (such as the SMA types) are secured by screws. Connection and disconnection is moderately convenient. Quicker mating and remating are possible with non-screw devices such as the ST connector.[14] This is a keyed and spring-loaded assembly (as illustrated in Fig. 8-26), which attaches to a coupling bushing much like a coaxial BNC connector. The spring is attached to the ferrule in such a way that the ends of the connector make contact, but the force between the fibers is determined by the spring tension (not by the tightness of the end caps or other securing mechanism).

The concept of an overlap connector is illustrated in Fig. 8-27. This structure is similar to the V-groove splice. Polished fibers lie in rounded grooves in mirror-image sections. A polishing tool is required not only to hold the fiber while it is being prepared but to control the fiber length accurately. The grooved sections, made of a compliant plastic, are clamped together to make the connection. When pressure is applied, the resilient materials conform to the shape of the fiber. This causes both fibers to become aligned among a common central axis. Axial alignment is enhanced in this manner. While engaged, the connector halves are kept apart (one above the other) until the fiber ends nearly meet. The two sections are then pressed together and locked by a clamping mechanism. The clamp can be spread to disengage the connector.

A lensed connector is shown schematically in Fig. 8-28. The expanding beam radiating from the transmitting fiber is collimated by a lens. The fiber-to-lens distance is equal to the focal length, as required for collimation. (See Chapter 2 for a review of ray optics, if needed.) An identical arrangement exists at the receiver. This configuration is an imaging system with unity magnification, regardless of the spacing between the lenses. Because the beam is enlarged at the connecting plane, the sensitivity to lateral offset is reduced compared to a butt joint. The lateral losses given by Eq. (8-1) and plotted in Fig. 8-3 now apply to the enlarged beam diameter.

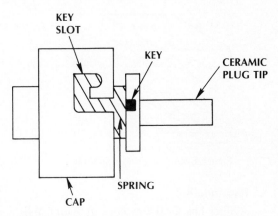

Figure 8-26 The ST type connector is keyed and spring loaded.

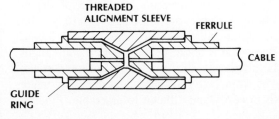

Figure 8-25 Tapered-sleeve connector. Caps fit over the ferrules, rest against the guide rings, and screw onto the threaded sleeve to secure the connection.

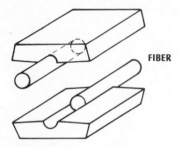

Figure 8-27   Overlap connector.

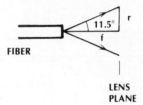

Figure 8-29   Design of a lensed coupler.

Example 8-4

A fiber has a 50-μm core diameter and an 0.2 NA value. The beam is expanded to a 2-mm diameter. Design the lens arrangement and compute the allowable lateral offset for a 0.5-dB loss.

Solution:

The beam diverges at an angle $\sin^{-1}$ NA = 11.5°, as sketched in Fig. 8-29. The beam radius and the distance from the fiber to the lens are related by $r/f = \tan 11.5°$. Setting $r = 1$ mm yields $f = 4.9$ mm. We choose a lens with a focal length of 4.9 mm and a diameter a little larger than 2 mm. It is placed 4.9 mm from the fiber. According to Fig. 8-3, a loss of 0.5 dB occurs if $d/2a = 0.09$. Since $2a = 2$ mm, then $d = 0.18$ mm = 180 μm. This tolerance should be compared with the result obtained in Example 8-1, in which the allowed offset was only 4.5 μm for a 0.5-dB loss.

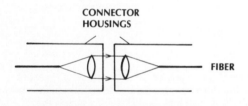

Figure 8-28   Lensed connector.

In addition to reduced sensitivity to lateral offset, lensed connectors allow larger gaps than butt joints. Because the beam is collimated, the gap can be significant before losses become large. This property can be quite useful. Not only can the separation tolerance be loosened, but a flat glass plate can cover each connector, protecting it from the environment. It might have occurred to you that the highly polished fiber ends in a connector need to be treated carefully when not engaged to keep them from being scratched or from gathering dirt. This is indeed the case. A permanent glass cover can be part of a lensed connector to reduce the probability of damage. The lens separation cannot be arbitrarily large because off-axis rays do not enter the receiving fiber at the same angle that they left the transmitting fiber unless the lens separation is twice the focal length. Figure 8-30 illustrates the change in ray direction for a gap greater than $2f$. The deviation increases as the gap enlarges. Of course, when the deviation causes the ray to exceed the fiber's acceptance angle, the ray is no longer coupled to the receiving fiber.

Lensed connectors have several disadvantages. Losses are more sensitive to angular misalignments than they are in butt joints. Because it is not difficult to obtain good angular alignment, this problem is not too serious. The complexity of lensed connectors make them costly and difficult to assemble. Finally,

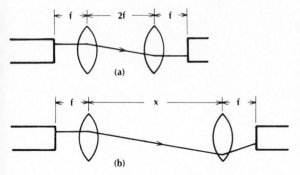

(a)

(b)

Figure 8-30 Lens separation is limited by the change in receiving angles of off-axis rays. (a) The separation is 2*f* and the emitted ray angles are preserved. (b) When the separation is greater than 2*f*, the emitted and received ray angles differ. The ray angle at the receiver may now fall outside the fiber's acceptance cone, contributing to coupler inefficiency.

because of reflections from the two lenses and the two cover plates (if present) the fixed losses of lensed connectors may exceed those of butt couplers. Antireflection coatings on the boundary surfaces can reduce the reflection losses.

The GRIN rod lens was introduced in Section 2-2. It can substitute for a conventional lens in a connector, as illustrated in Fig. 8-31. Each connector section consists of a fiber attached to a solid, quarter-pitch GRIN lens. This type of connector has the same advantage as the spherical-lens connector: good tolerance to both lateral offset and end separation. It can also be covered with a protective glass plate to reduce damage and minimize additional losses from scratches and dirt. This connector might be easier to assemble and maintain than its spherical counterpart.

Multichannel connectors can be constructed. The simplest example is a two-channel connector, which is convenient for duplex systems in which information is carried in one direction in one fiber and in the opposite direction in a second fiber. The overlap design (in Fig. 8-27) can accommodate two

fibers if it contains sections with two parallel grooves rather than one. The overlap concept could be extended to more than two channels if there were additional grooves.

Multichannel connectors may use the straight or tapered-sleeve approaches in what might be called the *bayonet* style. In one possible design (see Fig. 8-32), each fiber has its own ferrule. The ferrules for one of the multichannel cables are attached to a plug and the ferrules for the other cable are attached to a receptacle. The ends of the plug ferrules extend from the plug body itself, while the receptacle ferrules are recessed within a sleeve in the receptacle. When the plug and receptacle are mated, the protruding ferrule tips are guided into alignment by the sleeve. The ferrules are pressed into the plug and receptacle by springs pushing against the ferrule guide rings. The bayonet style of coupler can be modified to produce connectors accepting a circular (rather than linear) array of fibers.

Typical connector losses range from about 0.5 dB to 3 dB, a bit larger than splice losses. Matching fluid could improve the efficiency, but it is often not acceptable because of inconvenience, evaporation or leakage away from the joint, loss of transparency with time, and a tendency to trap small particles of foreign matter in the joint. Molded plastic connectors have lower prices and less mechanical precision (and thus higher losses) than metal connectors.

GRIN ROD
LENSES

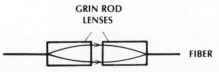

FIBER

Figure 8-31 GRIN-rod lensed connector.

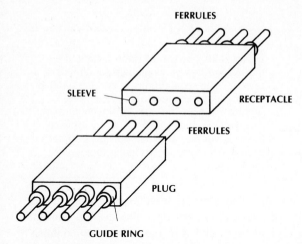

FERRULES

SLEEVE

RECEPTACLE

FERRULES

PLUG

GUIDE RING

Figure 8-32 Multichannel connector.

Why do connectors not have losses as low as splices? The most likely reason is the difficulty of achieving the required lateral alignment tolerance by relying on purely mechanical arrangement (i.e., without resorting to active alignment). In fact, some connectors can be tuned by rotating the ferrules until maximum transmission occurs before tightening the end caps.

Another factor that makes connector losses larger than those of splices is that the fiber ends do not make contact in some designs. This is a safety factor built into connectors so that overtightening of the securing caps does not create excessive forces on the finely polished fiber ends. A spring-loaded connector (for example, the one drawn in Fig. 8-26) is an effective solution to the end-gap problem. In a splice, the fibers can usually be gently moved toward each other until they touch before being secured. The splice joint may even be directly viewable, so the alignment can be inspected visually.

8-5 SOURCE COUPLING

Coupling from the light source to the fiber can be very inefficient. We define the coupling efficiency as

$$\eta = \frac{P_f}{P_s} \qquad (8\text{-}10)$$

where P_f is the power in the fiber and P_s is the power emitted by the source. Expressed in decibels, the coupling loss is $L = -10 \log \eta$. Several mechanisms contribute to the inefficiency. They include reflection loss, area-mismatch loss, packing-fraction loss, and numerical-aperture loss. These problems are illustrated in Fig. 8-33 and described next.

Reflection Loss

If an air gap exists between the emitting surface and the fiber, then power is reflected at the boundary according to Eq. (3-28) for normal incidence. The equation is adequate for the small acceptance angles usually found. In Section 3-5 the loss at an air-to-glass interface was calculated to be just under 0.2 dB. If the source is in contact with the fiber, or if a matching fluid fills the gap, then this loss disappears. The 0.2-dB loss is so small that

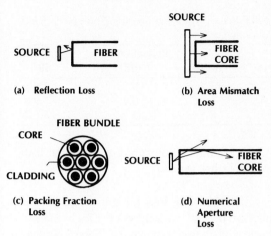

Figure 8-33   Source coupling losses.

matching fluid may not be advisable except, perhaps, to provide structural support for the fiber. The fiber end face is prepared just as it is for a splice to eliminate scattering from uneven surfaces.

Area-Mismatch Loss

If the source area is larger than the area of the fiber core, then some of the power is lost (see Fig. 8-33). The reduction in efficiency is the ratio of the core area to the source area, A_c/A_s. If the source is smaller than the core, then this loss disappears.

Packing-Fraction Loss

Sometimes a bundle of fibers (pictured in Fig. 8-33) is used with a single emitter. Many fibers are packed together with their claddings in contact. This structure has several applications. The large bundle can match a large area source, eliminating area-mismatch loss. Because large sources can emit more light than smaller ones, more power can be coupled into a bundle than into a single fiber. Bundles also

provide redundancy. Reception is not terminated if one of the fibers breaks, although the power level is reduced. The fiber bundle can also be used to distribute information to several locations by splitting the bundle and routing the separate fibers along different paths. Fiber bundles incur a coupling loss we have not yet mentioned. Light from the source that strikes the cladding or the air space between fibers is lost. The coupling efficiency is reduced by the *packing fraction pf*, which is the ratio of the sum of the core areas to the area of the bundle. Packing fractions of 0.4–0.75 are typical. Large packing fractions are obtained by minimizing the cladding thickness. The possibility of crosstalk between thinly cladded fibers is unimportant because every fiber is carrying the same information.

Numerical-Aperture Loss

The losses described in this section apply only to multimode fibers.

As first explained in Section 4-4, rays of light that are incident at angles larger than the waveguide's acceptance angle are not efficiently transmitted. When coupling light from a source into a fiber, the losses owing to this effect can be very significant. The efficiency for this phenomenon is given by[15]

$$\eta = \text{NA}^2 \qquad (8\text{-}11)$$

for a step-index fiber excited by a Lambertian source, such as a surface-emitting LED. The Lambertian power distribution was discussed in Section 6-2 and was sketched in Fig. 6-10. The numerical aperture of an SI fiber is given by $\text{NA} = \sqrt{n_1^2 - n_2^2}$ as determined in Section 4-4. Equation (8-11), which is plotted in Fig. 8-34, applies when an air gap (or matching fluid) separates the source and fiber or when the two are in direct contact. This follows

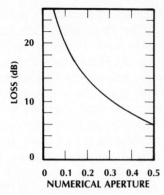

Figure 8-34 Coupling loss from a Lambertian source into a step-index fiber.

TABLE 8-1. Typical Lambertian Source—SI Fiber Coupling Losses

NA	NA Loss (dB)	Reflection Loss (dB)	Total Loss (dB)
0.24	12.4	0.2	12.6
0.41	7.7	0.2	7.9
0.48	6.4	0.2	6.6

from the knowledge that the relationship between the ray angle in the fiber and the corresponding ray direction inside the source is independent of the material between the two. The argument is illustrated in Fig. 8-35.

Example 8-5

Compute the coupling losses for SI fibers having the characteristics in Table 5-1 when excited by a surface-emitting LED.

Solution:

Assuming no area mismatch, the NA loss is calculated from Eq.(8-11) and the reflection loss from Eq. (3-28). The results, expressed in decibels, appear in Table 8-1. The reflection loss, which only applies when an air gap is present, is negligible compared to the numerical-aperture losses.

The coupling loss for a low-NA fiber is substantial. The advantage of a high-quality, low-loss fiber over one having larger loss may disappear if the path is short and the NA of the high-loss fiber is greater. An example will illustrate this point.

Example 8-6

A Lambertian source radiates 2 mW (3 dBm). How much power is coupled into the SI glass and plastic fibers listed in Table 5-2? How much power remains in the fibers after 10 and 100 m?

Solution:

The results of this problem are given in Table 8-2. The NA and attenuation data are from Table 5-2. The coupling loss is found in Table 8-1. The source power of 3 dBm is reduced by the 12.6-dB glass-fiber coupling loss to $3 - 12.6 = -9.6$ dBm. Similarly, the plastic fiber couples $3 - 6.6 = -3.6$ dBm. The power in the plastic fiber is about 6 dB more than in the glass fiber. Ten meters of the glass fiber contribute 0.05 dB of loss, and the same length of plastic introduces 2 dB of loss. Adding these to the coupled power yields -9.7 and -5.6 dBm in the glass and plastic fibers, respectively. For a 10-m path, there is 4.1 dB more power in the high-loss plastic than in the more efficient

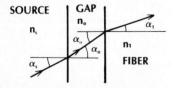

Figure 8-35 Ray progression from the source to the fiber. According to Snell's law, $n_s \sin \alpha_s = n_0 \sin \alpha_0 = n_1 \sin \alpha_1$, so that the ray angle within the fiber (α_1) depends only on the original ray direction inside the source (α_s).

TABLE 8-2. Power Transmission for a Glass and a Plastic Fiber Excited by a 2-mW Lambertian LED

Fiber	NA	Attenuation (dB/km)	Coupling Loss (dB)	Coupled Power (dBm)	Output Power 10 m (dBm)	Output Power 100 m (dBm)
Glass	0.24	5	12.6	−9.6	−9.7	−10.1
Plastic	0.48	200	6.6	−3.6	−5.6	−23.6

glass fiber. For the 100-m path, the attenuation in the plastic fiber (20 dB over this length) reduces the power below that in the glass by more than 13 dB. The power levels at the end of 100 m are 97.7 μW in the glass (corresponding to −10.1 dBm) and 4.36 μW in the plastic (corresponding to −23.6 dBm).

Evaluation of the coupling efficiency is complicated by several factors. A diverging source will usually excite cladding modes. These modes propagate a short distance through the fiber before attenuating significantly. The coupled power measured near the input point will include these modes, and a measurement far from the input will not. Since cladding modes were not included in obtaining the efficiency in Eq. (8-11), that equation predicts lower efficiencies than are actually obtained for short paths. The area-mismatch loss is also reduced for short paths if parts of the source radiate directly into the cladding, exciting cladding modes. Another factor that improves the short distance coupling efficiency compared to the long fiber efficiency is the existence of *leaky modes*, which are characterized by skew rays. Skew rays do not pass through the fiber axis but circulate around it in a helical fashion. Leaky modes are attenuated but may persist for fairly long distances in some fibers.

In Section 5-2 we saw that GRIN fibers had numerical apertures that dropped from a maximum along the axis to zero at the edge of the core. Because of this, light is coupled less and less efficiently as the excitation point

moves away from the axis. Therefore, power is coupled less efficiently into a GRIN fiber than into a SI fiber. For the parabolic index fiber the efficiency is[16]

$$\eta = \frac{NA^2}{2} \qquad (8\text{-}12)$$

for a Lambertian emitter, where the axial NA is used. This is just half the efficiency of the comparable SI fiber. Thus, coupling into a SI fiber is 3 dB *better* than coupling to a comparable parabolic-index fiber.

This last equation applies to the case where the LED's emitting surface is the same size as the fiber core. If the emitting surface is smaller, however, most of the light will be collected over parts of the fiber near its axis where the NA is larger. Thus, a greater fraction of the emitter's light rays will fall within the local acceptance angle of the fiber and the overall collection efficiency will improve. Under these conditions, the efficiency is given by[17]

$$\eta = NA^2[1 - 0.5(a_e/a_f)^2] \qquad (8\text{-}13)$$

where a_e is the radius of the emitting surface, a_f is the radius of the fiber core, and the axial NA is used. As expected, this equation simplifies to the previous one when the emitter and fiber are the same size.

Edge-emitting LEDs and laser diodes radiate beams that are more compact than the Lambertian distribution, improving the coupling efficiency. For LEDs the improvement can be several decibels and for laser diodes even more. We can model a narrow power distribution by the expression $\cos^m \theta$, where m is an integer and θ is the viewing angle measured with respect to the normal to the emitting surface. Figure 8-36 illustrates the power patterns, showing how higher values of m correspond to narrower beams.

Measured power patterns can be compared with curves like those drawn in the figure to determine the appropriate values of m for a particular source. The SI coupling efficiency is given by[18]

$$\eta = 1 - (1 - NA^2)^{(m+1)/2} \qquad (8\text{-}14)$$

This result can also be written as $\eta = 1 - (\cos \alpha_o)^{m+1}$, where α_o is the fiber's acceptance angle. For $m = 1$ (the Lambertian distribution), this reduces directly to Eq. (8-11). For small values of NA, the efficiency equation reduces to

$$\eta = \frac{(m + 1)NA^2}{2} \qquad (8\text{-}15)$$

This approximate result is appropriate for the case where $(m - 1)NA^2 < 0.4$. The improvement in efficiency for narrow beams ($m \gg 1$) is evident from this expression. For coupling into a parabolic-index fiber, Eq. (8-15) is multiplied by $1 - 0.5(a_e/a_f)^2$.

As we have noted several times, the main cause of coupling inefficiency in multi-mode fibers is the wide angular distribution of light from common emitters and the restricted acceptance angles of fibers. A reasonable way to reduce the angular spread from the source is to use a lens, as drawn in Fig. 8-37. The angular change was given by Eq. (2-9). The small-angle approximation (Eq. (2-10)) shows that the angular spread is reduced by the magnification of the imaging system. Large magnifications can reduce the beam divergence considerably, with a corresponding increase in coupling efficiency. It may even be possible to reduce the beam divergence so that nearly all the source rays are within the acceptance angle. Of course, the magnified image of the source must remain smaller than the fiber core to avoid area-mismatch loss. It is clear that reducing the beam spread by magnifying the source will work only when the core is larger than the source. When this is not the case, butt coupling will be more efficient.

For a large source having area-mismatch losses, a lens can be used to reduce the cross-sectional area of the beam to match the fiber core. Demagnification ($M < 1$) causes an increase in beam divergence, again according to Eq. (2-10). This method is not appropriate for LEDs. It can be useful for coupling from lasers (such as gas lasers) which have very small beam divergence.

Edge emitters produce beam patterns that are not symmetrical. A cylindrical lens, placed as shown in Fig. 8-38(a), reduces the

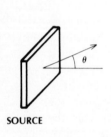

SOURCE

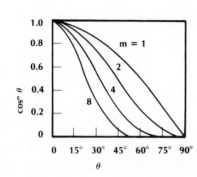

Figure 8-36 Power distribution models.

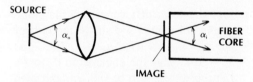

Figure 8-37 Reducing the angular spread of a source. $\alpha_i \simeq \alpha_0/M$, where M is the linear magnification.

nonsymmetry by narrowing the ray angles only in the plane of greatest beam divergence. The cylindrical lens, drawn as a circle in the figure, is often simply a short length of glass fiber placed at right angles to the transmitting fiber. The beam emerging from the cylindrical lens can excite the fiber directly. A spherical lens further reduces the beam spread, if required, as drawn in Fig. 8-38(b). The conventional spherical lens may be replaced by an equivalent GRIN rod lens, as shown in Fig. 8-38(c).

Commercial sources are often available with short fiber pigtails attached, ready for splicing or connecting to the transmitting fiber. The details of the coupling between the

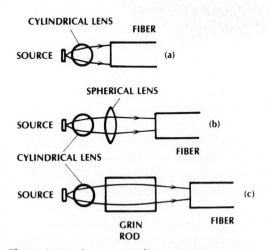

Figure 8-38 Source coupling using (a) a cylindrical lens, (b) a combination cylindrical and spherical lens, and (c) a combination cylindrical and GRIN rod lens.

source and the pigtail might be unknown to the purchaser. The system designer needs to know only the power emerging from the pigtail and the core size and type (SI or GRIN) of the pigtail itself. When buying a source, the purchaser should check whether the output power provided by the manufacturer is the power directly out of the emitter or is the power that emerges from a pigtailed fiber. The latter power is considerably lower in most cases because of low coupling efficiency.

Single-Mode Fiber Coupling

Coupling into a single-mode fiber would be very efficient if the incoming wave had a Gaussian distribution matching that of the propagating HE_{11} mode. This requires that the spot sizes (discussed in Sections 2-5 and 5-4) of the two waves be equal. Figure 2-27 illustrates an arrangement for coupling a collimated laser beam into the fiber. The beam is focused to reduce the spot size to that of the fiber. Similarly, a lens can be placed between a laser diode and a single-mode fiber to match the spot size of the emitter to that of the fiber. Alternatively, a lens can be formed on the end of the fiber for the same purpose. The fiber can also be tapered (by heating and pulling) to enhance further the matching of the two waves. Coupling efficiencies as low as a few decibels are possible by using these techniques.

8-6 SUMMARY

Producing efficient splices, connectors, and source couplers requires a great deal of care and attention. Fiber ends must be precisely prepared and positioning and alignment must be very accurate. With the proper precautions, splice losses of a few tenths of a decibel or less can be obtained. Good connectors provide less than 1 dB of loss.

The efficiency of source couplers depends on the radiation pattern of the source and the NA of the fiber. Surface-emitting LEDs have losses of more than 12 dB when coupled to fibers having NA < 0.24. Lenses can improve the efficiency when the LED is smaller than the fiber core. Laser diodes and edge-emitting LEDs, radiating narrower beams than do surface-emitting LEDs, have better coupling efficiencies. The losses are still considerable when coupling to small fibers having low NA values.

PROBLEMS

8-1. Derive Eq. (8-1) by computing the area of overlap of the two fiber cores. Perform the evaluation by integration.

8-2. Use Eq. (8-1) to confirm at least four points on Fig. 8-3.

8-3. At what value of fractional lateral offset is the loss equal to 10 dB for the multimode SI fiber?

8-4. Repeat Problem 8-3 for a single-mode fiber. Assume that $V = 2.4$.

8-5. A single-mode fiber contains a beam whose spot size is 5 μm. What lateral offset will produce 0.5 dB of loss? Assume that $V = 2.4$.

8-6. Compute the normalized intensity of the beam at the core-cladding interface of a SI single-mode fiber if $V = 2.4$. Assume that the propagating beam is Gaussian.

8-7. Plot the loss (dB) versus angular misalignment for a multimode SI fiber if its NA $= 0.1$ and air fills the groove formed by the fibers. The misalignment angle should range from 0 to 10°.

8-8. Plot the loss (dB) versus angular misalignment angle for a single-mode fiber having $n_1 = 1.47$, $n_2 = 1.468$, and $V = 2.4$. Vary the angle from 0° to 4°. Let the wavelength be 0.8 μm and then repeat the problem for a 1.3-μm wavelength.

8-9. Derive Eq. (8-4) for the loss caused by angular misalignment.

8-10. Plot the spot size versus wavelength for a SI fiber whose single-mode cutoff wavelength is 1250 nm and whose NA is 0.1. The wavelength range should be 1250–1600 nm. Also compute and plot the lateral offset loss versus wavelength if the offset is 1 μm.

8-11. Plot the loss (dB) versus end spacing for the single-mode fiber described in Problem 8-8. Let the wavelength be 0.8 and then 1.3 μm. Vary the spacing from 0 to 500 μm.

8-12. A SI fiber has a 100-μm core diameter and a value of 0.28 NA. Design a lensed connector following the procedure described in Section 8-4. The lens diameter is 3 mm. Compute the allowable lateral offset for a 0.8-dB loss.

8-13. A large-core multimode SI fiber is excited by a surface-emitting LED. The fiber's NA is 0.2. The LED's output power is 5 mW. The fiber's loss is 4 dB/km. Compute the power in the fiber at 1 m, 1 km, and 10 km. Repeat this problem if the fiber now has NA $= 0.5$ and a loss of 500 dB/km.

8-14. Derive Eq. (8-15) from Eq. (8-14). (*Hint*: Use the binomial formula.) If an error of 10% is allowed, determine the maximum value of NA that allows use of the approximate result [Eq. (8-15)]. Evaluate the result for values of m equal to 1, 2, 4, 6, 8, 10, and 20. Compute the exact and approximate values of coupling efficiency at the maximum values of NA just determined.

8-15. A source has a half-power radiation angle of 32.8° as measured from the normal to the emitting surface. Compute the coupling efficiency into a multimode SI fiber having a NA of 0.2.

8-16. A source has a half-power radiation angle of 45° as measured from the normal to the emitting surface. Compute the coupling efficiency into a multimode SI fiber having a NA of 0.2.

REFERENCES

1. Haruhiko Tsuchiya, Hiroshi Nakagome, Nobuo Shimizu, and Seiji Ohara. "Double Eccentric Connectors for Optical Fibers," *Appl. Opt.* 16, no. 5 (May 1977): 1323–31.

2. Dietrich Marcuse, Detlef Gloge, and Enrique A. J. Marcatili. "Guiding Properties of Fibers." In *Optical Fiber Telecommunications*, edited by Stewart E. Miller and Alan G. Chynoweth. New York: Academic Press, 1979, pp. 71–72.

3. Tsuchiya et al. "Double Eccentric Connectors for Optical Fibers." pp. 1324–25.

4. Marcuse et. al. "Guiding Properties of Fibers." pp. 71–72.

5. Tsuchiya et al. "Double Eccentric Connectors for Optical Fibers." p. 1324.

6. Dietrich Marcuse. "Loss Analysis of Single-Mode Fiber Splices," *Bell Syst. Tech.* J. 56. no. 5 (May 1977): 703–18.

7. Tsuchiya et al. "Double Eccentric Connectors for Optical Fibers." p. 1326.

8. John Joseph Esposito. "Optical Connectors, Couplers, and Switches." In *Handbook of Fiber Optics: Theory and Applications,* edited by Helmut F. Wolf. New York: Garland, 1979. pp. 241–303.

9. Detlef Gloge, Allen H. Cherin, Calvin M. Miller, and Peter W. Smith. "Fiber Splicing." In *Optical Fiber Telecommunications,* edited by Stewart E. Miller and Alan G. Chynoweth. New York: Academic Press, 1979, pp. 456–61.

10. Jack F. Dalgleish. "Splices, Connectors, and Power Couplers for Field and Office Use." *Proc. IEEE* 68, no. 10 (Oct. 1980): 1226–32.

11. Gloge et. al. "Fiber Splicing." pp. 461–82.

12. W. John Carlsen. "An Elastic-Tube Fiber Splice," *Laser Focus* 16, no. 4 (April 1980): 58–62.

13. T. Leslie Williford, Jr., Kenneth W. Jackson, and Christian Scholly. "Interconnection for Lightguide Fibers," *The Western Electric Engineer* 24, no. 1 (Winter 1980): 86–95.

14. AT&T manufacturer's literature, New York, 1986.

15. Michael K. Barnoski. "Coupling Components for Optical Fiber Waveguides." In *Fundamentals of Optical Fiber Communications,* 2nd ed., edited by Michael K. Barnoski. New York: Academic Press, 1981, pp. 147–86.

16. Ibid.

17. R.H. Saul, T.P. Lee, and C.A. Burrus. "Light-Emitting Diode Device Design." In *Semiconductors and Semimetals*, vol. 22, pt. C, edited by W.T. Tsang. New York: Academic Press, 1985.

18. Barnowski, op. cit.

Chapter 9

Distribution Networks and Fiber Components

Up to now we have been considering, at least implicitly, only point-to-point unidirectional links. However, the versatility of fiber optics makes possible the design of bidirectional systems, which propagate signals on a single fiber in both directions simultaneously. Also, distribution of information over fibers to multiple terminals is important and practical. Multiterminal architectures have many applications. The most important might be the *local area network* (LAN), an interconnection of numerous input and output devices that is located within a restricted region.[1] Examples of a restricted region are a single building or a campus containing several buildings.

An office LAN includes CRT terminals located throughout the business premises. From each terminal an employee can access a variety of equipment and services, such as electronic data files, word processor, videotext service, computer, computer printer, or copying machine. Computers themselves can be linked by a LAN. Facilities for video teleconferencing can also be included. In a fiber optic LAN, fibers carry the information between the connected devices. Advantages over wires include improved security, smaller size, lower weight, and broader bandwidth. In another application, LANs installed in manufacturing plants monitor and control operations. We might classify the fibered city (described in Section 1-5) as an extended LAN.

Fibers are suitable for communicating between a number of field positions in a tactical command system. The low weight of fiber cables allows quick installation of the network. Security of fiber communications is a big advantage in this multiterminal application.

In this chapter we study basic system configurations and components for distributing and controlling information over fiber cables in ways that are not as restrictive as the single-optical-channel, unidirectional link connecting a single transmitter to a single receiver.

9-1 DISTRIBUTION NETWORKS

A directional coupler forms the basis of many distribution networks.[2] Figure 9-1 illustrates a four-port directional coupler. Later we will describe couplers with more ports. The directions of allowed power flow are indicated by the arrows in the figure. For a description of coupler characteristics, we will assume that power P_1 is incident on port 1 of the coupler.

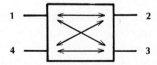

Figure 9-1 Four-port directional coupler.

This power will divide between ports 2 and 3 according to the desired splitting ratio. Ideally, no power will reach port 4, the isolated port. Without loss of generality, we can assume that the power emerging from port 2 (P_2) is equal to, or greater than, the power emerging from port 3 (P_3). We then define the following characteristic coupler losses (in dB):

1. *Throughput loss*

$$L_{THP} = -10 \log \frac{P_2}{P_1} \qquad (9\text{-}1)$$

specifies the amount of transmission loss between the input port and the favored port (port 2).

2. *Tap loss*

$$L_{TAP} = -10 \log \frac{P_3}{P_1} \qquad (9\text{-}2)$$

specifies the transmission loss between the input port and the tap (port 3).

3. *Directionality*

$$L_D = -10 \log \frac{P_4}{P_1} \qquad (9\text{-}3)$$

specifies the loss between the input port and the port we wish to isolate (port 4).

4. *Excess loss*

$$L_E = -10 \log \frac{P_2 + P_3}{P_1} \qquad (9\text{-}4)$$

specifies the power lost within the coupler. It includes radiation, scattering, absorption, and coupling to the isolated port.

In an ideal coupler, no power reaches port 4 ($L_D = \infty$). Additionally, no power is lost, so the total power emerging from ports 2 and 3 equals the input power ($P_2 + P_3 = P_1$), making the excess loss zero. Good directional couplers have excess losses less than 1 dB and directionality greater than 40 dB.

The *splitting ratio* is P_2/P_3, the ratio of the powers at the two output ports. Couplers are often described by their tap loss. For example, a 10-dB coupler is one that has a 10-dB tap loss. Table 9-1 lists values of throughput loss, tap loss, and splitting ratio for several ideal couplers.

TABLE 9-1. Characteristics of Several Ideal Four-Port Directional Couplers

Coupler Description	L_{TAP} (dB)	L_{THP} (dB)	Splitting Ratio
3 dB	3	3	1:1
6 dB	6	1.25	3:1
10 dB	10	0.46	9:1
12 dB	12	0.28	15:1

For the lossless coupler, $P_2 = P_1 - P_3$, so the throughput loss [Eq. (9-1)] can be written as

$$L_{THP} = -10 \log(1 - 10^{-L_{TAP}/10}) \qquad (9-5)$$

This result provides the relationship between the tap loss and the throughput loss.

In the following example, we illustrate how excess loss changes the throughput and the tap losses.

Example 9-1

A coupler has an excess loss of 1 dB and a splitting ratio of 1:1. How much of the input power reaches the two output terminals?

Solution:

Using $P_2 = P_3$ and $L_E = 1$ dB in Eq. (9-4) yields $P_2/P_1 = P_3/P_1 = 0.397$. This corresponds to a 4-dB throughput loss and a 4-dB tap loss. These losses exceed the losses of an ideal coupler having the same splitting ratio (3 dB according to Table 9-1) by 1 dB, the excess loss itself.

If L'_{THP} and L'_{TAP} are the losses of an ideal directional coupler having a given splitting ratio, then the losses of an actual coupler having the same splitting ratio, but excess loss L_E, are

$$L_{THP} = L'_{THP} + L_E \qquad (9-6a)$$

$$L_{TAP} = L'_{TAP} + L_E \qquad (9-6b)$$

The ideal losses are simply increased by the excess loss. These are the losses that one would actually measure by comparing the output powers at ports 2 and 3 to the input power entering port 1. Because the losses in Eq. (9-6) are the actual losses found when inserting the coupler into the system, they are often called the *insertion losses*.

As indicated by the arrows in Fig. 9-1, the coupler is bidirectional. Any of the four ports can serve as the input. Possible couplings (with the favored port listed just after the input port) are 1 to 2 and 3, 2 to 1 and 4, 3 to 4 and 1, and 4 to 3 and 2. Directional couplers are normally constructed symmetrically, so the characteristic losses have the same values regardless of which port is chosen as the input.

Duplexing Network

In the most straightforward scheme for transmitting and receiving at both ends of a point-to-point link, two fibers are used. One carries information in one direction, and the other carries signals in the opposite direction. A full-duplex system (one permitting simultaneous transmission in both directions along the same fiber) conserves fiber, a particularly significant advantage for long links. Figure 9-2 illustrates the full-duplex architecture with a directional coupler at each terminal. In this application, perfect 3-dB couplers would provide 6 dB of loss between the transmitter and receiver. Excess loss and the connector loss at each port would lower the received power even farther.

Tee Network

The *tee* network, drawn in Fig. 9-3, interconnects many terminals. Each terminal contains

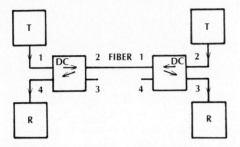

Figure 9-2   Full-duplex communications system. T, transmitter; R, receiver; DC, directional coupler. The unused ports are also shown.

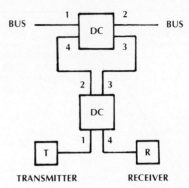

Figure 9-4   Tee coupler using two directional couplers.

a transmitter and a receiver. A trunk fiber, also known as a *bus*, or *data bus*, carries the information between taps. Taps are provided by tee couplers. The tee coupler shown in Fig. 9-4 permits bidirectional information flow in the bus fiber. In the figure, two directional couplers constitute the tee coupler. Terminals 1 and N are connected to the bus by a single directional coupler.

  A network with many terminals requires a large splitting ratio (throughput power $\gg$ tapped power) for the tee couplers. This ensures that signals reaching receivers far from the transmitter will have sufficient strength to be properly detected. Consider the total loss between terminals 1 and N, assuming that the directional couplers that attach to the bus fiber each have throughput loss L_{THP} and tap loss L_{TAP}. The signal must pass through $N - 1$ directional couplers before reaching the coupler at the receiver. The receiver connects to the

tap port of this coupler, so the total distribution loss is

$$L = (N - 1)L_{\text{THP}} + L_{\text{TAP}} \qquad (9\text{-}7)$$

We conclude that the total loss, in decibels, increases linearly with the number of terminals.

  In an actual system, we need to account for the losses in the connectors used to assemble the network. Each coupler input and output port requires a connector, so there are $2N$ connectors in the path between terminals 1 and N. A loss of L_C dB per connector adds a loss of $2NL_C$ to Eq. (9-7). The total distribution loss is now

$$L = (N - 1)L_{\text{THP}} + L_{\text{TAP}} + 2NL_C \qquad (9\text{-}8)$$

  Figure 9-5 shows a few examples of the distribution loss. The bottom part of the figure applies for ideal couplers (no excess loss and no connector loss). The top curves assume a 1-dB excess loss for each coupler and a 1-dB loss for each connector. As indicated by the figure, losses become prohibitively large when connecting many terminals.

  In addition to loss, tee networks have other characteristics worth mentioning. These characteristics involve special receiver re-

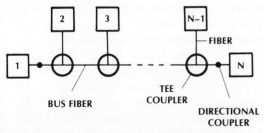

Figure 9-3   Tee network interconnecting N terminals.

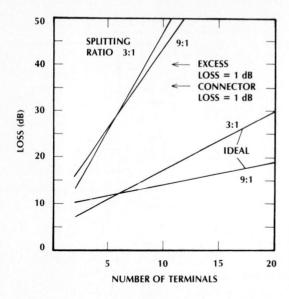

Figure 9-5 Distribution loss in a tee network. For the two bottom curves, the couplers are ideal. For the two top curves, the excess loss is 1 dB per coupler and the connectors have 1 dB of loss.

quirements, susceptibility to damage, and ease of adding new terminals. We briefly discuss these topics next.

A terminal in a tee network will receive more power from an adjacent terminal than from a distant one. Therefore, the receiver must be able to process signals having a wide range of power levels. In other words, the receiver must have a large *dynamic range*.

Localized damage to a tee network does not shut off all communications. A break in the bus fiber divides the system into two parts, with information flow intact on each side of the break. Damage to one of the tee couplers divides the system and eliminates contact with the tapped terminal. Damage to a terminal merely disconnects that terminal, leaving the rest of the system in operation.

New terminals can be added to a tee network simply by cutting the bus fiber and inserting a tee coupler.

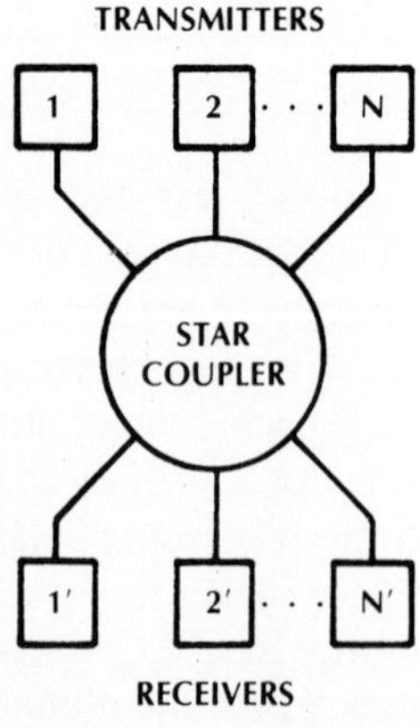

Star Network

An alternative to the tee, for multiterminal networks, is the *star* configuration, drawn in Fig. 9-6. In this scheme, a *transmissive star coupler* interconnects N terminals. The coupler has $2N$ ports. It may be viewed as a directional coupler with more than four ports. The star coupler distributes power equally to each of the receiver ports from any one of the transmitter ports, as illustrated in Fig. 9-7. An ideal star divides the input power N ways without loss. The transmission efficiency for

Figure 9-6 Star network.

INPUT PORTS

1   2 · · · N

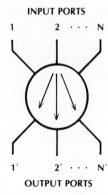

1'   2' · · · N'

OUTPUT PORTS

Figure 9-7 A transmissive star coupler distributes power from any input port to all the output ports.

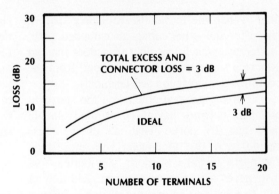

Figure 9-8 Distribution losses in a star network.

each port is then $1/N$ and the corresponding insertion loss (in decibels) is

$$L_{IN} = -10 \log \frac{1}{N} \qquad (9\text{-}9)$$

If we include two connectors, each having loss L_C, and the star excess loss L_E, then the total distribution loss associated with the star coupler is

$$L = -10 \log \left(\frac{1}{N}\right) + L_E + 2L_C \qquad (9\text{-}10)$$

Figure 9-8 illustrates these last two equations. We can note the differences between the losses of the star and tee networks by comparing Figs. 9-5 and 9-8. Generally, the star network provides significantly higher efficiency when more than five terminals are interconnected. This occurs because the logarithmic loss variation of the star configuration increases much slower with N than does the linear change of the tee. For every new terminal added to a tee system, the signal must pass through two more connectors. In a star, an added terminal does not increase the number of connectors that a signal must pass through on its path from transmitter to receiver.

Example 9-2

Compare the added loss when a network increases from 10 to 11 terminals for a tee and a star system. Assume a $9:1$ splitting ratio and a 1-dB excess loss for the tee coupler. Use connectors having a 1-dB loss for both systems.

Solution:

The added loss for the tee is the throughput loss of one directional coupler plus the loss of two connectors. Table 9-1 and Eq. (9-6a) show that $L_{THP} = 1.46$ dB for a 9:1 coupler having 1 dB of excess loss. The added loss for the tee is then $1.46 + 2 = 3.46$ dB. By adding just one terminal, the received power diminishes to less than half its previous value. For the star, the loss changes from $-10 \log (1/10) = 10$ dB to $-10 \log (1/11) = 10.4$ dB, an increase of only 0.4 dB.

For systems with just a few terminals, the tee losses may be acceptable, particularly if the connector loss L_C is minimized by carefully splicing the coupler ports to the fiber bus. For a large number of terminals (more than 10), tee losses prohibit a practical design. Why consider the tee at all? The tee saves fiber. When the terminals are widely spaced, one after the other along an extended path, the tee uses much less fiber than does the star (where a separate cable must extend from the centralized coupler to each terminal).

For the greatest efficiency, the star coupler in a network having N terminals should have just $2N$ ports. That is, all the ports should be in use. A coupler with more than $2N$ ports introduces more distribution loss than necessary. For this reason, addition of new terminals to an existing system requires a new star coupler (one with more ports). In the preceding example, we assumed that the new star coupler would have no more excess loss than the old one. This is a reasonable assumption for an addition of just two ports, although the excess loss of practical devices does increase with the number of ports. The excess loss may vary from about 1 dB for 16 ports ($N = 8$) to 3 dB for 128 ports ($N = 64$).

In a star network, damage to a branch cable connecting a terminal to the coupler merely interrupts service to that terminal. However, destruction of the star coupler itself terminates all data flow.

Ring Network

Fibers can also connect numerous terminals in the *ring* network illustrated by Fig. 9-9. The ring is actually a serial connection of independent point-to-point links. Each ring node contains an optical transmitter and receiver. The node's function is that of an active regenerator. After the receiver detects the delivered message and the station reads it, the data are regenerated and then retransmitted to the next station.

In a ring, the power from any one optical transmitter travels to a single receiver. There is no sharing of the optical power by several stations, as is the case in the tee and star networks. For this reason, the ring can interconnect more terminals than any of the other networks described in this section. That is, the ring is not limited by power distribution losses as are the tee and star. Of course,

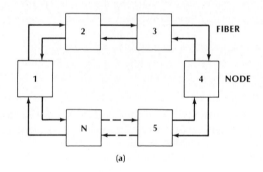

(a)

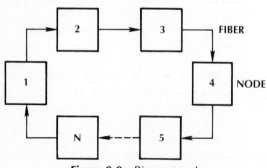

Figure 9-9   Ring network.

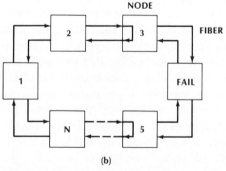

(b)

Figure 9-10   Counter-rotating rings. (a) Basic network and (b) reconfigured network when node 4 has failed.

ding from one side of a short section of each fiber and then butting the two polished surfaces together. The cores must be within several micrometers for good coupling in this arrangement. As is the case with the tapered coupler, this device depends upon evanescent coupling for its success.

The tapered (or polished) single-mode coupler is of such importance that a little more should be said about its operation. By using Fig. 9-1 as a reference for the ports and assuming an input at port 1, the coupling to ports 2 and 3 are given by

$$P_2/P_1 = \cos^2(\Delta\beta L)$$
$$P_3/P_1 = \sin^2(\Delta\beta L)$$
(9-11)

where $\Delta\beta$ is the *coupling coefficient* (given in radians per meter) between the two waveguides and L is the length of fiber over which coupling exists. As seen from these equations, the input power divides between the two output ports with no loss. In a good practical coupler, the excess loss may be only a few tenths of a dB. As seen from the preceding equations, all the power appears at port 3 when the length of the interaction region is

$$L_c = \pi/2\Delta\beta$$
(9-12)

The resulting length is called the *coupling length*. Figure 9-14 is a plot of the coupled power as a function of the length of the coupling region. Note that any desired splitting ratio can be obtained by suitably adjusting the length of the coupling region. Also notice that the fractional coupling repeats itself as the interaction length increases.

An offset butt joint can be used to form a four-port directional coupler in the manner illustrated in Fig. 9-15.[4] With an input at port 1, the favored port (port 2 in the figure) collects an amount of power determined by the offset. The lateral misalignment curve (Fig.

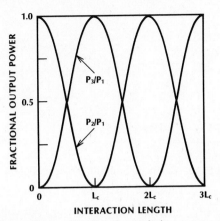

Figure 9-14 Fractional coupled power as a function of the length of the coupling region.

8-3) predicts the offset coupling loss for SI fibers. A portion of the incident light travels from the joint to the tap (port 3) along a planar curved plastic waveguide. The waveguide, and grooves for accurately positioning the fibers, can be produced by a thick film photolithographic process.

For conventional optic systems, a *beamsplitter* (partial reflector) serves as a simple directional coupler. A *beamsplitting plate,* pictured in Fig. 9-16(a), consists of a thin partially reflective coating (either dielectric or metallic) on a transparent substrate. The thickness and composition of the coating de-

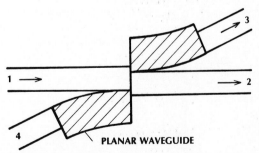

Figure 9-15 Offset butt-joint directional coupler. Power couples between the trunk fibers (1 and 2) and the tap fibers (3 and 4) by way of planar dielectric waveguides.

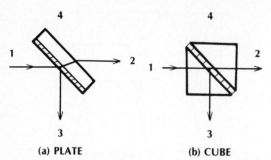

(a) PLATE   (b) CUBE

Figure 9-16 Beamsplitting directional couplers.

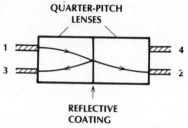

Figure 9-18 Directional coupler.

termines the splitting ratio. The beamsplitting plate displaces the transmitted beam laterally with respect to the incident beam. The *beamsplitting cube,* shown in Fig. 9-16(b), removes the displacement. The cube consists of two prisms separated by a partially reflective coating.

A beamsplitter cannot be used by itself when dividing power among fibers. The space occupied by the splitter is equivalent to a gap. As discussed in Section 8-1, gaps between connecting fibers produce large losses because the diverging rays emitted by the input fiber miss the receiving fiber. Collimating the rays incident on the beamsplitter and refocusing the divided light onto the receiving fibers solves this problem. Figure 9-17 illustrates a

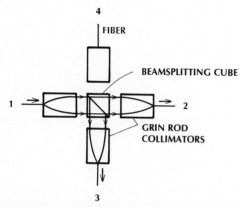

Figure 9-17 Directional coupler using four GRIN rod collimating lenses attached to the fibers.

beamsplitter-type directional coupler by using GRIN rod lenses for collimating and refocusing.[5] The beamsplitting cube aligns ports 1 and 2 (and ports 3 and 4). These ports would be offset if a beamsplitting plate were used. The coupler in Fig. 9-17 would also work if conventional spherical lenses replaced the GRIN lenses.

A variation of the beamsplitting coupler appears in Fig. 9-18.[6] Two quarter-pitch GRIN rod lenses, separated by a partially reflective coating, make up the coupler. The connecting fibers are offset from the axes of the lenses. Consider an input at port 1. The combined lenses image light from port 1 onto the fiber at port 2. Light reflected by the coating is imaged onto port 3. None of the light reaches port 4. Inputs at the other ports are similarly distributed.

Beamsplitting couplers are *amplitude-division* devices. They distribute light by dividing the amplitude of the incident wave into the desired proportions. Couplers can also be produced by *wavefront division,* dividing the wavefront into several parts and directing the separated waves to the desired ports.[7] Figure 9-19 illustrates a coupler operating on this principle. The input light, from port 1, diverges. The upper half of the wave is imaged onto the fiber at port 2 by concave reflector M_1. The lower half of the wave is imaged onto the fiber at port 3 by concave reflector M_2. As drawn, the splitting ratio is 1–1. Other ratios are obtained by enlarging one of

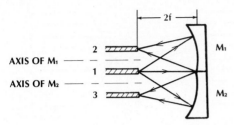

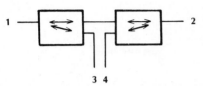

Figure 9-19   Wavefront-dividing directional coupler.

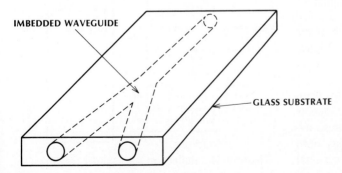

Figure 9-20   Connecting two three-port couplers to obtain a four-port directional coupler.

the reflectors, so that it intercepts more of the wavefront than the other reflector.

The fibers in Fig. 9-19 are placed near the center of curvature and slightly off the axis of each reflector. A spherical reflector's focal length f is half its radius of curvature, so the fibers are at a distance $2f$ from the lenslike mirror. According to the imaging equations [Eqs. (2-5) and (2-6)], this placement produces a focused image with unit magnification. One-to-one imaging ensures that there will be no increase in the beam divergence, so all the light incident on an output fiber will be accepted.

An input at port 2 of the wavefront-dividing device couples only to port 1. Simi-

larly, an input at port 3 couples only to port 1. The coupler in Fig. 9-19 has just three ports. A duplexing system (see Fig. 9-2) only requires three-port couplers. Connecting two three-port couplers, as in Fig. 9-20, produces a four-port directional coupler if needed.

Couplers can also be fabricated in the integrated optic format. In one such implementation, waveguides are diffused into a glass substrate by using ion exchange techniques.[8] These waveguides are circular, as indicated in Fig. 9-21, conforming to the structure of optical fibers. This configuration simplifies the connection of these couplers to the rest of the fiber system. Core sizes and numerical apertures are made to match those of the fibers to which the couplers will be attached. Both single-mode and multimode couplers have been constructed. For example, core sizes in the couplers are approximately 9, 50, or 62.5 μm to match the most widely used single-mode and graded-index multimode fibers.

The coupling is formed by branching the imbedded waveguides in a Y junction as illustrated on Fig. 9-21. One-by-two (1 × 2) couplers are the basic building blocks of this structure, but many more terminals can be connected by cascading several of them together. Figure 9-22 shows how a 1 × 8 coupler can be constructed by cascading several 1 × 2 couplers in a tree configuration.

Figure 9-4 showed how two four-port

Figure 9-21   Integrated optic branching coupler.

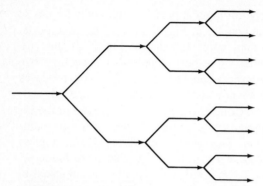

Figure 9-22 Cascaded 1 × 2 couplers to produce a 1 × 8 coupler.

directional couplers could be combined to obtain a tee coupler for bidirectional transmission along a single fiber bus. Any of the four-port couplers described in this section can be used for this purpose.

9-3 STAR COUPLERS

The fused biconically tapered technique can be extended to produce multimode fiber couplers having more than four ports.[9] An 8 × 8 *transmission star coupler* and an eight-port *reflection star coupler* are illustrated in Fig. 9-23. Individual multimode fibers are wound around one another and fused while under tension.

For the transmission star, power put into any port on one side of the coupler emerges from all the ports on the other side, divided equally. Ideally, ports on the same side of the coupler are isolated from each other. Figure 9-7 illustrates how the transmission star coupler interconnects terminals.

The reflection star couples light from any one port to all the ports. It interconnects terminals as shown in Fig. 9-24. Because every fiber connected to the star carries both transmitted and received data, a directional

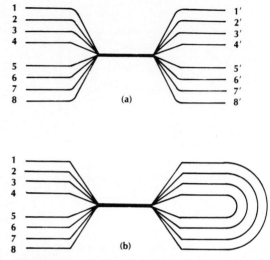

Figure 9-23 Star couplers. (a) Transmission star. (b) Reflection star.

coupler is needed to separate the two signals at each terminal.

Fusion of more than two single-mode fibers to produce multi-terminal star couplers does not work well because of the need to couple between the individual evanescent fields of the many fibers. For single-mode sys-

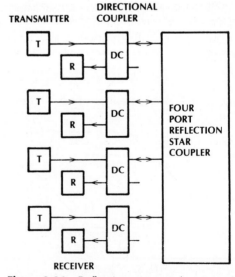

Figure 9-24 Reflection star coupler network.

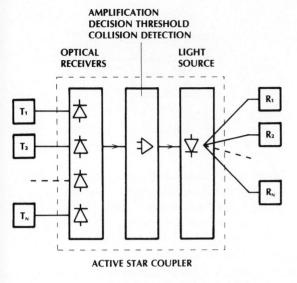

Figure 9-25   Active star network.

tems, star couplers are made by cascading 1×2 port fused couplers. The scheme, illustrated in Fig. 9-22, works for any type of coupler construction.

The star couplers just described are passive devices. They feature reliability and low cost compared with active devices. However, active stars can be very useful in implementing LANs.[10] A schematic of an active star network appears in Fig. 9-25. The active star acts as a repeater. It receives a signal from any transmitter, converts it from optic to electrical form, and amplifies the resulting current. This current drives a light source, reproducing the optic signal. The light source divides its power equally among all the receiving stations. One method for achieving the power division is sketched in Fig. 9-26. Output fibers, which are fused together and conically tapered, share the light emitted by the source.

The active star can include provisions for detecting collisions between data packets transmitted simultaneously by different terminals. If collisions occur, then the repeater signals the stations to take corrective actions. Active stars add flexibility to a distribution network because of their regenerating and collision detecting properties.

9-4 SWITCHES

Fiber optic switches reroute the optic signals. Switches are useful in, for instance, distribution networks, measuring equipment, and experiments. We will describe two devices: a two-position switch and a bypass switch. These examples illustrate some of the general features of fiber switches.

Figure 9-27 shows a two-position switch. An input at port 1 can be switched to either port 2 or port 3. For the following definitions, assume the switch is set for cou-

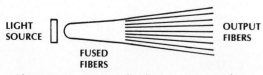

Figure 9-26   Power-dividing source coupling.

Figure 9-27   Two-position switch.

pling to port 2. The insertion loss (IL) (in decibels) is

$$L_{IL} = -10 \log \frac{P_2}{P_1} \qquad (9\text{-}13)$$

where P_1 is the power going into port 1 and P_2 is the power emerging from port 2. Insertion loss depends on fiber alignment, just like the loss of a simple connector. Losses of less than 1.5 dB can be obtained with good mechanical switches. In addition to having low insertion loss, a good switch will have the same value of insertion loss for all switch positions.

Crosstalk (CT) is a measure of how well the uncoupled port is isolated. It is given by

$$L_{CT} = -10 \log \frac{P_3}{P_1} \qquad (9\text{-}14)$$

where P_3 is the power emerging from port 3. Crosstalk depends on the particular design of the switch, but values of 40–60 dB are typical.

Reproducibility (achieving the same insertion loss each time the switch is returned to the same position) may be more important than the value of the insertion loss itself. A good switch will reproduce the insertion loss within about 0.1 dB.

Switching speed (how fast the switch can change from one position to the other) is a crucial factor in some applications. Switching can be done electromechanically. In this type of device, an energized electromagnet attracts a magnetic material to which an optic device is attached. Mirrors, lenses, and prisms (even

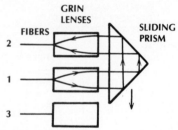

Figure 9-28   Sliding-prism, two-position switch.

fibers themselves) can be moved in this manner. When the electromagnet is turned off, a spring pulls the magnetic holder back to its rest position. Switching times on the order of a few milliseconds can be obtained with electromechanical switches.

The two-position switch drawn in Fig. 9-28 consists of a sliding prism and quarter-pitch lenses attached to each fiber.[11] In the position shown, light couples between ports 1 and 2. Let us follow the progress of an input at port 1. A GRIN lens collimates the diverging beam emitted by the fiber. The right-angled prism deflects the light by total-internal reflection at its two slanting surfaces. A GRIN lens refocuses the collimated beam onto fiber 2. To direct light from port 1 to port 3, the prism moves in the direction shown in the figure, aligning the beam between fibers 1 and 3. Collimating lenses are required to eliminate insertion loss caused by beam spreading and to ensure that all rays strike the prism's reflecting surfaces beyond the critical angle. The right-angled prism not only reflects the light but it translates the beam parallel to itself, thereby effectively angularly aligning the input and output fibers.

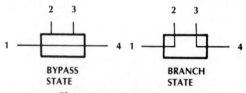

Figure 9-29   Bypass switch.

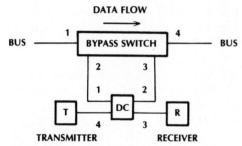

Figure 9-30 Bypass switch incorporated into a tap for a tee (or ring) network.

Figure 9-29 illustrates the functions of a bypass switch. In the bypass state, ports 1 and 4 are coupled. Ports 2 and 3 are isolated. In the branch state, ports 1 and 2 are coupled and ports 3 and 4 are coupled. The bypass switch can be incorporated into a tee (or ring) network by attaching it to the data bus in the way indicated in Fig. 9-30. The terminal shown can be bypassed or included in the network, as desired. A station that is not transmitting or receiving can be bypassed.

A repeater can be connected to a data bus, replacing the terminal in Fig. 9-30. If the repeater needs repair, then it can be bypassed without shutting down the entire network. The switch provides a fail-safe feature in this application. A second repeater (also attached to the bus by a bypass switch) can be switched into the network, taking the place of the malfunctioning repeater. This strategy, using re-

dundancy, improves the network reliability at the expense of increased system complexity.

A practical electromechanical bypass switch is illustrated in Fig. 9-31.[12] In the bypass state, light passes directly between ports 1 and 4. In the branch state, mirrors couple light between ports 1 and 2 and ports 3 and 4. GRIN lenses collimate the beams to minimize the insertion loss. An iron bar with mirrors attached to its ends is raised when the electromagnet is activated. This produces the branch state. When the magnet is turned off, a spring pulls the iron bar out of the optic path, returning the switch to the bypass condition.

9-5 FIBER OPTICAL ISOLATOR

Laser diodes are particularly sensitive to light energy reflected back from the rest of the system. The reflected light increases the noise in the emitted beam, degrading system performance. The returning photons arrive back inside the laser cavity where they are amplified and generally take part in laser action. They compete with the photons already in the cavity for the excited atomic states. Because the returning photons are unlikely to be in phase with the wave existing in the laser cavity, they tend to force the diode to restart its oscillation. The new oscillation is in phase with the

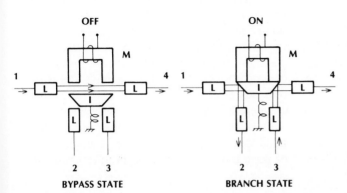

Figure 9-31 Bypass switch.
L, GRIN lens; M, electromagnet;
I, iron bar with mirrored endfaces.

returned beam of light. The result is that the laser diode periodically and randomly shifts the phase of its output radiation, adding to the system noise.

Reflections close to the transmitter cause the major disruptions. Reflections occurring farther away are attenuated by the fiber and the connectors (and any other components in the path), becoming small when arriving back at the laser diode. Thus, most of the precautions to be mentioned here apply only to portions of the system located near the transmitter. Reflections are minimized in a number of ways in practical systems. Components, such as the fiber or any lenses in the system, can have their ends antireflection coated to reduce the amplitude of any return. Fiber ends can be rounded so that reflected rays do not propagate back to the transmitter but are diverted out of range of the allowed propagating modes. Connectors and couplers are specifically designed to minimize the amount of reflected light. The measure of effectiveness in controlling reflections is the *return loss*. Expressed in decibels, it is defined as

$$L_R = -10 \log \frac{P_r}{P_i} \qquad (9\text{-}15)$$

where P_i is the power incident on the component and P_r is the power reflected. Return losses of 30 or 40 dB are representative of well-designed components.

Example 9-3

A laser diode feeding a glass fiber may be separated from it by a small air-gap. Compute the return loss at the air-to-fiber interface.

Solution:

Assuming an index of refraction of 1.5 for the glass fiber, we find from Eq. (3-28)

that 4% of the light is reflected. The return loss is then $L_R = -10 \log 0.04 = 13.98$ dB.

An optical isolator will ensure a low level of return to the laser diode. An optical *isolator* is a one-way transmission line. That is, it will allow propagation in only one direction along the fiber. The basic structure of such a device is shown in Fig. 9-32(a). It consists of two linear polarizers and a 45° Faraday rotator. A beam of light incident from the left is vertically polarized by polarizer P_L as indicated in Fig. 9-32(c). The resultant wave passes through the rotator. Generally, a Faraday rotator rotates the direction of linear polarization of an incident beam by an amount determined by the properties of the device. To construct an isolator, the rotation angle must be 45°. Thus, the beam emerging from the rotator is linearly polarized at 45° to the vertical. The polarizer on the right P_R, aligned at 45°, allows passage of this wave. Transmission from left-to-right occurs in this manner. Now consider a beam traveling from right to left as in Fig. 9-32(d). This beam is polarized at 45° by P_R as it enters the isolator. It is rotated another 45° by the Faraday device so that it is horizontally polarized as it exits the left side of the rotator as shown on the figure. Polarizer P_L will block passage of this horizontally polarized beam. We conclude that no light can travel from right to left through the isolator. It should be emphasized that Faraday rotation is nonreciprocal. When viewed along the direction of wave travel, the rotation always takes place in a direction determined by the direction of wave travel. In this example the rotation was clockwise for the forward wave and counterclockwise for the reverse wave. A reciprocal device would rotate the beam in the same direction, irrespective of the direction of wave travel. Assuming right-to-left rotation

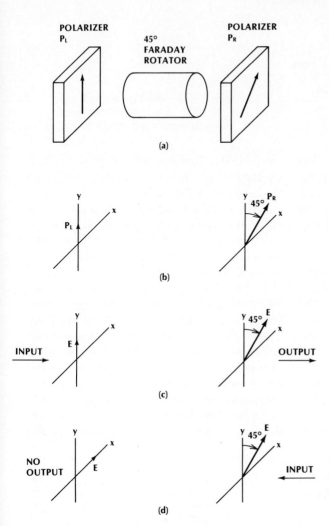

Figure 9-32   Farady rotation isolator. (a) Components, (b) allowed polarization states, (c) electric field for left-to-right propagation, and (d) electric field directions for right-to-left propagation.

is still counterclockwise as shown on the figure, the 45° beam coming from the right would be rotated back to a vertical polarization state (counterclockwise again as viewed along its direction of travel) and we would no longer have an isolator.

The angle of rotation (in radians) in a Faraday rotator is given by

$$\theta = VHL \qquad (9\text{-}16)$$

where V is the *Verdet constant*, a measure of the strength of the Faraday effect, H is the strength of the applied magnetic field, and L is the interaction length. For silica glass, V is 4.68×10^{-6} rad/A.

A transparent material having a large Faraday effect is yttrium iron garnet, $Y_3Fe_5O_{12}$. It is commonly referred to as YIG. It produces Faraday rotation when a magnetic field is applied to a single crystal along the direction of

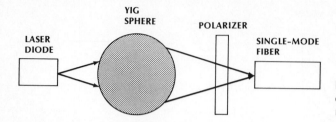

Figure 9-33 Isolator-coupled laser diode. The circuit for providing the magnetic field is not shown.

wave travel. As already emphasized, the biggest need for an isolator in a fiber link is for protection of the laser diode from reflected light. An isolator that combines this function along with source coupling is sketched in Fig. 9-33.[13] The YIG sphere acts as a lens to focus light from the source onto the fiber and provides the required Faraday rotation. An initial polarizer is not required because the laser diode output is already polarized. Reflected light returning at 90° to the laser polarization will not be coupled with that light beam. The magnetic field is applied by a permanent magnet.

An ideal isolator would have no loss in the forward (or allowed) direction and infinite loss in the reverse (or forbidden) direction. In practice, because of reflections at the component interfaces and imperfections in the polarizers and in the rotator, the isolator performance is not ideal. Forward losses of around 1 dB and reverse losses of about 30 dB are representative.

9-6 WAVELENGTH-DIVISION MULTIPLEXING

Optic beams with different wavelengths propagate without interfering with one another, so several channels of information (each having a different carrier wavelength) can be transmitted simultaneously over a single fiber. This scheme, called *wavelength-division multiplex-*

ing (WDM), increases the information-carrying capacity of a fiber. In Chapters 3, 4, and 5 we determined the capacity limits owing to material and waveguide dispersion and modal distortion. These limits apply to the information carried at any one wavelength. Increasing the number of carriers increases the capacity proportionately.

An *optic multiplexer* couples light from the individual sources to the transmitting fiber, as shown in Fig. 9-34. At the receiving station an *optic demultiplexer* separates the different carriers before photodetection of the individual signals (see Fig. 9-34). Generally, multiplexers/demultiplexers have fibers at their input and output ports. It is also possible to replace the input fibers in a multiplexer with optic sources directly integrated into the

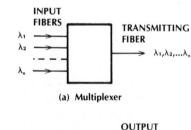

(a) Multiplexer

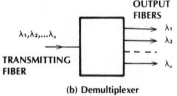

(b) Demultiplexer

Figure 9-34 Schematic of (a) an optical multiplexer and (b) an optical demultiplexer.

device. Similarly, photodetectors can replace the output fibers in a demultiplexer. Often the same device can perform as a multiplexer or demultiplexer.

Insertion loss and crosstalk are the important properties of multiplexers/demultiplexers. Insertion loss is the attenuation of a wave traveling from the input port to the desired output port. For example, referring to Fig. 9-34(a), the insertion loss for channel 1 is the fraction of the input power at wavelength λ_1 that reaches the transmitting fiber. A multiplexer/demultiplexer has *uniformity* when the insertion loss is nearly the same for each channel. Crosstalk is the wave attenuation measured at an unintended port. For example, referring to Fig. 9-34(b), the crosstalk is the fraction of the input power at λ_1 that reaches the output fiber assigned to λ_2. Crosstalk is chiefly a problem at the receiver, where mixing of two or more channels can seriously interfere with the desired signal.

Figure 9-35 illustrates loss curves for a demultiplexer. The curves are generated by applying a narrow-band (nearly single wavelength) beam at the input and measuring the power transmitted to each of the output channels. Laser diodes (or light from a monochromator) provide suitable coherence. The source wavelength is changed and the measurement repeated until the spectral range of interest is covered. The result is a plot of the wavelength dependence of both the insertion loss and crosstalk.

Example 9-4

Estimate the insertion loss and crosstalk for the demultiplexer in Fig. 9-35 if the input wavelength is 805 nm and the source linewidth is 1 nm.

Solution:

Because the linewidth is so small, it can be ignored. According to the figure, the power reaching channel 1 is 5 dB below the input power. The power in channels 2 and 3 is more than 60 dB below the input power. Thus, the insertion loss is 5 dB and the crosstalk is greater than 60 dB.

In Example 9-4 the source spectral width was negligible, a condition obtained in practice by using laser diodes. The linewidths of LEDs are so large (20–100 nm) that they cannot be neglected.

Example 9-5

Estimate the insertion loss and crosstalk for the demultiplexer in Fig. 9-35 if the input wavelengths are centered at 800,

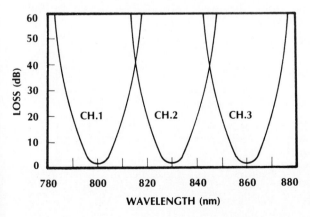

Figure 9-35 Demultiplexer loss. The power measured in output channels 1, 2, and 3 is shown.

830, and 860 nm and the source line-widths are 30 nm.

Solution:

The 800-nm source radiates between 785 and 815 nm. Over this range, the insertion loss varies from a maximum of 40 dB (at the band edges) to a minimum of 2 dB (at the center wavelength). The net insertion loss depends on the precise way that the power is distributed over the 30-nm range. Because the loss curve in Fig. 9-35 is fairly flat around 800 nm, and because most of the source power lies close to 800 nm, the total loss will be closer to 2 dB than to 40 dB. A reasonable estimate of the insertion loss would be 5 or 6 dB. The crosstalk from channel 1 into channel 2 has a maximum of 40 dB at 815 nm. Radiation below 815 nm is coupled more weakly into channel 2. The total crosstalk power will be more than 40 dB below the input level. The crosstalk from channel 1 to channel 3 is much greater than 60 dB.

If the source linewidth in Example 9-5 were larger than 30 nm, then the insertion loss would increase and the crosstalk would get worse. This illustrates the advantages of narrow-band sources for WDM systems. Additionally, the smaller the source linewidth, the greater the number of channels that can be squeezed within the low-loss range 800 to 900 nm. Of course, if more channels are desired, multiplexers/demultiplexers capable of combining and separating very closely spaced carriers must be designed. Figure 9-36 illustrates a three-channel WDM system. In its simplest form, this network is unidirectional. It can, however, operate in both directions if the wavelength separation devices are bidirectional. Later in this section we will see how such devices can be constructed. When operating bidirectionally,

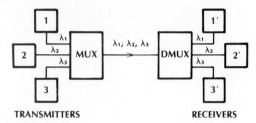

TRANSMITTERS   **RECEIVERS**

Figure 9-36   Three-channel wavelength-division-multiplexed network. MUX, multiplexer; DMUX, demultiplexer.

directional couplers must be included at each terminal to separate the transmitted and received waves.

Previously installed systems, designed to operate with a single carrier, can be upgraded by WDM. Only the terminal equipment need be changed. The original fibers can remain in place.

A particularly attractive combination for WDM operates with one channel at 1.3 μm and one at 1.55 μm. The large wavelength spacing simplifies the design of the multiplexer. The similarity in fiber characteristics (attenuation and bandwidth) at these two wavelengths makes the system practical. An additional channel operating in the region 800–900 nm could be added, but the larger loss and dispersion in this range would limit the length or bandwidth of the entire link. Wavelength division can be used to produce a fully duplexed network, as shown in Fig. 9-37.

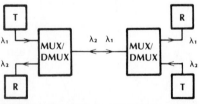

Figure 9-37   Full-duplex network. T, transmitter; R, receiver; MUX/DMUX, bidirectional multiplexer.

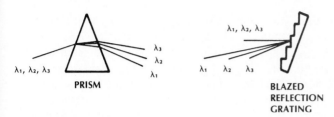

Figure 9-38 Angular dispersion.

Several multiplexer designs are based on either of two mechanisms: angular dispersion or optic filtering. Two devices exhibiting angular dispersion are the prism and the reflecting diffraction grating. Figure 9-38 shows how these bidirectional elements separate (or combine) beams of different wavelengths. The grating may be metal coated to enhance its reflectance. Optic filters (drawn in Fig. 9-39) consist of thin layers of transparent materials of different refractive indices. Interference within the thin films causes the filter to pass certain wavelengths and reflect others. In the scheme shown in the figure, two filters in series separate (or combine) three wavelengths.

Multiplexers/demultiplexers often incorporate lenses to capture the diverging rays emitted by the input fiber, to direct these rays onto the combining/separating elements, and to refocus the light onto the output fiber. Without the lenses, the gap loss between the input and output fiber would be excessive. Lenses perform another needed function. They collimate the beam striking the wavelength selective element. This is necessary because angularly dispersive components and optic filters are sensitive to the angle of incidence. Diverging incident rays would also di-

verge at the output of the selective element, each wavelength occupying a range of angles. This in turn would decrease the possibility of spatially separating the individual wavelengths present.

To illustrate some of the available construction techniques, two multiplexers will be described. First, consider the grating multiplexer drawn in Fig. 9-40.[14] For simplicity, only the central rays associated with each fiber are shown. However, the beams leaving any fiber are diverging and the beams entering any fiber are converging. The beams are collimated in the space between the lens and the grating. The upper fiber is the system's transmission line. When used as a demultiplexer, wavelengths λ_1, λ_2, and λ_3 enter the quarter-pitch GRIN rod lens from the transmission line. The lens collimates the beam before the rays strike the grating. The grating spatially separates the three wavelengths, after which the lens focuses the angularly displaced collimated beams onto the three output fibers. This device is bidirectional. When used as a multiplexer, the direction of ray travel is reversed. Inputs at the lower three fibers are combined by the grating and focused onto the transmission line (the upper fiber).

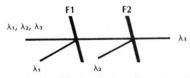

Figure 9-39 Optical filtering. Filter F1 reflects λ_1 and transmits λ_2 and λ_3, and filter F2 reflects λ_2 and transmits λ_3.

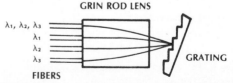

Figure 9-40 Grating multiplexer/demultiplexer. Only the fiber's axial rays are drawn.

The second multiplexer, drawn in Fig. 9-41, uses a GRIN rod lens and optic filters.[15] Filter F1 passes wavelength λ_1 and reflects λ_2, and filter F2 passes λ_2 and reflects λ_1. As in the grating multiplexer in Fig. 9-40, the lens collimates the light emerging from the fibers and, after the rays are reflected by filter F2 or the mirror, focuses the beams onto the output fibers. When used as a demultiplexer, wavelengths λ_1 and λ_2 enter the lens from the upper fiber. Filter F2 reflects the power at λ_1. This beam travels toward the lower fiber, where it passes through filter F1 and enters channel 1. Meanwhile, filter F2 passes the power at λ_2. This beam is reflected by the tilted mirror, which directs the rays toward channel 2 for collection. Filter F1, which reflects λ_2, improves the crosstalk by minimizing the amount of power at λ_2 that reaches channel 1. When operating as a multiplexer, λ_1 enters the lens from channel 1 and λ_2 enters from channel 2. Filter F2, the mirror, and the lens combine the two wavelengths onto the transmitting fiber. Filter F1 is not required by the multiplexer.

Although many variations of the grating and filter multiplexer/demultiplexer are possible, the examples in the last two figures indicate a few of the basic elements of multiplexer construction and design. The devices described are quite simple, using the same lens

to couple to the input and output ports. Some multiplexer designs use separate lenses for each port. The advantage of the latter construction is that each lens can be used axially. That is, fibers are attached to the axis of the lens, rather than displaced from it as in Figs. 9-40 and 9-41. Lenses have smaller aberration losses when used axially.

An alternative multiplexer for single-mode systems is based on the single-mode fused tapered coupler described in Section 9-2. Because the coupling coefficient ($\Delta\beta$) in Eq. (9-11) depends on the wavelength, the coupling length (L_c) is different for different wavelengths. The tap ratio, P_3/P_1, is sketched in Fig. 9-42 for two different wavelengths for a particular coupler. Concentrate attention on a coupler having interaction length $3L_{c1}$ (three times the coupling length corresponding to λ_1). For this coupler, an input at λ_1 will be entirely coupled to the tap port, port 3. On the other hand, an input at λ_2 has zero coupling to the tap port, so it will emerge entirely at the throughput port, port 2. In other words, the coupler acts as a demultiplexer, separating the two wavelengths incident on port 1. Conversely, an input of λ_1 at port 3 and an input

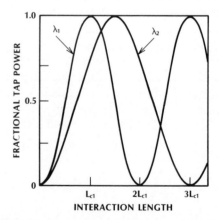

Figure 9-42 Fractional tap power (P_3/P_1) dependence on wavelength and length of the coupling region for a single-mode fused coupler.

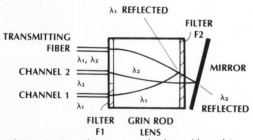

Figure 9-41 Filter-type multiplexer/demultiplexer. F1 passes λ_1 and reflects λ_2; F2 passes λ_2 and reflects λ_1.

of λ_2 at port 2 will both exit at port 1. Thus, the coupler acts as a multiplexer also. The fused fiber multiplexer is useful for WDM networks and for combining the pump and signal beams in an erbium-doped amplifier of the type described in Section 6-7.

9-7 SUMMARY

Fiber optic technology can be applied to bidirectional and multiterminal communications networks by using the basic techniques presented in this chapter. The multiterminal capability increases the attractiveness of fiber optics.

Several alternative network architectures are available. These include the tee, star, ring, and a hybrid combination of basic architectures. Many system components (directional couplers, transmission star couplers, reflection star couplers, switches, and multiplexers/demultiplexers) are also available. The system designer must understand the many alternative strategies to choose the best one.

A comprehensive design procedure is not practical because of the variety of possible multiterminal applications. However, a few generalizations can be made. The designer must consider the quantity of fiber needed and the system losses. When a number of terminals are being connected, a tee network generally requires less fiber than a star network. For systems covering a large area, the savings produced by minimizing the total fiber length can be significant. The physical location and spacing of the terminals dictates the amount of fiber necessary and the optimum network configuration.

The losses in a tee network increase with the number of terminals faster than do the losses in a star network. The star is superior in this respect. Additionally, a star system delivers nearly the same power to each terminal. The received power does not change when a different terminal transmits. In a tee network, the power arriving at each receiver differs. A receiver located close to the transmitter (transmitter and receiver at adjacent taps along the data bus) receives more power than one located several taps away. If a different terminal transmits, then the power levels change because the receivers are now located differently relative to the transmitter. Therefore, receivers in a tee network must operate over a large range of input power levels; that is, they must have a wide dynamic range. The losses in a tee network may be excessive if the system contains more than just a few terminals.

In a point-to-point link, amplifying repeaters make up for the attenuation of long lengths of fiber. In multiterminal networks, distribution losses often overshadow fiber attenuation losses. In such cases, repeaters can again be used to overcome the distribution losses, thereby increasing the number of allowable terminals. Repeaters increase system complexity and cost, but they solve power level problems.

Multiplexers increase the amount of information that can be transmitted along a single fiber by allowing simultaneous propagation of several optic carriers. Multiplexing terminals are much more complex than single-carrier terminals. The added system capacity can easily compensate for the increased terminal complexity.

Fibers were first commercially applied on a large scale to unidirectional, point-to-point telephone links. They are much more versatile than originally thought, appearing now in complex systems that require elaborate distribution of the optic signals. Application of multiterminal networks include local-area networks and fibered cities.

PROBLEMS

9-1. An ideal four-port directional coupler has a 4:1 splitting ratio.
(a) What fraction of the input power goes to each of the ports?
(b) Compute the throughput loss, the tap loss, the directionality, and the excess loss.

9-2. A four-port directional coupler has a 4:1 splitting ratio and an excess loss equal to 2 dB. The coupler's directionality is 40 dB.
(a) What fraction of the input power goes to each of the ports?
(b) Compute the throughput loss and the tap loss.
(c) Compute the loss (in decibels) owing to radiation, scattering, and absorption in the coupler.

9-3. Consider the full-duplex network shown in Fig. 9-2. Ideal 3-dB couplers and ideal connectors and fibers are used. Compute the total loss (in decibels) from transmitter to receiver.

9-4. Repeat Problem 9-3 if the directional couplers each have 1.5 dB of excess loss and all connectors (one at each directional coupler port, one at the transmitter, and one at the receiver) have 0.8 dB of loss. The fiber has a 4-dB loss.

9-5. A five-terminal tee network is structured like the one shown in Fig. 9-3. The tee couplers are like the one shown in Fig. 9-4. Assume ideal 3-dB couplers, ideal fibers, and lossless connectors.
(a) Draw the entire network.
(b) Compute the transmission loss to each of the receivers when terminal 1 is the transmitter.

9-6. Repeat Problem 9-5 when ideal 10-dB directional couplers are used. Compare the results of this problem with Problem 9-5. Which coupler (3 dB or 10 dB) is best?

9-7. Repeat Problem 9-5 when the directional couplers each have 1.5 dB of excess loss, all connectors have 0.8 dB of loss, the fiber loss is 35 dB/km, there is 100 m between terminals, and splices produce a 0.2-dB loss. (The number of connectors and/or splices is up to you to specify. Be sure to include them in your drawing of the network.)

9-8. A star network (like that in Fig. 9-6) connects five terminals. Assume ideal couplers, connectors, and fibers.
(a) Sketch the network.
(b) If terminal 1 transmits, compute the total transmission loss (in decibels) to each of the receivers.

9-9. Repeat Problem 9-8 when the excess loss of the star coupler is 2 dB, connector losses are 0.8 dB, splice losses are 0.2 dB, and the fiber loss is 35 dB/km. Terminals 1, 2, 3, and 4 are 100 m from the star coupler. Terminal 5 is 20 m from the star coupler. (Include all the connectors and splices you think you need and note them on your sketch.)

9-10. Plot the loss versus number of terminals (3–20 terminals) for a star network if the star has a 3-dB excess loss and each port of the star has a connector whose loss is 0.8 dB. Neglect fiber losses and transmitter and receiver connector losses. The result will be the loss associated with the star coupler.

9-11. A two-position switch (as in Fig. 9-27) operates by mechanically positioning a flexible input fiber so that it lines up with either of the two output fibers, as needed. The fibers are SI multimode and have 100-μm core diameters. Compute the maximum allowable lateral misalignment if the loss must be less than 1.5 dB.

9-12. Repeat Problem 9-11 if a GRIN lens terminates each of the fibers, expanding the beam to 1 mm.

9-13. Sketch a four-channel, full-duplex WDM network. Choose specific multiplexer/demultiplexer and directional coupler schemes and clearly draw the paths taken by the four different wavelengths.

9-14. Repeat Problem 9-5 if terminal 2 is the transmitter.

9-15. Repeat Problem 9-6 if terminal 2 is the transmitter.

9-16. Suppose you wish to interconnect 6 terminals.
  (a) How many transmitters are required if independent point-to-point links are established between all the terminals?
  (b) How many transmitters are required if a multifiber bundle network (as in Fig. 9-12) is used?

9-17. A laser diode illuminates a 2.5-km length of fiber. The total link loss over the 2.5 km is 4 dB. Power is reflected back toward the laser by the end of the fiber, which is cleaved so as to have a smooth, polished surface. The fiber end radiates into air. Compute the level of the reflected light power when it returns to the LD. That is, how many decibels down from the emitted signal is the reflected signal?

9-18. An isolator is used to reduce the reflected power in the Problem 9-17 by a factor of 50. What is the reverse loss of this isolator?

9-19. What is the level of returned power (in dB) reflected into the transmitter from a connector with an end gap between the two fibers being attached. Assume the connector is very close to the emitting source so that there are no significant losses between the two of them.

9-20. An isolator is made by using the Faraday rotation available from a silica glass fiber.
  (a) If the fiber is 100 m long, what magnitude of magnetic field is required?
  (b) If the fiber is 1 m long, what is the strength of the required magnetic field?

9-21. Equation (9-11) gives the power distribution for an ideal single-mode tapered directional coupler. Prove that the total coupled power equals the input power. Excess loss would subtract from the total coupled power in a real coupler.

9-22. Consider an ideal tapered single-mode directional coupler having a 1:1 splitting ratio. The coupling coefficient is $\Delta\beta = 1$ rad/mm.
  (a) What length is the coupling region?
  (b) Repeat for a 10:1 splitting ratio.

9-23. An ideal tapered and fused single-mode coupler is used as a two-wavelength demultiplexer with inputs of λ_1 and λ_2 at port 1. At wavelength λ_1 the coupling coefficient has a value of 2 rad/mm. This wavelength is coupled entirely to port 3. Wavelength λ_2 is entirely transmitted to port 2.
  (a) Compute the length of the coupling region.
  (b) Compute the value of the coupling coefficient at λ_2.
  (c) Compute the value of the coupling length at λ_2.

REFERENCES

1. Donald G. Baker. *Local-Area Networks with Fiber-Optic Applications.* Englewood Cliffs, N.J.: Prentice Hall, 1986.
2. Two general references covering distribution networks follow:
   John Joseph Esposito. "Optical Connectors, Couplers, and Switches." In *Handbook of Fiber Optics: Theory and Applications,* edited by Helmut F. Wolf. New York: Garland, 1979, pp. 241–303.
   Michael K. Barnoski. "Design Considerations for Multiterminal Networks." In *Fundamentals of Optical Fiber Communications,* 2nd ed., edited by Michael K. Barnoski, New York: Academic Press, 1981, pp. 329–51.
3. B. S. Kawasaki and K. O. Hill. "Low-Loss Access Coupler for Multimode Optical Fiber Distribution Networks," *Appl. Opt.* 16, no. 7 (June 1977): 1794–95.
4. F. Auracher and H. -H. Witte. "New Planar Optical Coupler for a Data Bus System with Single Multimode Fibers," *Appl. Opt.* 16, no. 8 (Aug. 1977): 2195–97.
5. W. J. Tomlinson. "Applications of GRIN-Rod Lenses in Optical Fiber Communications Systems," *Appl. Opt.* 19, no. 7 (April 1980): 1127–38.
6. Ibid.
7. Narinder S. Kapany. "A Family of Kaptron Fiber Optics Communications Couplers." In *Proceedings of the Third International Fiber Optics and Communications Exposition* San Francisco, Calif.: Information Gatekeepers, Inc., 1980.
8. A. Beguin, T. Dumas, M. J. Hackert, R. Jansen, and C. Nissim. "Fabrication and Performance of Low Loss Optical Components Made by Ion Exchange in Glass," *J. Lightwave Tech.* 6, no. 10 (Oct. 1988): pp. 1483–87.
9. Michael Barnoski. "Design Considerations for Multiterminal Networks," pp. 334–40.
10. Eric G. Rawson, Ronald V. Schmidt, Robert E. Norton, M. Douglas Bailey, Lawrence C. Stewart, and Hallam G. Murray. "Fibernet II: An Active Star-Configured Fiber-Optic Local Computer Network with Data Collision Sensing." In *Digest of the Topical Meeting on Optical Fiber Communication.* Phoenix, Ariz.: Optical Society of America, 1982, pp. 22–23.
11. Tomlinson. "Applications of GRIN-Rod Lenses," pp. 1123–33.
12. M. Nunoshita and Y. Nomura. "Optical Bypass Switch for Fiber-Optic Data Bus Systems," *Appl. Opt.* 19, no. 15 (Aug. 1980): 2574–77.
13. T. Sughie and M. Saruwatri. "Nonreciprocal Circuit for Laser-Diode-to-Single-Mode Fiber Coupling Employing a YIG Sphere." *Electron. Lett.* 18 (1982): p. 1026.
14. Tomlinson. "Applications of GRIN-Rod Lenses," pp. 1133–34, 1137–38.
15. F. Tanaka, S. Kishi, and T. Tsutsomi. "Fiber-Optic Multifunction Devices Using a Single GRIN-Rod Lens for WDM Transmission Systems," *Appl. Opt.* 21, no. 19 (Oct. 1982): 3423–29.

Chapter 10

Modulation

Throughout the preceding chapters we have been evaluating fiber links on the basis of their information capacities (as measured by the highest modulation frequency for analog systems and the upper bound on the data rate for digital systems). In this chapter we probe into various analog and digital formats suitable for fiber networks. We also describe specific circuit techniques for modulating laser diodes and LEDs. This discussion expands the introductory remarks concerning source modulation made in Chapter 6. We also discuss time-division multiplexing, a method of combining several channels of information onto a fiber by using a single optic carrier.

10-1 LIGHT-EMITTING-DIODE MODULATION AND CIRCUITS

It would be impractical to describe even a small fraction of the circuits utilized, or pro-

posed, for modulation of LEDs. Instead, we present basic modulation requirements and strategies and illustrate them with a few specific circuits.

Analog Modulation

Figure 6-7 illustrates the basic requirements for analog modulation of a LED. The total modulating current and the resulting optic power, illustrated in Fig. 10-1, are given by

$$i = I_{dc} + I_{SP} \sin \omega t \qquad (10\text{-}1)$$

and

$$P = P_{dc} + P_{SP} \sin \omega t \qquad (10\text{-}2)$$

In these equations the first term is the dc bias and the second represents the information sig-

239

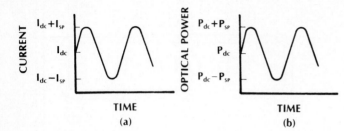

Figure 10-1 (a) LED driving current. (b) Resulting output power

nal. We will use a sinusoidal waveform (represented either by the sine or cosine function) to determine the network performance.

The *modulation factor* m' is the peak current excursion relative to the average current, divided by the average current. That is,

$$m' = \frac{I_{SP}}{I_{dc}} \qquad (10\text{-}3a)$$

Since the total peak and minimum currents are $I_{dc} + I_{SP}$ and $I_{dc} - I_{SP}$ respectively, the signal amplitude I_{SP} can have its largest value if the dc bias is half the maximum permissible diode current. Setting $I_{SP} = I_{dc}$ for this case produces a peak current of $2I_{dc}$, a minimum current of zero, and a unity modulation factor.

We define the *optic modulation factor* in terms of the optic power. Thus

$$m = \frac{P_{SP}}{P_{dc}} \qquad (10\text{-}3b)$$

allowing us to write the optic power as

$$P = P_{dc}(1 + m \sin \omega t) \qquad (10\text{-}4)$$

Combining Eqs. (10-3a), (10-3b), and (6-7) yields

$$m = \frac{m'}{\sqrt{1 + \omega^2 \tau^2}}$$

showing how the optic modulation factor decreases with modulation frequency. For $\omega\tau \ll 1$ (modulation well below the LED 3-dB bandwidth), $m = m'$.

A variety of LED analog modulation circuits exist. We will describe a simple one, the transistor amplifier drawn in Fig. 10-2.[1] The transistor's collector current i_C is the LED driving current. When used as a conventional amplifier, a load resistor replaces the LED in the circuit and a capacitor bypasses resistor R_e. We can understand the modulator's operation with the help of the transistor characteristic in Fig. 10-3. The supply voltage V_{dc}, together with resistors R_a and R_b, provide the dc base current I_B (capital letters denote dc quantities in this analysis). I_B forward biases the base-emitter junction, turning the transistor on (i.e., causing flow of collector current). The resulting collector current is $I_C = \beta I_B$, where β is the transistor's current-amplifica-

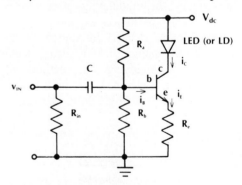

Figure 10-2 Analog modulator.

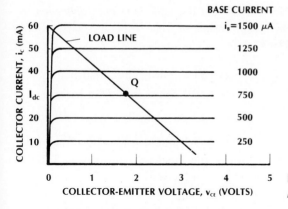

BASE CURRENT

$i_g = 1500 \ \mu A$

LOAD LINE

1250

1000

750

500

250

Figure 10-3   Transistor characteristic.
$\beta = \Delta i_c / \Delta i_b = 40$.

tion factor. I_C is the LED bias current, labeled I_{dc} in Eq. (10-1). With no input signal, the transistor operates at point Q in Fig. 10-3. This illustrates class A amplification, defined as the condition in which the Q point is well above the collector current cutoff. Cutoff occurs when the base current drops to zero.

The signal voltage v_{IN} produces a time-varying base current that adds to I_B. The time-varying collector current is an amplified replica of the ac base current. The Q point is chosen so that the total base current neither shuts the transistor off during a negative swing nor drives the transistor to saturation during a positive swing. The resistor R_e stabilizes the operating point.

An example will illustrate the design of the analog modulator. We will use a silicon transistor with the characteristics in Fig. 10-3. Also, we will let $V_{dc} = 5$ V, $R_a = 2$ kΩ, $R_b = 5$ kΩ, $R_{IN} = 50 \ \Omega$, and $R_e = 60 \ \Omega$. From Fig. 10-3 we note that β is about 40. The voltage V_o across the forward-biased base-emitter junction is about 0.6 V for silicon (0.2 V for germanium). The equivalent parallel resistance of R_a and R_b is

$$R_1 = \frac{R_a R_b}{R_a + R_b} \qquad (10\text{-}5)$$

These two resistors divide the supply voltage, producing the equivalent voltage

$$V_1 = \frac{R_b}{R_a + R_b} V_{dc} = 3.57 \text{ V} \qquad (10\text{-}6)$$

in series with R_1. The dc collector current is

$$I_C = \frac{\beta (V_1 - V_0)}{R_1 + (1 + \beta)R_e} = 30.5 \text{ mA} \qquad (10\text{-}7)$$

and the dc base current is $I_B = I_C/\beta = 763 \ \mu$A. The load line is determined by the loop equation

$$i_C R_e + v_{CE} + v_d = V_{dc} \qquad (10\text{-}8)$$

where v_d is the diode voltage and the small base current flowing through R_e is ignored (i.e., we assume $i_E \simeq i_C$). The diode voltage is nearly constant for forward currents of more than a few milliamperes. We will take it to be 1.4 V in this example. The equation of the load line now reduces to

$$i_C R_e + v_{CE} = 3.6$$

We can easily find the coordinates of several points on this line. When $v_{CE} = 0$, then $i_C =$

3.6/60 = 60 mA, giving the coordinates of the upper point on the load line. At the Q point $i_c = I_{dc} = 31$ mA, so that

$$v_{CE} = 3.6 - 0.031(60) = 1.7 \text{ V}$$

Now we have the coordinates at the Q point and can draw the load line on the transistor curve in Fig. 10-3. From the figure we can see that the base current cannot exceed about 1400 μA without saturating the collector current. This corresponds to a maximum collector current of 55 mA. Therefore, the peak diode *signal* current can be as large as $I_{SP} = 55 - 31 = 24$ mA, resulting in a modulation factor $m' = 24/31 \simeq 0.80$. This circuit can operate at about 80% modulation.

Analog modulators must produce optic power variations that resemble the input voltage (or current) waveforms as closely as possible. Deviations occur if the source's power-current characteristic is not a perfectly straight line. Junction heating is the principal cause of LED nonlinearities.

We can investigate the LED nonlinearity by modeling its output characteristic by

$$P = P_{dc} + a_1 i_s + a_2 i_s^2 \qquad (10\text{-}9)$$

where i_s is the signal current and P_{dc} is the constant power produced by the dc current. The last term expresses the LED departure from linearity. Additional terms, involving higher powers of the current, can be added if greater accuracy is desired.

A single sinusoidal input, $i_s = I \sin \omega t$, produces optic power

$$P = P_{dc} + 0.5\, a_2 I^2 + a_1 I \sin \omega t$$
$$- 0.5\, a_2 I^2 \cos 2\omega t \qquad (10\text{-}10)$$

when inserted into Eq. (10-9). The last term, oscillating at twice the signal frequency, is the offending second harmonic distortion.

We define the *total harmonic distortion*, THD, in terms of the receiver's electrical power.

$$\text{THD} = \frac{\text{electrical power in the harmonics}}{\text{electrical power in the fundamental}}$$
$$(10\text{-}11)$$

Because the electrical power is proportional to the square of the incident optic power, THD can be written in the form

$$\text{THD} = \frac{(\text{optic power in the harmonics})^2}{(\text{optic power in the fundamental})^2}$$
$$(10\text{-}12)$$

When expressed in decibels,

$$\text{THD}_{dB} = -10 \log \text{THD} \qquad (10\text{-}13)$$

For the single sinusoidal input, we use Eq. (10-10) to find THD = $0.25\, (a_2 I/a_1)^2$. The amount of nonlinearity varies greatly among the LEDs available. Distortions of 30–60 dB below the signal level are representative.

An input current, $i_s = I_1 \sin \omega_1 t + I_2 \sin \omega_2 t$, containing two frequencies, produces output power

$$P = P_{dc} + 0.5\, a_2(I_1^2 + I_2^2)$$
$$+ a_1(I_1 \sin \omega_1 t + I_2 \sin \omega_2 t)$$
$$- 0.5\, a_2(I_1^2 \cos 2\omega_1 t + I_2^2 \cos 2\omega_2 t)$$
$$+ a_2 I_1 I_2[\cos (\omega_1 - \omega_2)t$$
$$- \cos (\omega_1 + \omega_2)t] \qquad (10\text{-}14)$$

In addition to harmonics, the power spectrum contains combinations (sum and difference, in this example) of the input frequencies. These combinations illustrate *intermodulation distortion*. In a multichannel system, such as a cable television distribution network with numerous carrier frequencies, intermodulation couples power between channels. Obviously, this effect must be minimized to prevent mixing of the pictures from the different channels.

This discussion on nonlinearity, including Eqs. (10-9)–(10-14), applies to laser diodes when operated at currents above their threshold.

The light source produces most of the nonlinearity in a fiber system because photodetectors have extremely good linearity and very linear transistor transmitting and receiving circuits can be designed. The fiber itself adds a negligible amount of distortion to the analog waveform. The distributed-feedback (DFB) laser diode (described in Section 6-6) has particularly good linearity, allowing its use in cable television signal transmission. Good DFB lasers have total harmonic distortion and intermodulation distortion better than 70 dB.

Digital Modulation

Unlike analog modulators, digital LED drivers need not provide dc bias currents. The digital circuit simply turns the LED on or off. In the off state, the LED emission should be low, creating a large on-to-off power ratio. In the on state, it is desirable that the driving current be independent of the magnitude of the input signal. Then the output power will be identical for every pulse, even if successive input signals vary somewhat.

The circuits in Fig. 10-4 illustrate two concepts for meeting the preceding requirements. For the series circuit, an opened switch permits no current, shutting off the LED. Closing the switch produces current

$$I = \frac{V_{dc} - v_d}{R} \qquad (10\text{-}15)$$

where v_d is the diode's forward voltage drop. The resistor R and the supply voltage deter-

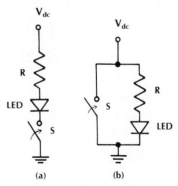

Figure 10-4 Schematic of (a) series-switched and (b) parallel-switched digital modulators.

mine the current for a given diode. An ideal switch (one having negligible resistance and, thus, negligible voltage drop when closed) does not affect the current amplitude. The resistor R functions as a limiter, protecting the diode from excessive currents. The parallel-switched modulator in Fig. 10-4(b) works similarly to the series circuit. Closing the switch shuts the LED off by bypassing the current to ground. Opening the switch sends all the current through the branch containing the LED, turning it on.

In practical circuits, transistors often provide the switching mechanism. Figure 10-5 illustrates a transistorized series-switched modulator.[2] The transistor characteristic in Fig. 10-6 shows that the collector current is small when the base current is zero (corresponding to an opened switch) and the collector-emitter voltage is small (≤ 0.3 V) when the base current is large (this condition corresponds to a closed switch). The current in the on condition is

$$I_C = \frac{V_{dc} - v_d - 0.3}{R} \qquad (10\text{-}16)$$

This is close to the ideal result in Eq. (10-15). The transistor not only produces switching but also provides amplification. A small input cur-

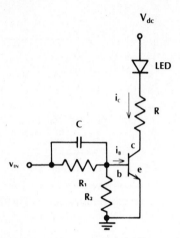

Figure 10-5 Transistor-switched LED digital modulator.

rent (~1 mA) controls the larger current (50–100 mA) required by the LED. R_1 and R_2 in Fig. 10-5 are chosen to impedance match the signal source to the transistor. The input capacitor C increases the speed of the circuit, if needed. This modulator can operate up to 30 MHz.

Example 10-1

For the series-switched modulator, find the diode current when fully on and the required base current to produce this condition. Use the transistor characteristic in Fig. 10-6 and let $V_{dc} = 5$ V and $R = 45$ Ω. The LED forward biased voltage drop is 1.4 V.

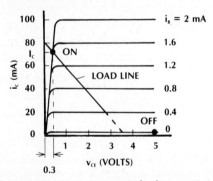

Figure 10-6 Transistor switch characteristics.

Solution:

Drawing the load line will help solve this example. The load line equation is

$$i_C R + v_{CE} + v_d = V_{dc} \qquad (10\text{-}17)$$

When $v_{CE} = 0$, we obtain $i_C = 80$ mA. When $i_C = 0$, we find $v_{CE} = 3.6$ V. Connecting these two points yields the load line drawn in Fig. 10-6. This line correctly predicts the operating point except at low current levels, where the diode voltage (which we took as 1.4 V) drops to zero. When the base current is zero (the off state), the collector current is nearly zero and the entire supply voltage appears across the transistor. In this condition $v_{CE} = V_{dc} = 5$ V, as noted in Fig. 10-6. In the on state, the base current should be so large that slight input amplitude variations will not affect the collector current. Figure 10-6 shows that this condition will occur if $i_B \geq 1.6$ mA. The collector current *saturates* (does not increase if the base current increases) for base currents satisfying this condition. The collector current in the on state, calculated from Eq. (10-16), is

$$I_C = \frac{5 - 1.4 - 0.3}{45} = 73 \text{ mA}$$

As long as $i_B \geq 1.6$ mA, the driving current (and the optic power) will be the same for all input pulses.

10-2 LASER-DIODE MODULATION AND CIRCUITS

Laser diodes present more problems to the circuit designer than LEDs. The troubles arise from.

1. The existence of a threshold current
2. The threshold current's age dependence
3. The threshold current's temperature dependence
4. The emission wavelength's temperature dependence

Digital systems generally operate just below threshold in the off state. The dc current is $I_{dc} \simeq I_{TH}$, as illustrated in Fig. 6-23. Operation near threshold (rather than at zero current) minimizes turn-on delay. Analog systems require a bias current in addition to the threshold current to achieve linear operation, as indicated in Fig. 6-24. The increase in threshold current owing to aging or a temperature rise causes a decrease in the output power if the current remains fixed.

Carrier wavelength changes are on the order of 0.2 nm/°C.[3] This corresponds to a frequency shift of 89 GHz/°C at 0.82 μm. For some applications the shift is insignificant, and for others it may be very important. For links operating near the minimum dispersion wavelength, a shift away from the optimum wavelength decreases the system's bandwidth. Wavelength-multiplexed systems also require a high degree of carrier-wavelength stability to minimize crosstalk between adjacent channels.

The laser's temperature dependence can be overcome by cooling the diode. Strategies include adequate heat-sinking and thermoelectric cooling (described briefly in Section 6-5). Threshold variations can be corrected by increasing (or decreasing) the dc current to compensate for the temperature- or age-induced changes in the laser's characteristics. This last solution, accomplished automatically by feedback control, does not solve the temperature-dependent wavelength shift, however.

The circuits that follow illustrate basic techniques for analog and digital modulation

of laser diodes. Although they do not include the complexity of feedback control, these circuits are practical when temperature and aging are unimportant. This is often the case, for example, in laboratory tests in which age is not a factor and in which the operator can monitor the laser output and adjust the drive current manually to maintain the desired power level. Finally, these circuits can be the modulator sections of complex networks that do contain feedback control.

Analog Modulation

The circuit in Fig. 10-2 (which we previously described for LED modulation) is suitable for analog modulation of a laser diode. In the preceding section we found that the transistor provided a dc current of 31 mA. Laser diodes usually need larger currents. A typical LD might have a 75-mA threshold and require a 25-mA bias above threshold, so a total dc current of 100 mA is needed. The additional current can be supplied by a high-impedance dc current source connected directly to the diode (at the transistor's collector terminal in Fig. 10-2). An inductor placed in series with this source decouples the ac and dc circuits.

As with LED systems, linearity must be considered for high-quality laser-diode analog links. Junction heating causes deviations from linearity in the power-current characteristic above threshold. Distortions 30 dB or more below the signal can be expected from good laser diodes.

Digital Modulation

The circuit in Fig. 10-7, using an n-channel GaAs MESFET, is suitable for high-speed digital modulation.[4] Rates better than 1 Gbps can be achieved. The circuit shown is an example

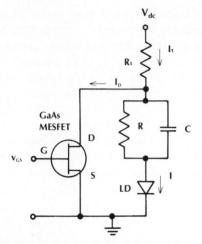

Figure 10-7   Laser diode digital modulator.

However, the modulation circuit should ensure that the diode's drive current (and thus the transmitted optic power) is the same for each on pulse. The circuit in Fig. 10-7 achieves this. When v_{GS} is high, the drain current I_D is so small that it does not affect the value of the laser current I. Almost all the current supplied by V_{dc} flows through the diode. Under these conditions, the supply voltage and resistors R_1 and R determine the laser current. That is, the diode's on current does not depend on the signal voltage v_{GS} as long as this voltage is above some minimum level.

of a parallel-switched modulator. The MESFET gate voltage v_{GS} (either zero or negative) controls the flow of current in the circuit. When v_{GS} is small, the resistance of the transistor's drain-to-source channel is low, whereas a large negative gate voltage produces a high channel resistance. In the modulator's off state the gate voltage is low, permitting a fraction of the current in resistor R_1 to bypass the branch containing the laser and flow through the transistor. The gate voltage is adjusted so that the diode's current is at its threshold value. An increased (more negative) gate voltage turns the laser on. Now most of the supplied current passes through the diode because of the high resistance presented by the MESFET.

The diode voltage (typically less than 2 V) is smaller than the drain-to-source voltage v_{DS} required for operation of the MESFET. The resistor R, in series with the diode, ensures that v_{DS} is large enough in both the on and off states. The capacitor C improves the switching speed of the circuit.

Linearity of the laser diode characteristic is not a problem for digital modulators.

10-3 ANALOG MODULATION FORMATS

In Section 10-1 we investigated the simplest type of analog modulation, transmission of a single sinusoidal current variation. The analysis illustrates *optic baseband transmission,* in which the signal is carried on a light beam modulated at the baseband frequencies of the information. For example, a baseband optic communications link carrying a single voice channel would contain modulation frequencies from a few tens of hertz up to 4 kHz. Because the optic power varies in proportion to the input current, the term *intensity modulation* (IM) applies. Intensity modulation differs from amplitude modulation (AM), which is commonly used with radio frequency carriers. In AM the amplitude of the carrier (rather than its power) varies in proportion to the information waveform. Fiber systems almost always use some form of intensity modulation. An exception, frequency modulation of the optic source, will be discussed in Section 10-5.

Analog formats other than baseband IM exist. For comparison purposes, and to simplify the notation, we will first rewrite Eqs. (10-1) and (10-2) as

$$i = I_o + I_s \cos \omega_m t \qquad (10\text{-}18)$$

$$P = P_o + P_s \cos \omega_m t \qquad (10\text{-}19)$$

I_o is the total dc current and ω_m is the modulation frequency. P_o is the average optic power. These expressions apply for both LEDs and laser diodes, as will all the equations in this section. In all cases, the current I_o places the operating point at the appropriate place along the linear portion of the source's power-current characteristic.

AM/IM Subcarrier Modulation

Conventional amplitude modulation places the message on a carrier whose frequency is much greater than any of the frequencies contained in the baseband. The resulting waveform has a spectrum that surrounds the carrier frequency. In essence, AM shifts the baseband to a new region of the electromagnetic spectrum. AM radio stations broadcast at different carrier frequencies, so they can be individually received by using filters tuned to the assigned carrier. After reception, the modulated signals are electronically returned (demodulated) to the original baseband frequencies.

Amplitude modulation of a single sinusoid can be written as

$$i = I_s(1 + m \cos \omega_m t) \cos \omega_{sc} t \qquad (10\text{-}20)$$

where ω_{sc} is the carrier frequency and, to maintain an undistorted signal, $m \le 1$. For 100% modulation, $m = 1$. The spectrum of this signal appears in Fig. 10-8. We can add a dc current I_o to the current in Eq. (10-20) and drive an optic source with the result, producing intensity modulation of a light beam by an amplitude modulated signal. This is AM/IM

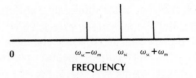

Figure 10-8 Spectrum of an amplitude-modulated wave.

modulation. Figure 10-9 illustrates the waveforms. AM/IM modulation generates optic power

$$P = P_o + P_s(1 + m \cos \omega_m t) \cos \omega_{sc} t \qquad (10\text{-}21)$$

The optic carrier oscillates very quickly. The radio-frequency subcarrier is slower, and the information waveform is even slower. The detected current, being proportional to the optic power, has the same form as Eq. (10-21). That is, the detected current is still in the AM format. The receiving circuit demodulates this current.

Frequency-Division Multiplexing

By using subcarrier modulation, several messages can simultaneously travel along the

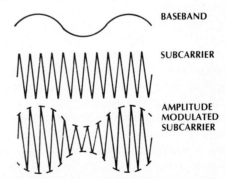

Figure 10-9 AM waveforms. The optic power and the detected current have the same waveshape as the modulated subcarrier.

fiber. Each message must modulate a different subcarrier, and the subcarriers must be so far apart that the spectra of adjacent channels do not overlap. Overlapping spectra produce crosstalk. Referring to Fig. 10-8, we see that each channel occupies a bandwidth equal to twice the highest modulation frequency; that is, half of the band is above the subcarrier and half is below. Therefore, subcarriers separated by twice the maximum expected modulation frequency will prevent overlap. The different channels of information are separated at the receiver (by filters) after photodetection. There is a limit, of course, to the number of channels added. New subcarriers can be added only if their frequencies are less than the fiber's bandwidth.

Simultaneous transmission of several messages by using different radio-frequency subcarriers is frequency-division *multiplexing* (FDM). FDM differs from wavelength-division multiplexing, described in Section 9-5, in which different optic carriers were used to distinguish between channels. Both of these multiplexing strategies increase the number of messages transmitted. In fact, the two techniques can be combined. The resulting system would contain several sources, each emitting at a different wavelength and each intensity modulated by a frequency-multiplexed modulated current.

Several problems accompany FDM. Inevitable nonlinearities in the source's power-current characteristic cause coupling (crosstalk) between channels. Nonlinearities elsewhere in the system (transmitter circuit, photodetector, receiving circuit) must be evaluated also and minimized to reduce crosstalk distortion. Furthermore, if several channels modulate a source, then the current due to each channel must be small so the total current does not drive the emitter beyond its linear range. The circuit designer must ensure that the peak current expected, when combining several channels, remains less than

the light source's rated limit. Otherwise, the source could be damaged or destroyed. Lowering the driving current in each channel reduces the amount of transmitted power containing each of the desired messages. The signal quality will be degraded just as much by reducing the transmitted power in each channel as it is by losses in couplers, connectors, the fiber, or any other component. Despite the difficulties just mentioned, AM FDM cable television fiber transmission has been very successful. AM has the advantage over FM and digital transmission in that no conversion is necessary from the AM format of the original signal at the transmitter and to the AM format of the signal required by conventional TV receivers. Also, AM signals occupy less bandwidth than either FM or digital signals. As the cost of digital conversion equipment lowers, however, it is expected that digital transmission will increase in television transmission systems. Characteristics (and advantages) of digital transmission are presented in the next section.

FM/IM Subcarrier Modulation

In conventional FM systems, operating at radio frequencies, the transmitted information is contained in the *phase* of the carrier wave. The current may be described generally by

$$i = I_s \cos \left[\omega_{sc} t + \theta(t) \right] \qquad (10\text{-}22)$$

where the message resides in the time variation of the phase angle θ. If the modulation is a single sinusoid oscillating at frequency $f_m = \omega_m / 2\pi$, the FM current takes the form

$$i = I_s \cos \left(\omega_{sc} t + \beta \sin \omega_m t \right) \qquad (10\text{-}23)$$

where β is the *modulation index*. The spectrum of a FM signal occupies a region that surrounds the carrier $f_{sc} = \omega_{sc}/2\pi$. The spectrum has (approximately) a total bandwidth

$$B_T = 2\Delta f + 2B \qquad (10\text{-}24)$$

In this expression, B is the baseband bandwidth (equal to f_m for the single sinusoid) and Δf is the maximum *frequency deviation*. It is given by

$$\Delta f = \beta f_m \qquad (10\text{-}25)$$

where f_m is the highest modulation frequency in the message. Normally, the baseband bandwidth equals the highest modulation frequency, so that

$$B_T = 2f_m(1 + \beta) \qquad (10\text{-}26)$$

For small values of the modulation index ($\beta \ll 1$), the total system bandwidth is just $2f_m$, the same as the bandwidth of an AM system. For larger values of β, however, the FM spectrum exceeds that of the comparable AM channel. Since the modulation index can be much bigger than one, the FM spectrum can far exceed that required for AM.

Adding a dc current to either Eq. (10-22) or (10-23) and intensity modulating an optic source with the result produces FM/IM subcarrier modulation. For the single sine wave, the optic power varies as

$$P = P_o + P_s \cos(\omega_{sc}t + \beta \sin \omega_m t) \quad (10\text{-}27)$$

FM waveforms are sketched in Fig. 10-10. The detected current has the same form as the optic power. Conventional FM demodulation circuits retrieve the information embedded in the phase of the detected current.

As first discussed in Section 10-1, nonlinearities in fiber optic analog-modulated systems distort the transmitted signal waveforms. The signal degradation severely limits the ap-

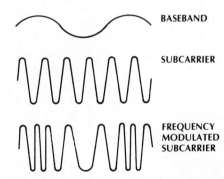

Figure 10-10 Subcarrier-FM waveforms. The optic power and the detected current have the same waveshape as the modulated subcarrier.

plication of fibers (using analog techniques) if good signal reproducibility is required. Commercial television picture transmission is just such an application. The effects of nonlinearities can be minimized by using FM/IM subcarrier modulation rather than baseband intensity modulation or AM/IM subcarrier modulation. The reason is that the information is extracted from the phase of the FM/IM waveform and not from its amplitude. High-quality television signals can be transmitted over fibers in the FM/IM format if a bandwidth of about 10 MHz is used.

Several FM channels can be simultaneously transmitted by frequency-division multiplexing, just as we described for subcarrier amplitude modulation. However, because the FM bandwidth is larger than the AM bandwidth, fewer FM messages can be fitted within the fiber's limited range of frequencies. The FM subcarriers must be separated by the bandwidth B_T given by Eq. (10-26).

10-4 DIGITAL MODULATION FORMATS

In Section 1-2 we noted how analog messages can be coded for digital transmission. In an example, we showed that sampling and coding of a 4-kHz voice message produced a 64-kbps pulse stream. A fiber system needs only a 4-kHz bandwidth for analog baseband voice transmission. According to the equation preceding Eq. (3-20), an RZ pulse train requires an electrical bandwidth equal to the data rate. Thus the digital 64-kbps signal needs a system with a 64-kHz bandwidth. For video transmission, representative bandwidths are 6 MHz for baseband analog transmission and 81 MHz for digital RZ signaling at an 81-Mbps data rate. Clearly, digital systems require much more bandwidth than analog systems.

Why, then, choose optic digital links over analog ones? A few of the reasons follow:

1. LEDs and laser diodes can be switched rapidly, giving them large bandwidths. Fibers and photodetectors also have large bandwidths. Thus, fiber optic systems can operate at data rates comparable to those needed for video and other broadband applications.

2. Analog fiber signals are degraded by nonlinearities in the LD or LED power-current characteristic. Digital signals are less affected by these nonlinearities because only two (or maybe three) power levels are normally used, and one of these levels is zero. Unlike analog transmission, the waveshape need not be accurately preserved. The receiver determines only the existence of pulses in each bit interval, not the pulse shape.

3. Digital systems can use error-checking codes and redundant information transmission to minimize errors.

4. Digital optic links are compatible with nonoptic digital systems. For example, a network connecting microprocessors contains only digitized signals. The network could be tied together with a combination of wire and fiber links. In this example, only digital transmission makes sense. In any application that generates data in digital form, a digital link is preferable to an analog one.

5. Digital pulses can be easily regenerated at repeaters. Digital repeaters reshape incoming pulses and amplify them, overcoming both attenuation and distortion. Very long fiber links (several thousands of kilometers long) can be constructed by using repeaters. Analog signals can be amplified by repeaters, but their waveforms cannot be easily restored. For long systems requiring repeaters, digital transmission is highly favored.

6. Generally, digital systems produce better quality signals than analog ones. If desired, signal quality can be traded for increased path length. Improved quality and longer transmission paths are the major rewards for providing a large-bandwidth digital link.

In the remainder of this section we describe a few digital coding schemes compatible with fiber optic transmission.

Pulse-Code Modulation

In Chapter 3 we discussed both non-return-to-zero (NRZ) and return-to-zero (RZ) codes. Both of these two-level unipolar formats (illustrated together in Fig. 10-11) are examples of *pulse-code modulation* (PCM). When viewing these pictures, remember that the waveshapes shown represent the average power in the extremely fast oscillations of an optic carrier.

Example 10-2

At 0.82 μm, how many oscillations occur within a 1-ns pulse?

Solution:

The frequency is $f = c/\lambda = 3.66 \times$

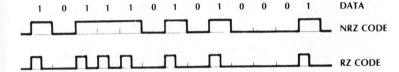

Figure 10-11 Non-return-to-zero and return-to-zero coding formats.

10^{14} Hz and the period is $T = 1/f = 2.73 \times 10^{-15}$ s. During the 10^{-9}-s pulse duration there are $10^{-9}/2.73 \times 10^{-15} = 365,853$ optic cycles.

Because optic PCM involves turning the carrier on and off, it goes by the name *on-off keying* (OOK).

The spectrum of an NRZ pulse train contains a large and important dc component. Its value in any short time period depends on the data. A series of 1s produces a larger value than a series of alternating 0s and 1s or a succession of 0s. In the receiver, the dc signal current partially determines the operating point of the amplifiers. A changing dc level changes the operating point, resulting in an undesirable variation (drift) in the receiver's characteristics. A disadvantage of the NRZ code is the need for dc coupling.

For the RZ code, ac capacitive coupling in the receiver blocks the dc spectral component, minimizing drift. Transitions between levels reveal the presence of 1s or 0s. Compatibility with ac coupling is an advantage of RZ over NRZ formats. Of course, as noted by comparing Eqs. (3-20) and (3-21), a fiber link with a fixed pulse spread (and thus a fixed bandwidth) can transmit NRZ signals at twice the rate of RZ signals. Put another way, for a

fixed data rate the RZ format requires twice as much transmission bandwidth as the NRZ code. In this case the NRZ format has the advantage.

In many instances, the receiver must know the rate at which data bits arrive (this is the *clock rate*). For the NRZ code, a series of alternating 1s and 0s reveals the clock rate, but a succession of all 1s or all 0s masks it. For the RZ code, the clock rate can be measured when a succession of 1s appears, but not when any other data pattern occurs.

The clock rate can be recovered from the train of data pulses by using the *Manchester* coding scheme, illustrated in Fig. 10-12. In this format, the signal polarity reverses in the center of each bit interval. The direction of this transition determines the logical state. Changes from high to low level indicate 1s and from low to high level indicate 0s. The numerous transitions permit the receiver to recover the clock, regardless of the distribution of 1s and 0s in the data. Because the data are contained in transitions between levels, ac coupling is appropriate. To summarize, Manchester coding offers the benefits of clock recovery and ac coupling. Its bandwidth requirement is the same as for RZ and double that of NRZ.

If clock recovery is not important, then

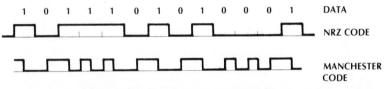

Figure 10-12 Manchester coding format.

RZ coding with ac coupling may be the best choice. This format does present a problem, however, if the receiving circuit contains *automatic gain control* (AGC). If a stream of 0s occurs, the AGC circuit increases the gain. In this condition, the next logical 1 to appear will be amplified much more than desired. Generally, each pulse will be amplified by an amount determined by the stream of data preceding it. This unstable operation makes it difficult for the receiver to correctly recognize the data bits.

Bipolar encoding, the three-level scheme illustrated in Fig. 10-13, solves the stability problem. This code provides a pulse whenever the data change. In the embodiment shown in the figure, the transmitter switches to full power for half a bit interval whenever a 0 follows a 1. It then returns to the half-power level and remains there until a 1 appears, at which time the power drops to zero for half a bit interval. The power then returns again to the median level. This code only transmits changes in the NRZ data stream. It is an *edge encoder*. The figure clearly shows that the dc (average) power level will not change, regardless of the data pattern. This characteristic occurs because high and low pulses always alternate. Stable operation results, even with an AGC receiver, because a fixed reference level is maintained.

The bipolar transmitter, although possessing three levels, is still a binary system. Only 0s and 1s are conveyed to the receiver.

Figure 10-14(a) shows a strategy for designing a three-level transmitter. No current flows in the circuit when both switches are

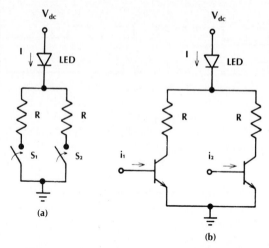

Figure 10-14 Producing a three-level LED output: (a) schematic; (b) practical circuit realization.

open, producing the zero-power first level. Closing switch S_1 causes LED current

$$I = \frac{V_{dc} - v_d}{R} \tag{10-28}$$

to flow, where v_d is the diode's voltage when forward biased. This current, normally half the LEDs allowed maximum, provides the second level at half the maximum optic power. Closing both switches changes the total resistance in series with the LED to $R/2$, almost doubling the second-level current. The LED now emits maximum power, creating the third level.

The circuit in Fig. 10-14(a) can be realized by using transistor switches, as illustrated in Fig. 10-14(b). The transistors are off (cor-

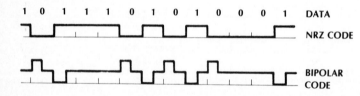

Figure 10-13 Bipolar coding.

responding to the open switch position) when their base currents (i_1 and i_2) are zero and on (closed switch position) when their base currents are positive. As described in Section 10-1, the state of a transistor switch (open or closed) depends on the input base current. Including the small voltage drop ($v_{CE} \simeq 0.3$ V) across the transistor when it is in the closed state, the diode current at level two becomes

$$I = \frac{V_{dc} - v_d - v_{CE}}{R} \qquad (10\text{-}29)$$

Prebias can be added to the three-level circuit by placing a resistor in parallel with the resistor-switch branches. When both switches are open, an amount of current dependent on the value of the added resistor will flow through the diode. A laser diode can replace the LED in Fig. 10-14, using prebias to set the first-level current at threshold.

Other Digital Formats

In addition to the codes already presented, numerous other digital schemes commonly found in electrical communications systems can be implemented by fiber optic links. We will briefly describe a few of them.

In *pulse-position modulation* (PPM) an analog waveform is periodically measured and the amplitude information of each sample is transmitted by a single, short optic pulse. Every pulse has the same height, regardless of the strength of the individual sample. The position of the pulse within the time slot allocated to each pulse conveys the amplitude information. The time slot is long compared to the pulse duration. The pulse delay relative to some reference point in the slot is proportional to the amplitude of the sample. Unlike PCM, in which the receiver decides whether a pulse occurs in any bit interval, the PPM re-

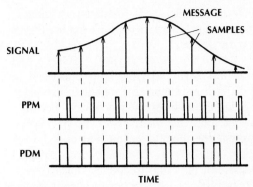

Figure 10-15   Pulse-position and pulse-duration modulation.

ceiver determines when the pulse arrives. Figure 10-15 illustrates pulse-position modulation.

Pulse-duration modulating (PDM) resembles pulse-position modulation. A pulse is transmitted for every data bit, but now the pulse *duration* is proportional to the sampled amplitude. An example of a PDM pulse train is drawn in Fig. 10-15 for comparison with PPM.

On-off keying of a subcarrier, shown in Fig. 10-16, is a possibility for fiber communi-

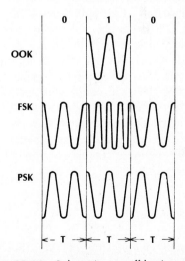

Figure 10-16   Subcarrier on-off keying, frequency-shift keying, and phase-shift keying.

cations. A single radio-frequency oscillator, turned on for binary 1s and off for 0s, drives the optic source.

Frequency-shift keying (FSK) and *phase-shift keying* (PSK), both illustrated in Fig. 10-16, are binary formats that also place the data onto radio-frequency waves. For fiber transmission, the waveshapes in the figure represent the modulating current and the resulting optic power. The oscillations pictured are those of the radio-frequency subcarriers.

In FSK, the subcarrier frequency determines the logical state. For example, frequency f_1 corresponds to a 1 and frequency f_2 to a 0. In PSK, the *phase* of the subcarrier determines the state. Positive polarity represents a 1 and negative polarity a 0.

Subcarrier OOK, FSK, and PSK move the modulating signal spectrum from low frequencies to a region surrounding the subcarrier. Benefits of digital subcarrier codes include the capability for frequency-division multiplexing, similar to the scheme described in Section 10-3 for simultaneous transmission of multiple analog messages.

The complexity of the transmitter and receiver depends on the coding scheme. PPM, PDM, subcarrier OOK, subcarrier FSK, and subcarrier PSK formats generally require more intricate designs than PCM. For this reason, PCM is an attractive choice for many fiber systems. In some cases, compatibility with existing electrical communications links dictates use of one of the other codes. In other situations, system performance may be improved (fewer errors and more sensitive receiver) with the more complex equipment.

Time-Division Multiplexing

Time-division multiplexing (TDM) allows a number of digitized messages to timeshare the same transmission line. Unlike WDM systems, which propagate several messages simultaneously, TDM interleaves bits or groups of bits (words or characters) belonging to different messages prior to transmission. At the receiver, the process reverses. Pulses belonging to the individual messages are separated and routed to their appropriate destinations. Figure 10-17 illustrates TDM. In practice, electronic (or optic) switches replace the mechanical ones drawn in the figure. The system in Fig. 10-17 interleaves four-bit words belonging to N channels (N messages). The switch S_T sequentially samples each channel and then starts over again, producing a single *frame* of N four-bit words for each cycle.

The telephone system provides an excellent practical example of TDM. Recall from Section 1-2 that voice messages are sampled 8000 times a second and an eight-bit word represents the amplitude of each sample. Eight bits can describe $2^8 = 256$ unique levels. That is, the amplitude of the voice waveform is quantized into 256 levels. The data rate for a single voice message is $8(8000) = 64,000$ bps. As we have seen, fibers can easily transmit at much higher rates. Sending only 64 kbps would be wasteful. Time-division multiplexing results in better utilization of the fiber's available bandwidth.

Let us look at the T1 (24-channel) multiplexing arrangement. Multiplexing involves combining 24 eight-bit words; that is, each frame contains $8(24) = 192$ message bits. An additional bit is added at the front of each frame to mark its beginning, so each frame actually contains 193 bits. Because the frame rate equals the sampling rate, there are 8000 frames per second. The total pulse rate that must be transmitted is then $193(8000) = 1.544$ Mbps. As we determined earlier, this rather moderate rate is easily handled by optic fibers. In fact, TDM fiber telephone lines often operate at higher rates, such as T3 (44.7 Mbps), T3C (91 Mbps), and above (up to several Gbps).

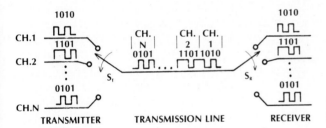

Figure 10-17 Time-division multiplexing. Switches S_T and S_R are synchronized.

10-5 OPTIC HETERODYNE RECEIVERS

As we know from the discussion in Chapter 7, photodetectors produce currents proportional to the incident optic power. Detectors respond to fluctuations in the light intensity, a characteristic independent of the light's phase or frequency. Thus optic detectors do not reproduce variations in the frequency or phase of the oscillating lightwave. Because of this, frequency modulation of an optic source is ineffective for communications using the direct detection methods described so far. However, optic frequency modulation systems are possible using heterodyne detection.[5,6,7,8]

Heterodyne Detection

In *heterodyne detection* (also called *coherent detection*), a beam of light (the *local oscillator*) mixes with the modulated wave at the entrance to the photodetector, as shown in Fig. 10-18. The heterodyne detector converts phase changes in the optic carrier to phase changes in the optic intensity. The latter variations are reproduced in the detected current waveform, making possible the reception and demodulation of frequency-modulated optic carriers. Heterodyne receivers are also effective for detecting intensity-modulated digital signals.

A simple analysis shows how the heterodyne scheme permits detection of the modulation. The electric fields of the transmitted signal and the local oscillator beams, respectively, are

$$E_{\text{SIG}} = E_S \cos [\omega_c t + \theta(t)] \qquad (10\text{-}30)$$

$$E_{\text{LO}} = E_L \cos [(\omega_c + \omega_{\text{IF}})t] \qquad (10\text{-}31)$$

where ω_c is the optic carrier frequency and $\theta(t)$ contains the frequency-modulated message. For modulation by a single sine wave, $\theta(t) = \beta \sin \omega_m t$ and β is the modulation index. Equation (10-30) can also represent an OOK signal. In this case, θ is constant and the signal amplitude E_s takes one of two values during each bit interval, depending on whether a 0 or a 1 is being transmitted. For FSK modulation, $\theta(t)$ is either $\omega_1 t$ or $\omega_2 t$. The local oscillator frequency $\omega_{\text{LO}} = \omega_c + \omega_{\text{IF}}$ is offset from the carrier frequency by the *intermediate frequency* ω_{IF}. The IF frequency is normally in the radio-frequency range. It may be a few tens or hundreds of megahertz. A *homodyne* detection system results if there is no offset, that is, if $\omega_{\text{IF}} = 0$.

The detected current is proportional to the intensity I (the square of the total electric field) of the incident light beam. Thus

$$I = (E_{\text{SIG}} + E_{\text{LO}})^2 \qquad (10\text{-}32)$$

Substitution of Eqs. (10-30) and (10-31) into this one and simplifying the results yields

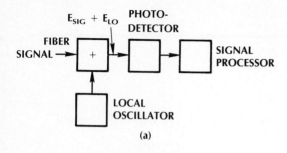

(a)

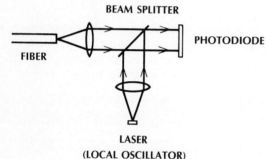

(b)

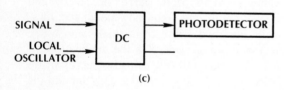

(c)

Figure 10-18   Optical heterodyne detection. (a) Receiver block diagram. Combining the signal and local-oscillator beams with (b) beamsplitter and (c) single-mode fiber directional coupler.

$$I = 0.5\, E_s^2 \{1 + \cos\,[2\omega_c t + 2\theta(t)]\}$$
$$+ \; 0.5\, E_L^2 \{1 + \cos\,[2(\omega_c + \omega_{IF})t]\}$$
$$+ \; E_L E_s \{\cos\,[\omega_{IF} t - \theta(t)]$$
$$+ \; \cos\,[2\omega_c t + \theta(t) + \omega_{IF} t]\}   \quad (10\text{-}33)$$

Three terms in this expression oscillate near frequency $2\omega_c$. This frequency (twice the optical carrier frequency) is much greater than the frequency response of the detector, so that all intensity components near this frequency are eliminated from the receiver. The elimination of the high-frequency terms is further explained by noting that the detected current is proportional to the average optical intensity, where the average is taken over a time interval long compared to the optical period but short

compared to the period of the IF frequency. The average intensity is then

$$I = 0.5(E_L^2 + E_s^2) + E_s E_L \cos\,[\omega_{IF} t - \theta(t)]$$
$$(10\text{-}34)$$

The corresponding optical power (which is proportional to the intensity) is then

$$P = P_L + P_s + 2\sqrt{P_L P_s}\, \cos\,[\omega_{IF} t - \theta(t)]$$
$$(10\text{-}35)$$

where P_L and P_s are the powers in the local oscillator and signal beams, respectively. You might wish to verify that the proportionality factor is correct in this result. This is done by taking the special case of no signal beam (E_s, and thus P_S, are zero). Then Eq. (10-35) should reduce to $P = P_L$, which it does. Note the preservation of the optical phase variation in the IF term.

The current, $i = \eta e P / hf$, includes a dc term

$$i_{dc} = \frac{\eta e}{hf}(P_L + P_S) \qquad (10\text{-}36)$$

and one at the IF frequency

$$i_{IF} = \frac{2\eta e}{hf}\sqrt{P_S P_L}\cos\left[\omega_{IF}t - \theta(t)\right] \qquad (10\text{-}37)$$

The dc current is generally filtered out and the IF current is amplified. Conventional FM electronic demodulators then extract the information contained in $\theta(t)$. If the system is on-off keyed (rather than frequency-modulated), then the phase θ remains fixed and P_S contains the information. Notice that the information-bearing IF current increases with local oscillator power. In effect, the LO acts as a signal amplifier, increasing the sensitivity of the receiver.

In the preceding analysis we wrote the equations assuming perfectly monochromatic light emitters. In practice, linearly polarized, single-longitudinal-mode, single-transverse-mode laser diodes suffice.

Heterodyne detection depends on interference between the local oscillator and signal light beams. The fields will not interfere unless they are identically polarized. This explains the source's linear polarization requirement. Unfortunately, most fibers do not maintain a wave's state of polarization. During travel, the direction of polarization may rotate and the linear state may change to some other polarization. Also, environmental factors (such as small temperature shifts and vibrations) may cause random variations in the state of polarization. Specially constructed po-

larization-maintaining single-mode fibers are required for practical heterodyne systems.

Figure 10-18 illustrated two techniques for combining the signal and local oscillator beams, a beam splitter and a single-mode fiber directional coupler. Regardless of the type of combiner used, it must maintain the polarization state of both beams. Only specially designed fiber couplers satisfy this requirement.

The frequency offset between the local oscillator and the transmitter can be finely tuned by using the temperature-dependent wavelength property of laser diodes. Two identical diodes, operated at slightly different temperatures, will oscillate at different frequencies. Once set, the laser temperature must be maintained within a very small fraction of a degree Celsius to keep the IF frequency from changing too much.

Example 10-3

Suppose that a laser diode has a wavelength change of 20 GHz/°C. What temperature fluctuations are allowed if the change in frequency offset must be less than 100 MHz?

Solution:

The allowed change is $(0.1\ \text{GHz})/(20\ \text{GHz/°C}) = 0.005°C$. A system with a nominal IF frequency of 1 GHz might find the 100-MHz variation unobjectionable.

Laser-Diode Frequency Modulation

The oscillation frequency of a single-mode laser diode depends on the instantaneous amplitude of the injected current. We can explain this result as follows: The current determines both the carrier density and the temperature in the semiconductor's active layer. In turn, these two factors determine the layer's refractive index. As shown earlier by Eq. (3-24), the resonant frequency of a cavity depends on its refractive index. Thus, the resonant frequency (which is also the output frequency) changes when the current does. In this way, modulation of the drive current produces frequency modulation of the emitting diode. We may view the result as *refractive-index* modulation.

The circuit for frequency modulating a laser diode, drawn in Fig. 10-19, appears to be very similar to an intensity-modulation circuit. A dc current biases the diode to the middle of the linear region of its power-current characteristics. The ac modulating current must be small (maybe just a few milliamperes) to minimize the undesired intensity modulation that occurs. Electronic limiters in the receiver further reduce the amplitude variations before demodulation of the signal. The ac current produces frequency modulation of the optic carrier. For single sine-wave modulation at frequency f_m, the FM frequency deviation is $\Delta f = \beta f_m$. The frequency deviation (or equivalently, the modulation index β) varies linearly with the peak amplitude of the ac current. A representative normalized frequency deviation for AlGaAs laser diodes is 200 MHz/mA at a modulation frequency of 300 MHz. To clarify this last statement, if the ac drive current oscillates at 300 MHz, then the frequency deviation increases by 200 MHz for each increase of 1 mA in the peak value of the ac current.

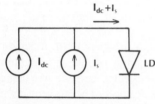

Figure 10-19 Frequency modulation of a laser diode. I_s is the modulating signal.

Example 10-4

Use the numerical value of 200 MHz/mA frequency deviation at a modulation frequency of 300 MHz to compute the modulation index when the peak ac current is 1 mA and 5 mA.

Solution:

At 1 mA, the frequency deviation is $\Delta f = 200$ MHz, and at 5 mA it is $\Delta f = 1000$ MHz. Using $\beta = \Delta f / f_m$, we find $\beta = 200/300 = 0.67$ at 1 mA and $\beta = 1000/300 = 3.33$ at 5 mA.

As an alternative to internal modulation by varying the source's drive current, the information can be applied to the carrier externally. External electro-optic and acousto-optic modulators can turn the light beam on and off for digital systems, and they can frequency modulate the optic carrier for FM systems. They can also intensity and polarization modulate the beam. All these formats can be detected and demodulated by a heterodyne receiver. Electro-optic and acousto-optic modulators can be constructed as bulk devices or as integrated optic components (as discussed briefly in Section 4-6).

In some fiber optic applications (primarily sensors) environmental changes directly phase modulate the optic carrier. The fiber optic gyroscope, sketched in Fig. 10-20, is a good example.[9] Identical light beams from a single LD propagate in both directions around a multiturn coil of single-mode fiber. If the coil is stationary, then the two beams remain identical. When the coil rotates around its axis, the beam traveling in the direction of rotation has its phase shifted relative to the counter-rotating beam. The phase shift is $\theta = (8\pi A N/\lambda c)\Omega$, where Ω is the rotation rate (rad/s), N is the number of turns on the coil, and A is the coil area. The shift can be measured by homodyne detection. Equation

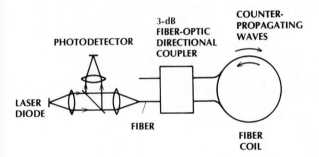

PHOTODETECTOR

3-dB
FIBER-OPTIC
DIRECTIONAL
COUPLER

COUNTER-
PROPAGATING
WAVES

LASER
DIODE

FIBER

FIBER
COIL

Figure 10-20   Fiber optic gyroscope.

(10-35) describes the power incident on the gyroscope's photodetector. If we set $\omega_{IF} = 0$ and let $P_L = P_S = P_o/2$ (where P_o is the power in each of the counterrotating beams), then

$$P = P_o(1 + \cos \theta) \qquad (10\text{-}38)$$

where θ is the rotationally induced phase difference. The detected current is

$$i = \frac{\eta e P_o}{hf}(1 + \cos \theta) \qquad (10\text{-}39)$$

so the amplitude of the photocurrent reveals the value of the phase shift and, consequently, the rotation rate.

There are many variations of the basic fiber gyroscope described in this section.

Optic-Frequency-Division Multiplexing

Several messages can be simultaneously transmitted along the fiber by *optic frequency-division multiplexing* (OFDM) combined with heterodyne detection. One scheme appears in Fig. 10-21. N identical laser diodes are temperature tuned to emit at slightly different frequencies, $\omega_1, \omega_2, \ldots, \omega_N$. The diodes are modulated with the desired messages. The output of each diode is coupled to a fiber and the multiple fibers are coupled (by fusion, for example) to the transmission fiber. At the receiver, light from a single local oscillator mixes with each transmitted beam, producing a different IF frequency for each channel. The IF frequencies, $\omega_{IF1}, \omega_{IF2}, \ldots, \omega_{IF3}$, are electronically sorted by filters. The speed of the photodetector and the receiving circuit determines the maximum allowable IF frequency.

In an alternate OFDM arrangement, a tunable local oscillator replaces the fixed one in Fig. 10-21. In addition, only one IF filter is needed. A given channel will be detected when its optical carrier frequency differs from that of the LO by an amount equal to the IF frequency. In this case, all other channels will produce IF frequencies outside the passband of the IF filter. To receive a different channel, the LO frequency is changed. In this system, a station can tune into any of the transmitted channels.

Optic frequency-division multiplexing and wavelength-division multiplexing are somewhat similar. They both use separate optic sources for each channel. Their differences are important, however. OFDM requires heterodyne detection, and WDM uses direct detection. Also, WDM systems sort the channels in the optic domain (before photodetection), and OFDM systems separate them electronically (after photodetection). Electrical separation at radio frequencies is much more selective than optic separation, so adjacent channels can be closer together in OFDM systems than in

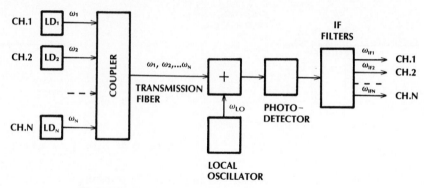

Figure 10-21 Optic frequency-division multiplexing.

WDM systems. Closer spacing allows transmission of more channels of information in the wavelength regions where fiber losses are low.

Example 10-5

Compute the allowed channel separation for an OFDM and a WDM system if the filters in both networks have a 1% fractional bandwidth. Assume optic operation in the short-wavelength window and IF frequencies as high as 1 GHz.

Solution:

First consider optic filtering. At 850 nm, a 1% passband represents 8.5 nm. Assuming that the spectral widths of the sources are much smaller, adjacent channels can be separated by 8.5 nm. By using the relationship obtained from Eq. (3-26), $\Delta f = c \Delta\lambda/\lambda^2$, the separation between adjacent WDM channels is 3.5×10^{12} Hz. On the other hand, a 1% passband at 1 GHz is 10^7 Hz, the minimum OFDM channel separation. At least in principle the OFDM technique allows much closer channel spacings than WDM. In practical situations, the temperature-induced frequency variations of the transmitter and local oscillator re-

quire that spacings larger than 10^7 Hz be maintained.

Advantages and Problems with Heterodyne Detection

Heterodyne detection has several attractions. It makes frequency-modulated optic links feasible. Also, heterodyne receivers are more sensitive than direct-detection receivers. We illustrate this characteristic in Section 11-2. A more sensitive receiver provides better quality reception and a capability for longer unrepeatered transmission paths. If repeaters are necessary, then their spacings can be greater with a sensitive heterodyne receiver than with a direct-detection receiver.

Increased complexity and cost is the price paid for the advantages of heterodyne systems. The laser diodes must be single-mode devices. The transmitting source and the local oscillator laser diode must be frequency stabilized so that the IF frequency does not drift. The signal and local oscillator beam alignment (in the mixer) is critical. The two wavefronts must be identical and must coincide. The spot sizes, polarizations, and propagation directions of the two beams must be the same.

10-6 SUMMARY

Early in this chapter we discussed a few simple modulating circuits. These circuits illustrate basic strategies. They can be used as described or can form the basis for more complex circuits.

We expanded the number of possible modulation formats over those studied in the preceding chapters. Tables 10-1 and 10-2 summarize the techniques introduced. Although not all-inclusive, these lists should be helpful in developing and analyzing most systems.

The various types of multiplexing schemes presented in this book are summarized as follows:

1. *Time-division multiplexing (TDM)*. Data bits corresponding to different messages

TABLE 10-1. Analog Modulation Formats

Name	Description	Comments
Baseband modulation		
Intensity modulation	Optic power varies in proportion to the baseband message	Simplest analog scheme
Optic frequency modulation	Direct frequency modulation of an optic carrier	Requires heterodyne detection
Subcarrier modulation		
AM/IM	Intensity modulation of an optic source by a lower-frequency, amplitude-modulated signal	Permits subcarrier frequency-division multiplexing
FM/IM	Intensity modulation of an optic source by a frequency-modulated signal	Permits subcarrier frequency-division multiplexing

TABLE 10-2. Digital Modulation Formats

Name	Comments
Pulse-code modulation	
Non-return-to-zero (NRZ)	Requires the least bandwidth for digital transmission
Return-to-zero (RZ)	Requires twice the bandwidth of NRZ systems
Manchester	Clock recovery is possible
Bipolar	Dc level remains constant
Pulse-position modulation	
Pulse-duration modulation	
Subcarrier modulation	
On-off keying (OOK)	Permits subcarrier FDM
Frequency-shift keying (FSK)	Permits subcarrier FDM
Phase-shift keying (PSK)	Permits subcarrier FDM

are interleaved to timeshare the fiber channel. A single optic source and a single photodetector are required if the message interleaving and separation are done when the signals are in electrical (rather than optic) form. TDM is appropriate for digital communications. It does not change the information capacity of the fiber: it merely distributes the allowed bits among several messages.

2. *Wavelength-division multiplexing (WDM)*. Several messages simultaneously travel along the fiber, each message carried at a different optic wavelength. Multiple sources, oscillating at different frequencies, are required. Message separation is performed in the optic domain, before detection. A separate photodetector is needed for each message. WDM accommodates both analog and digital signals. The information capacity of a fiber is increased (approximately) by a factor equal to the number of multiplexed messages for WDM. Essentially, each source and detector combination constitutes an independent channel.

3. *Subcarrier frequency-division multiplexing (SFDM)*. Messages are modulated onto different subcarriers and combined electrically. The combined signal modulates a single optic source. A single photodetector returns the signal to electrical form. At this point, electronic filters separate the messages. Subcarrier FDM can be used for both analog and digital signals. As with TDM, subcarrier FDM does not increase fiber capacity. The maximum subcarrier frequency cannot exceed the fiber's bandwidth. This scheme merely divides the available bandwidth among several messages.

4. *Optic frequency-division multiplexing (OFDM)*. Messages are modulated onto sources having slightly different wavelengths. Heterodyne detection, using a single photodetector, produces a signal current containing all the messages. Each message's spectrum surrounds a different intermediate frequency. Electrical filters then separate the messages. Optic FDM does increase the fiber's capacity. In practice, however, the speed of the photodetector will limit the maximum intermediate frequency, and this limits the number of messages that can be communicated.

It is sometimes said that digital modulation is more compatible with optic communications than analog modulation. The arguments include the general advantages of digital systems (improved signal quality, longer transmission paths, simpler repeaters), the relative ease of digital modulation (simply turn the source on and off), and the nonlinearity of optic sources (which degrades analog signals). Nonetheless, digital transmission of messages that originate in analog form (voice and video, for example) has its problems. The major disadvantage is the need to convert messages from analog to digital at the transmitter and from digital to analog at the receiver. The cost of the required conversion equipment may justify a completely analog system, particularly for short transmission paths. In any case, when the modulation method has not been predetermined, the system designer should consider both digital and analog formats.

PROBLEMS

10-1. An LED has its 3-dB bandwidth equal to 80 MHz. Its output optical power versus input current curve has a slope of 0.1 mW/mA. The input current consists of a 50-mA dc component and a 40-MHz sinusoid having a peak-to-peak current of 60 mA.

(a) Sketch a few cycles of the input current.

(b) Compute the modulation factor for the input current.

(c) Compute and plot the resulting optical power.

(d) Compute the optic modulation factor.

10-2. Repeat Problem 10-1 if the modulation frequency changes to (a) 80 MHz and (b) 120 MHz.

10-3. Design an analog modulator like that shown in Fig. 10-2 by using a silicon transistor. For this transistor, $\beta = 60$. Set the dc bias at 50 mA. $R_{in} = 50\ \Omega$, $R_a = 3\ k\Omega$, $R_b = 6\ k\Omega$, and the bias supply voltage is 10 V. When the light source is on, its voltage drop is 1.8 V.

(a) Determine the value of the emitter resistance R_e.

(b) Compute the collector-emitter voltage at the Q point.

(c) Determine the maximum modulation factor this circuit can provide.

10-4. An LED produces 50 dB of total harmonic distortion when driven by a sinusoidal current having a peak value of 2 mA. The slope of the LED power versus current curve at the dc bias point is 0.05 mW/mA. Compute the value of the nonlinearity coefficient [a_2 in Eq. (10-9)].

10-5. For the series-switched digital modulator (Fig. 10-5), find the LED current when the source is fully on and compute the base current required to produce this condition. Use the transistor characteristic shown in Fig. 10-6, and let the supply voltage be 8 V and $R = 60\ \Omega$. The LED forward voltage is 1.8 V when fully on.

10-6. Suppose that you want to transmit the entire commerical AM radio broadcast band over a single fiber. Draw a block diagram for accomplishing this. What system bandwidth is required?

Use AM/IM subcarrier modulation. The AM band has 107 channels having carriers ranging from 540 to 1600 kHz and with 10 kHz allowed for each channel (as described in Chapter 1).

10-7. Determine the bandwidth required to transmit 10 5-kHz radio stations over a fiber link using RZ-coded digital modulation.

10-8. Repeat Problem 10-6, substituting the entire commercial FM broadcast band for the AM radio band. Use FM/IM subcarrier modulation. The FM band extends from 88.1 to 200 kHz with a 200-kHz separation between carriers. There are 100 channels.

10-9. Draw a block diagram for transmitting all the VHF television channels over a single fiber. Use analog modulation. What system bandwidth is required?

10-10. Repeat Problem 10-9, using digital modulation.

10-11. The LED whose output characteristic is shown in Fig. 6-5 is used in a two-channel frequency-division-multiplexed system. The LED total harmonic distortion is measured as 25 dB when a sinusoidal current with a 50-mA peak is applied. The bias current is set at 50 mA. The AM modulation of each subcarrier is 50%. The intensity modulation is 100%. The subcarriers are at 1 MHz and 2 MHz and have equal peak currents. The information is a 1000-Hz tone for both channels. Write the equation of the input current and the radiated power when distortion is neglected and when it is not neglected. All the radiated power is detected by a photodiode whose responsivity is 0.5 A/W. Compute the receiver current and plot its spectrum. Is there any crosstalk in this system? How could you redesign the system to eliminate the crosstalk?

10-12. Draw the pulse trains for the signal 1001110001010 for NRZ, RZ, Manchester, and bipolar codes.

10-13. Design a three-level transmitter (like that in Fig. 10-14b). Use the transistor described in Fig. 10-6. Compute values for the components used. The three LED current drive levels are 0, 73, and 146 mA. The LED voltage drop is about 1.4 V when forward biased.

10-14. Consider a two-channel frequency-division multiplexed AM/IM subcarrier modulated optical signal, using a LED source. The two subcarriers are ω_{sc1} and ω_{sc2}. The modulation signal is the single frequency ω_{m1} and ω_{m2} for channels 1 and 2, respectively. The modulation coefficients and the peak amplitudes are the same for both channels.

(a) Using Eq. (10-9) as a model, compute the equation of the resulting optical power.

(b) Sketch the spectrum of the detected signal, showing the amplitudes of the frequency components.

(c) Discuss the possibilities of crosstalk between the two channels and measures that can be taken to minimize the crosstalk.

10-15. A 10-Gbps RZ digital system operates at a wavelength of 1550 nm. How many oscillations of the carrier occur in each pulse?

REFERENCES

1. P. W. Shumate and M. DiDomenico, Jr. "Lightwave Transmitters." In *Semiconductor Devices for Optical Communication,* edited by H. Kressel. Berlin: Springer-Verlag, 1980, pp. 161–200.

2. R. Adair. "CW Lasers and LEDs," Application Note A/N 101. New Brunswick, N.J.: Laser Diode Laboratories, Inc.

3. Ibid.

4. Shumate. "Lightwave Transmitters." pp. 182–188.

5. Francois Favre, Luc Jeunhomme, Irene Joindot, Michel Monerie, and Jean Claude Simon. "Progress towards Heterodyne-Type Single-Mode Fiber Communications Systems," *IEEE J. Quantum Electron.* 17, no. 6 (June 1981): 897–906.

6. Soichi Kobayashi, Yoshihisa Yamamoto, Minoru Ito, and Tatsuya Kimura. "Direct Frequency Modulation in AlGaAs Semiconductor Lasers," *IEEE J. Quantum Electron.* 18, no. 4 (April 1982): 582–95.

7. Shigeru Saito, Yoshihisa Yamamoto, and Tatsuya Kimura. "Optical FSK Heterodyne Detection Experiments Using Semiconductor Laser Transmitter and Local Oscillator," *IEEE J. Quantum Electron.* 17, no. 6 (June 1981): 935–41.

8. Yoshihisa Yamamoto and Tatsuya Kimura. "Coherent Optical Fiber Transmission Systems," *IEEE J. Quantum Electron.* 17, no. 6 (June 1981): 919–35.

9. Thomas G. Giallorenzi, Joseph A. Bucaro, Anthony Dandridge, G. H. Sigel, Jr., James H. Cole, Scott C. Rashleigh, and Richard G. Priest. "Optical Fiber Sensor Technology," *IEEE J.Quantum Electron.* 18, no. 4 (April 1982): 626–65.

Chapter 11

Noise and Detection

A variety of phenomena cause signals to degrade as they progress through a fiber communications link. We have already discussed how waveform distortion occurs in the fiber and how this limits the information capacity (and length) of the path. In addition, we have studied how signals attenuate owing to fiber, coupling, and distribution losses. Intuition leads us to believe that only so much attenuation can be tolerated before the power reaching the receiver becomes too small to detect accurately. On the other hand, we might decide that amplifiers can always boost the signal to the required level. The latter conclusion would be correct if it were not for another signal disturbing phenomenon, *noise*. Noise degrades the signal and is always present (as shown in Section 11-1). Signal amplification is always accompanied by an equal amount of noise amplification and the amplifier contributes additional noise of its own. For this reason, amplification cannot improve the ratio of signal power to noise power. As the received signal power diminishes toward the noise power, the signal becomes less and less discernible. In this way, attenuation ultimately limits the length of a fiber transmission system.

In this chapter we investigate the major sources of noise and show how to compute the noise power. The signal quality, given by its signal-to-noise ratio, can then be calculated. For digital communications, noise increases the probability of errors. For these systems, we also calculate the error rates.

Toward the end of this chapter we describe a few basic receiver circuit designs.

11-1 THERMAL AND SHOT NOISE

Two major causes of signal degradation occurring during reception are thermal noise and shot noise.

Thermal Noise

Thermal noise (also called *Johnson noise* and *Nyquist noise*) originates within the photodetector's load resistor R_L. Electrons within any resistor never remain stationary. Because of their thermal energy, they continually move, even with no voltage applied. The electron motion is random, so the net flow of charge could be toward one electrode or the other at any instant. Thus, a randomly varying current exists in the resistor, as pictured in Fig. 11-1. This is the thermal noise current i_{NT}. Its average value is zero. The average noise power generated within the resistor is $R_L \overline{i_{NT}^2}$, where $\overline{i_{NT}^2}$ is the mean-square value of the thermal noise current. (The bar indicates average value.) Figure 11-1 illustrates the square of the noise current and its mean (average) value. The noise current adds to the signal current generated by the photodetector. Figure 11-2 shows the results when constant optic power P illuminates the photodetector. Instead of remaining fixed at $i = \eta eP/hf$, the load current varies randomly around this value. When the incident power is so small that the signal current and the noise current have comparable

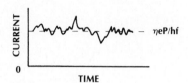

Figure 11-2   Receiver current when the optic power is constant, showing the signal degradation caused by thermal noise.

amplitudes, the presence of the signal is masked. Even with moderate amounts of optic power, the signal current might not be large enough (relative to the noise current) to achieve the desired clarity of reception.

The presence of thermal noise can be modeled by the equivalent circuit drawn in Fig. 11-3.[1] In this circuit, R_L is an ideal noiseless resistor. The noise is produced by a current source generating mean-square current

$$\overline{i_{NT}^2} = \frac{4kT\Delta f}{R_L} \qquad (11\text{-}1)$$

where k is the Boltzmann constant (given in Table 1-2), T is the absolute temperature (K), and Δf is the receiver's electrical bandwidth. Circuit elements in the receiver limit its bandwidth. At high frequencies, amplifiers cut off and capacitances short circuit signals. To process all of the desired message, the receiver's bandwidth must be at least as large as that of the information. The receiver's bandwidth should be limited, however, to minimize the noise. Low-noise receivers have bandwidths that range from a little more than, to twice, the information bandwidth. The larger bandwidths are sometimes needed to account for

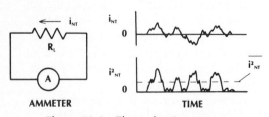

Figure 11-1   Thermal noise current.

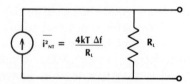

Figure 11-3   Thermal noise equivalent circuit.

bandlimiting of the signal by the transmitter and the fiber. A further discussion of receiver bandwidth appears in Chapter 12, which covers system design.

Equation (11-1) assumes that the thermal-noise spectrum is uniform over all frequencies. This is true up to around 10 GHz, making our model sufficient for analyzing most systems.

Shot Noise

The discrete nature of electrons causes a signal disturbance called *shot noise*. In photodetectors, either photoemissive tubes or semiconductor junction devices, incoming optic signals generate discrete charge carriers. Each carrier contributes a single pulse to the total current. We illustrate this for the vacuum photodiode in Fig. 11-4. The pulse starts when the electron escapes from the cathode and ends when the electron strikes the anode (where it disappears by recombining with a positive charge). Thus, the pulse duration equals the electron's transit time τ (the time it takes the electron to travel from the cathode to the anode). The exact pulse shape is relatively unimportant. The increase in current amplitude during transit arises from the electron's acceleration under the force of the electric field existing between the electrodes. The faster the electron, the greater the current.

Now consider what happens when an incoming wave having constant optic power P illuminates the cathode. We expect the photocurrent to be constant, as shown in Fig. 11-5. However, this constant current is made up of a large number of pulses of the form shown in Fig. 11-4. Although all the pulses are identical, they are generated at random times (as pictured in Fig. 11-5). Addition of identical, randomly delayed pulses does not produce a constant level. Instead, a ragged current results, having as its average value the current predicted for the noiseless case ($\eta eP/hf$). The deviations from the ideal current (caused by random generation of discrete charge carriers) is shot noise. In semiconductor photodiodes, shot noise arises from random generation and recombination of free electrons and holes.

Shot noise can be represented by an equivalent circuit consisting of a single current source, as drawn in Fig. 11-6. The mean-square shot noise current is[2]

$$\overline{i_{NS}^2} = 2eI\Delta f \qquad (11\text{-}2)$$

where e is the magnitude of the charge on an electron, I is the average detector current, and Δf is the receiver's bandwidth. The shot-noise

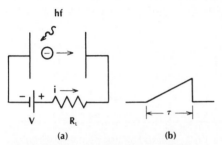

Figure 11-4   (a) Emission of a single photoelectron. (b) Resulting current pulse.

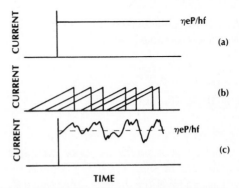

Figure 11-5   Shot noise. (a) Expected (ideal) photocurrent owing to constant optical power P. (b) Randomly produced current pulses created by the emitted electrons. (c) The sum of the current pulses (i.e., the total current).

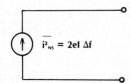

Figure 11-6 Shot-noise equivalent circuit.

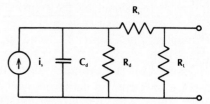

Figure 11-7 Junction photodiode equivalent circuit.

spectrum is uniform over all modulation frequencies of interest. Just like thermal noise, the shot-noise current depends on the system bandwidth, not on the location of the band. According to Eq. (11-2), shot noise increases with current. Thus, shot noise increases with an increase in the incident optic power. This differs from thermal noise, which is independent of the optic power level. In the next section we will determine how this behavior affects signal quality by computing the signal-to-noise ratio. Also, we will numerically evaluate the noise power as determined from Eqs. (11-1) and (11-2).

The current in Eq. (11-2) includes both the average current generated by the incident optic wave and the dark current I_D (introduced in Section 7-4). Then,

$$\overline{i_{NS}^2} = 2e(\overline{i_s} + I_D)\Delta f \qquad (11\text{-}3)$$

where i_s is the photocurrent and the bar indicates its average value.

11-2 SIGNAL-TO-NOISE RATIO

In Fig. 7-10 we showed the equivalent circuit of a junction photodiode. A more complete circuit appears in Fig. 11-7. For simplicity, we will assume that the diode's capacitance C_d and the transit time do not limit the signal. C_d can then be removed from the noise equivalent circuit. Its effect on noise is accounted for in the determination of the receiver's band-

width Δf. Semiconductor diodes have a small amount of series resistance R_s (maybe a few ohms) owing to conduction in the bulk p and n regions. This resistance will be neglected. Similarly, the diode has a resistance R_d in parallel with its equivalent current source. This is just the resistance of the depleted junction region. Since R_d is normally much larger than the load resistance R_L, it can be ignored.

With the preceding assumptions in mind, we can now combine the equivalent circuits of the diode and the thermal-noise and shot-noise sources. The result appears in Fig. 11-8. The signal-to-noise ratio can be computed from the circuit for a variety of circumstances. We will compute the SNR for the following situations:

1. *Constant incident optic power.* This corresponds to reception of a 1 in a binary PCM system. We will first consider use of a detector with no internal gain (such as a *pn* or PIN diode) and then show the improvement by using a detector with internal gain (such as an avalanche

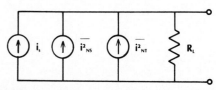

Figure 11-8 Photodetector receiving circuit, including the equivalent sources of thermal and shot noise.

photodiode) or by using heterodyne detection.

2. *Sinusoidally varying optic power*. This corresponds to an intensity-modulated analog signal.

Constant Power

In this case, the signal photocurrent has the constant value

$$i_s = \frac{\eta e P}{hf} \qquad (11\text{-}4)$$

where P is the incident optic power. The diode delivers average electrical signal power

$$\overline{P}_{ES} = i_s^2 R_L = \left(\frac{\eta e P}{hf}\right)^2 R_L \qquad (11\text{-}5)$$

to the load resistor.

The average shot-noise power delivered to the load is $i_{NS}^2 R_L$, which, using Eqs. (11-3) and (11-4), becomes

$$\overline{P}_{NS} = 2e\,\Delta f\left(\frac{\eta e P}{hf} + I_D\right)R_L \qquad (11\text{-}6)$$

We made the substitution $\bar{i}_s = \eta e P/hf$, because the current's instantaneous and average values are the same for the case of constant optic power.

The thermal-noise power delivered to the load is $\overline{i_{NT}^2}R_L$, which can be written as

$$\overline{P}_{NT} = 4kT\Delta f \qquad (11\text{-}7)$$

by using Eq. (11-1) for the current.

We are now in a better position than before to explicitly define signal-to-noise ratio. The signal-to-noise ratio is the average *signal* power divided by the average power owing to all noise sources. Combining Eqs. (11-5), (11-6), and (11-7), we obtain

$$\frac{S}{N} = \frac{(\eta e P/hf)^2 R_L}{2e R_L \Delta f\,(I_D + \eta e P/hf) + 4kT\Delta f} \qquad (11\text{-}8)$$

Let us investigate some special cases. Suppose that the average signal current $(\eta e P/hf)$ is much larger than the dark current. Then I_D can be dropped from Eq. (11-8). This situation occurs if the dark current is small and the optic power is not too low. Suppose also that the shot-noise power far exceeds the thermal power. Then the term $4kT\Delta f$ can be ignored. The optic power must be relatively high for this to happen. The signal-to-noise ratio then simplifies to

$$\frac{S}{N} = \frac{\eta P}{2hf\Delta f} \qquad (11\text{-}9)$$

In this case the SNR is *shot-noise limited* (also called *quantum limited*). This is the best result obtainable. In essence, by raising the optic power we have eliminated the effects of dark current and thermal noise. The quantum-limited signal-to-noise ratio can be rewritten in terms of the signal photocurrent by combining Eqs. (11-4) and (11-9) to obtain

$$\frac{S}{N} = \frac{i_s}{2e\,\Delta f} \qquad (11\text{-}10)$$

Unfortunately, we do not always have unlimited power. When the power is low, thermal noise usually dominates over shot noise. Then, Eq. (11-8) reduces to

$$\frac{S}{N} = \frac{R_L(\eta e P/hf)^2}{4kT\Delta f} \qquad (11\text{-}11)$$

This is the *thermal-noise-limited* result. It is usually much smaller than the quantum-limited SNR. Note that the SNR for this case can be improved by increasing the load resistance. As indicated in Table 7-3, however, this may reduce the receiver's bandwidth and dynamic range. We also see, from Eq. (11-11), that the SNR increases as the square of the incident optic power. We conclude that relatively small changes in system efficiency produce significant changes in the quality of the received signal in thermal-noise-limited systems.

Example 11-1

Suppose we have a system consisting of an LED emitting 10 mW at 0.85 μm, a fiber cable with 20 dB of loss, and a PIN photodetector of responsivity 0.5 A/W. The detector's dark current is 2 nA. The load resistance is 50 Ω, the receiver's bandwidth is 10 MHz, and its temperature is 300 K (27°C). The system losses, in addition to the fiber attenuation, include a 14-dB power reduction owing to source coupling and a 10-dB loss caused by various splices and connectors. Compute the received optic power, the detected signal current and power, the shot-noise and thermal-noise powers, and the signal-to-noise ratio.

Solution:

The total system loss is $14 + 20 + 10 = 44$ dB. By using Eq. (1-1) this converts to a transmission efficiency of $10^{-4.4} = 4 \times 10^{-5}$. The optic power reaching the receiver is then

$$P_R = 4 \times 10^{-5}(10)$$
$$= 4 \times 10^{-4} \text{ mW} = 0.4 \ \mu\text{W}$$

The photocurrent can be computed from Eq. (7-1) because the responsivity is given. Thus,

$$i_s = \rho P_R = 0.5(0.4)$$
$$= 0.2 \ \mu A = 200 \ nA$$

The dark current (only 2 nA) is small compared to the signal current, so it can be ignored in this example. The electrical signal power is

$$\overline{P}_{ES} = i_s^2 R_L = (0.2 \times 10^{-6})^2(50)$$
$$= 2 \times 10^{-12} \text{ W}$$

The shot-noise power, from Eq. (11-6), is

$$\overline{P}_{NS} = 2ei_s\Delta f R_L$$
$$= 2(1.6 \times 10^{-19})(0.2 \times 10^{-6})$$
$$\cdot (10^7)(50)$$
$$= 3.2 \times 10^{-17} \text{ W}$$

The thermal-noise power, from Eq. (11-7), is

$$\overline{P}_{NT} = 4(1.38 \times 10^{-23})(300)(10^7)$$
$$= 1.66 \times 10^{-13} \text{ W}$$

In this system the thermal noise is nearly four orders of magnitude greater than the shot noise. The thermal-noise-limited result [Eq. (11-11)] applies. We can compute the SNR from that equation or directly from

$$\frac{S}{N} = \frac{\overline{P}_{ES}}{\overline{P}_{NT}} = \frac{2 \times 10^{-12}}{1.66 \times 10^{-13}} = 12$$

Expressed in decibels, the SNR becomes $10 \log_{10} 12 = 10.8$ dB. For comparison, we can compute the quantum-limited SNR. From Eq. (11-10),

$$\frac{S}{N} = \frac{0.2 \times 10^{-6}}{2(1.6 \times 10^{-19})(10^7)} = 62,500$$

or 48 dB.

Example 11-2

In Example 11-1, decrease the system losses by 6 dB. (Perhaps a better fiber is used or the source coupling is improved.) Compute the new value of the SNR.

Solution:

The steps in the solution are the same as those followed in Example 11-1, so we will give the results very briefly. The 6-dB improvement corresponds to an increase in received optic power by a factor of 4. The signal photocurrent and the shot-noise power increase by this same factor, so $i_s = 0.8 \ \mu A$ and $\bar{P}_{NS} = 12.8 \times 10^{-17}$ W. The signal power flowing through R_L increases 16 times to $\bar{P}_{ES} = 32 \times 10^{-12}$ W. The thermal-noise power remains unchanged at $\bar{P}_{NT} = 1.66 \times 10^{-13}$ W, still far more than the shot-noise power. Then $S/N = \bar{P}_{ES}/\bar{P}_{NT} = 192$, 16 times that of the lossier system. In decibels, we find that $S/N = 22.8$ dB. Comparison with the preceding problem shows that a 6-dB increase in optic power produced a 12-dB improvement in the SNR.

Example 11-2 illustrates a general result for thermal-noise-limited systems. If the optic power increase by ΔP decibels, then the signal-to-noise ratio will increase by twice that amount ($2 \ \Delta P$ decibels). This follows from Eq. (11-11), which shows that the SNR is proportional to the square of the optic power. For shot-noise-limited systems, an increase in optic power of ΔP decibels produces only a ΔP decibel increase in SNR because, as indicated by Eq. (11-9), SNR is proportional to the optic power (not its square).

Examples 11-1 and 11-2 also illustrate the type of calculations that system designers make when determining whether the available

power is sufficient for the desired application. In the next section we show the digital error rates corresponding to the computed signal-to-noise ratios.

We can easily modify the SNR equations to include photodetectors with internal gain. If the gain is M, then the signal current increase by this factor. The signal power then increases by M^2. The shot-noise current is also amplified by the factor M. Its mean-square value, therefore, increases by M^2, as does the resulting shot-noise power. The thermal-noise current is not amplified because it is not generated inside the photodetector. With these modifications, Eq. (11-8) becomes

$$\frac{S}{N} = \frac{(M\eta eP/hf)^2 R_L}{M^2 2eR_L \Delta f (I_D + \eta eP/hf) + 4kT\Delta f}$$

$$(11\text{-}12)$$

If the gain is large enough, then the shot noise can far exceed the thermal noise, even for moderately low optic power levels. In this case (and assuming the dark current can be neglected), we find

$$\frac{S}{N} = \frac{\eta P}{2hf\Delta f} \qquad (11\text{-}13)$$

which is the ideal quantum-limited result first obtained in Eq. (11-9).

Example 11-3

The PIN detector in Example 11-1 is replaced by a detector that has a responsivity 160 times greater ($M = 160$). All other conditions are unchanged, so $0.4 \ \mu W$ is incident on the detector. Compute the SNR.

Solution:

The shot-noise power increases from 3.2×10^{-17} W to

$$\bar{P}_{NS} = (160)^2(3.2 \times 10^{-17})$$

$$= 8.19 \times 10^{-13} \text{ W}$$

The thermal-noise power remains at $\bar{P}_{NT} = 1.66 \times 10^{-13}$ W. Now the shot noise is about five times greater than the thermal noise and the system is nearly shot-noise limited. The signal power increases by M^2 over the value obtained without amplification. Thus

$$\bar{P}_{ES} = (160)^2(2 \times 10^{-12})$$

$$= 5.12 \times 10^{-8} \text{ W}$$

If we include the thermal noise,

$$\frac{S}{N} = \frac{\bar{P}_{ES}}{\bar{P}_{NS} + \bar{P}_{NT}}$$

$$= \frac{5.12 \times 10^{-8}}{8.19 \times 10^{-13} + 1.66 \times 10^{-13}}$$

$$= 52,000$$

or 47.2 dB. Neglecting the thermal noise we get the quantum-limited results $S/N = 62,500$, or 48 dB. The improvement over the thermal-noise-limited system ($S/N = 10.8$ dB) is evident. In this example, the system is within 1 dB of the ideal quantum limit.

Example 11-3 illustrates two facts: quantum-limited operation produces signals superior to those of thermal-noise-limited systems and ideal quantum-limited operation can be approached by using high-gain photodetectors.

Basically, the photodetector's gain increases the receiver's sensitivity. The improved sensitivity allows detection of low-level signals, such as those found at the ends of long fiber paths. Long path lengths are particularly beneficial when repeaters are used

because the repeaters can be spaced farther apart. This minimizes the number of repeaters required and allows flexibility in their physical placement.

Avalanche Photodiode Excess Noise

Although Eq. (11-12) holds pretty well for photomultipliers, it must be modified further for APDs.[3] Rather than increasing as M^2, the shot-noise power in an APD increases as M^n, where n lies between 2 and 3.[4] The shot-noise power is increased (relative to the signal power) by the *excess noise factor* $M^n/M^2 = M^{n-2}$ in an APD.[5] Now

$$\frac{S}{N} = \frac{(M\eta eP/hf)^2 R_L}{M^n 2eR_L\Delta f(I_D + \eta eP/hf) + 4kT\Delta f}$$

$$(11\text{-}14)$$

As before, increasing the gain beyond unity improves the SNR by making the thermal noise less dominant at low power levels. However, if the gain is so large that shot noise dominates, then we obtain (assuming that the dark current can be ignored),

$$\frac{S}{N} = \frac{1}{M^{n-2}} \frac{\eta P}{2hf\Delta f} = \frac{\text{quantum-limited SNR}}{\text{excess noise factor}}$$

This equation shows how the quantum-limited signal-to-noise ratio is diminished by the excess noise factor. Clearly, if M becomes too large then the signal quality will degrade. We conclude that an optimum value of M exists (somewhere between 1 and ∞), which produces a maximum value of SNR in Eq. (11-14). Fortunately, APD gain can be adjusted by varying the reverse bias voltage, as indicated earlier by Eq. (7-17).

Silicon APDs have less excess noise than

those made of either germanium or InGaAs. For example, typical silicon APDs produce excess noise factors close to 5 at current gains of about 100, and germanium APDs with gains of only 10 have excess noise factors of 7 and InGaAs APDs with gains of 20 have excess noise factors of 10.[6] Because Ge and InGaAs APDs are a little noisy, some long-wavelength fiber receivers use a good (low dark current) PIN photodiode followed by a low-noise preamplifier instead of an APD.

Noise-Equivalent Power

The *noise-equivalent power* (NEP), an alternative measure of a receiver's sensitivity, is related to the amount of optic power resulting in a signal-to-noise ratio of unity. For a simple illustration of its definition and determination, consider a thermal-limited PIN detector. Setting $S/N = 1$ in Eq. (11-11) and solving for the power yields

$$P_{\min} = \frac{hf}{\eta e} \sqrt{\frac{4kT\Delta f}{R_L}} \qquad (11\text{-}15)$$

This is the *minimum detectable power* if we take $S/N = 1$ as the criterion for detection. The noise-equivalent power is the minimum detectable power normalized by dividing by the square root of the system bandwidth. In the thermal-limited case

$$\text{NEP} = \frac{P_{\min}}{\sqrt{\Delta f}} = \frac{hf}{\eta e} \sqrt{\frac{4kT}{R_L}} \qquad (11\text{-}16)$$

The units of NEP are $\text{W/Hz}^{1/2}$.

The NEP can be computed in a similar manner for detectors having gain and for cases in which the dark current is not negligible. To do this, set $S/N = 1$ in the general expression of Eq. (11-14), while assuming the minimum power will be so low that $I_D \gg \eta e P/hf$. The result is

$$\text{NEP} = \frac{hf}{M\eta e} \sqrt{M^n 2eI_D + \frac{4kT}{R_L}} \qquad (11\text{-}17)$$

which can be put into the form

$$\text{NEP} = \frac{\sqrt{\bar{i}_{\text{NSD}}^2 + \bar{i}_{\text{NT}}^2}}{\rho \sqrt{\Delta f}} \qquad (11\text{-}18)$$

The detector responsivity is $\rho = M\eta e/hf$, $\bar{i}_{\text{NSD}}^2$ is the amplified shot noise owing to dark current alone, and $\bar{i}_{\text{NT}}^2$ is the thermal noise. From Eq. (11-18) we see that the NEP is equal to the root-mean-square (rms) noise current divided by the responsivity and the square root of the bandwidth.

Examination of Eq. (11-17) shows that thermal noise dominates for small load resistances and low detector gains. Small resistance, although producing low output voltage, is needed to achieve the large bandwidths required for high-frequency systems. (Table 7-3 summarized these conclusions.) Thus, for high-frequency operations, we can expect the thermal noise will dominate the value of NEP and the dark current will be insignificant. A relatively poor diode (large dark current) can be used. On the other hand, for large resistance and/or large gain, the dark-current noise exceeds the thermal noise. In this case we must choose a good diode (one with low dark current) to obtain the greatest receiver sensitivity.

Example 11-4
A PIN diode has responsivity 0.5 A/W at 0.85 μm and 2 nA of dark current. Compute the NEP as a function of the load resistance at a temperature of 300 K. What is the minimum detectable power if $R_L = 100\ \Omega$ and the bandwidth is 1 MHz?

Solution:

The rms thermal-noise current is

$$\sqrt{\overline{i_{NT}^2}} = \sqrt{\frac{4kT\Delta f}{R_L}}$$

$$= \sqrt{\frac{4(1.38 \times 10^{-23})(300)\Delta f}{R_L}}$$

$$= 129 \times 10^{-10}\sqrt{\frac{\Delta f}{R_L}}$$

The rms shot-noise current is

$$\sqrt{\overline{i_{NSD}^2}} = \sqrt{2eI_D\Delta f}$$

$$= \sqrt{2(1.6 \times 10^{-19})(2 \times 10^{-9})\Delta f}$$

$$= 2.53 \times 10^{-14}\sqrt{\Delta f}$$

Combining these results with a responsivity of 0.5 in Eq. (11-18) yields

$$NEP = \sqrt{2.56 \times 10^{-27} + \frac{6.62 \times 10^{-20}}{R_L}}$$

The first term, due to dark current, is negligible compared to the second term (thermally generated), until the load resistance exceeds 10^6 Ω. The plot of NEP versus load resistance, in Fig. 11-9, clearly shows the regions where shot noise or thermal noise dominates. For $R_L = 100$ Ω, NEP $= 2.57 \times 10^{-11}$ W/Hz$^{1/2}$. With 1-MHz bandwidth, we find

$$P_{min} = NEP\sqrt{\Delta f}$$

$$= 2.57 \times 10^{-11}(10^3) = 25.7 \text{ nW}$$

In some literature the NEP of the detector alone is given. The number given is the component of NEP owing to the diode's dark current. From Eqs. (11-17) and (11-18), we see that its value would be

$$NEP = \frac{hf}{M\eta e}\sqrt{M^n 2eI_D}$$

$$= \frac{\sqrt{\overline{i_{NSD}^2}/\Delta f}}{\rho} \qquad (11\text{-}19)$$

Analog Modulation SNR

If the optic signal is sinusoidally modulated (rather than constant), then we need only modify the general SNR result in Eq. (11-14) a bit. In the sinusoidal case, the optic power incident on the photodetector can be written as

$$P_i = P(1 + m\cos\omega t) \qquad (11\text{-}20)$$

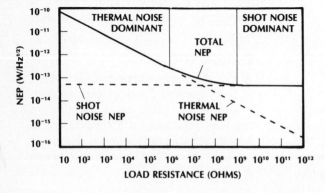

Figure 11-9 Noise equivalent power for a PIN diode having 2 nA of dark current and a 0.5-A/W responsivity at 300 K.

where m is the optic modulation factor and ω is the modulation frequency. The average incident power is P.

The power P_i produces photocurrent

$$i_s = \frac{\eta e P}{hf}(1 + m\cos\omega t) \qquad (11\text{-}21)$$

before internal amplification. The first term is the average current ($\bar{i}_s = \eta e P/hf$) and the second is the desired information signal. After amplification, the signal current increases to

$$i = \frac{M\eta e P}{hf} m\cos\omega t \qquad (11\text{-}22)$$

This current flows through the load resistor, delivering average electrical signal power $\bar{P}_{ES} = 0.5R_L i_p^2$, where i_p is the peak value of the signal current. Thus,

$$\bar{P}_{ES} = 0.5R_L\left(\frac{mM\eta e P}{hf}\right)^2 \qquad (11\text{-}23)$$

The noise power in the denominator of Eq. (11-14) is unchanged, since P correctly represents the average optic power in both the constant and the sinusoidal cases. Then

$$\frac{S}{N} = \frac{(m^2/2)(M\eta e P/hf)^2 R_L}{M^n 2e R_L \Delta f(I_D + \eta e P/hf) + 4kT\Delta f} \qquad (11\text{-}24)$$

This result differs only from the constant-power case by the factor $m^2/2$. The thermal-noise-limited and shot-noise-limited signal-to-noise ratios are, respectively,

$$\frac{S}{N} = \frac{m^2}{2}\frac{R_L(\eta e P/hf)^2}{4kT\Delta f}M^2$$

and

$$\frac{S}{N} = \frac{m^2/2}{M^{n-2}}\frac{\eta P}{2hf\Delta f}$$

To obtain the SNR from these equations when there is no gain, set $M = 1$. To obtain the SNR when there is no excess noise, set $n = 2$.

For 100% modulation, $m = 1$ and the SNR is a maximum. Since the SNR varies as the square of the modulation factor, making m as large as possible enhances reception of analog signals. However, as m increases, the source operates over a wider range of its power-current characteristic. Nonlinearities in this characteristic (causing signal distortion) may limit the useful range, which in turn limits the maximum allowable value of the modulation factor.

Broadband analog transmission (such as that needed for video) may require a signal-to-noise ratio of 40–60 dB.

Heterodyne SNR

In Section 10-5 we showed that a heterodyne detector produces average current [Eq. (10-36)]

$$i_{dc} = \frac{\eta e}{hf}(P_L + P_S)$$

and IF current [Eq. (10-37)]

$$i_{IF} = \frac{2\eta e}{hf}\sqrt{P_S P_L}\cos[\omega_{IF}t - \theta(t)]$$

where P_L and P_S are the optic powers in the local oscillator and signal beams. The local oscillator power P_L will always be constant. We will determine the SNR for the case in which the signal power P_S is also constant.

The average signal power delivered to the load resistor is $\bar{P}_{ES} = 0.5R_L(i_{IF,P})^2$, where

$i_{\text{IF,P}}$ is the peak value of the IF current. The signal power is then

$$\bar{P}_{\text{ES}} = 2R_L P_S P_L \left(\frac{\eta e}{hf}\right)^2 \quad (11\text{-}25)$$

The most striking result is the signal amplification provided by the local oscillator power. The larger the value of P_L, the greater the electrical signal power.

As before, the shot-noise power is $\bar{P}_{\text{NS}} = 2eR_L I \Delta f$, where I is the total average current and Δf is the bandwidth of the IF receiver. (The IF bandwidth typically needs to be twice the bandwidth of a baseband receiver because the modulation creates both upper and lower sidebands surrounding the IF frequency. That is, the spectrum of the IF signal has double the bandwidth of the baseband information. This is the same effect as that noted in Section 10-3 in a discussion of AM/IM modulation.) For the heterodyne receiver, I is the dark current plus the dc photocurrent. Thus,

$$\bar{P}_{\text{NS}} = 2eR_L \Delta f \left[I_D + \frac{\eta e P_L}{hf}\left(1 + \frac{P_S}{P_L}\right) \right] \quad (11\text{-}26)$$

The thermal noise remains at $\bar{P}_{\text{NT}} = 4kT\Delta f$, unaffected by the method of detection. Now,

mitted beam. If P_L is large, then Eq. (11-27) simplifies to

$$\frac{S}{N} = \frac{\eta P_S}{hf\Delta f} \quad (11\text{-}28)$$

This is the quantum-limited signal-to-noise ratio. Assuming an IF bandwidth equal to double the baseband bandwidth yields the same SNR as that in Eq. (11-9), which applies for direct detection. We conclude that heterodyne receivers can be very sensitive, approaching the ideal SNR (even at low signal levels). Heterodyne receivers do not introduce excess noise (like an APD), so they can closely approach the quantum-limited SNR.

11-3 ERROR RATES

In analog systems the transmitted waveshape is very important. Even small amounts of noise degrade the waveshape somewhat. The system designer must determine how much distortion is acceptable and set the signal-to-noise ratio high enough to ensure the required reproduction fidelity. On the other hand, the exact shape of a digital pulse need not be preserved. Digital receivers need determine only the presence (or absence) of pulses during

$$\frac{S}{N} = \frac{\bar{P}_{\text{ES}}}{\bar{P}_{\text{NS}} + \bar{P}_{\text{NT}}}$$

$$= \frac{2(\eta e/hf)^2 R_L P_S P_L}{2eR_L \Delta f [I_D + (\eta e P_L/hf)(1 + P_S/P_L)] + 4kT\Delta f} \quad (11\text{-}27)$$

Note that shot noise becomes dominant if P_L is large. This is normally the situation because the local oscillator is located at the receiver. P_L does not suffer propagation, distribution, and coupling losses as does the trans-

specified intervals. In this section, we will describe how noise introduces errors into this determination.

The bit-error rate, fractional number of detection errors, is a measure of digital system quality. If one error is made out of every 100 decisions, then the BER = 0.01. It follows that the chance of an error during any single bit interval just equals the BER. Thus, if BER = 0.01, then the *probability of error* P_e is 0.01. The two terms, bit-error rate and probability of error, are interchangeable.

Still another interpretation of the error rate can be made. Even though we cannot divide up an individual bit or an individual error, P_e can still be viewed as the number of errors per bit. For a data rate of R bps, the number of errors per second equals RP_e (the product of the bits per second and the errors per bit). If $P_e = 0.01$ and $R = 1$ Mbps, then there are 10,000 errors each second. This may well be intolerable. Changing P_e to 10^{-9} reduces the error rate to 0.001 errors per second, or 1 error every 1000 s (16.7 min). An error rate of 10^{-9} is sufficient for many applications.

Thermal-Noise-Limited
Error Rate

Figure 11-10 illustrates how thermal noise produces detection errors. The ideal (noiseless) received current is shown in Fig. 11-10(a). It is followed by the actual current [Fig. 11-10(b)] showing the effects of added noise and filtering. This current is sampled near the end of each bit interval (where the pulses are most likely to reach their maximum amplitudes) with the result appearing in Fig. 11-10(c). At this point, the amplitude of each sample is compared with a reference (or threshold) value. The threshold current is set somewhere between zero and the ideal current expected when a 1 arrives (i_s in the figure). If the sample exceeds the threshold, then it is further processed as a 1. If the sample is lower than the threshold, it is treated as a 0. Figure 11-10(d) shows the resulting data pattern.

Let us look a little closer at the reasons for the distorted pulse train in Fig. 11-10(b). When a 0 arrives, an ideal receiver produces no current. In reality, thermal noise and dark-current shot noise produce random currents. The noise current may be low on the average, yet still large enough during some bit intervals to exceed the threshold. In this case, an error occurs. When a 1 is received, the ideal current is constant [see level i_s in Fig. 11-10(a)]. In the actual receiver, noise may add out of phase with the desired current, causing the total to occasionally dip below the threshold level. Again, an error results. The figure illustrates errors in detecting both 1s and 0s.

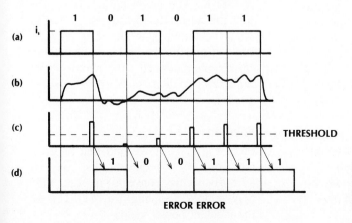

Figure 11-10 Detection errors.
(a) Ideal receiver current. (b) Actual current. (c) Sampled current.
(d) Resulting data pattern.

Clearly, the threshold cannot be too close to zero, for this would increase the number of errors when detecting 0s. It cannot be too close to the ideal level i_s either, because errors in detecting 1s would occur more frequently. As might be expected, the threshold level producing the fewest errors is half the ideal current received when a 1 arrives. (We set the threshold current equal to $0.5i_s$.) This is the optimum threshold if 1s and 0s are equally likely, a situation that exists for most messages. If the received power changes (e.g., owing to aging of the light source), then the threshold must be reset.

It is worth summarizing that the decision (0 or 1) in a thermal-noise-limited system is made by comparing the signal current amplitude with a predetermined threshold level.

Using a threshold of $0.5i_s$ results in the probability of error[7]

$$P_e = \frac{1}{2} - \frac{1}{2}\,\text{erf}\left(0.354\sqrt{\frac{S}{N}}\right) \qquad (11\text{-}29)$$

where erf is the *error function* (a well-known and tabulated quantity).[8] Table 11-1 lists values of the error function for arguments from 0 to 3. For arguments greater than 3, an equation giving approximate values is shown in the table. A plot of the error probability appears in Fig. 11-11. The signal-to-noise ratio used in determining P_e is the thermal-noise-limited value given by Eq. (11-11) for direct detection using a PIN detector. Equation (11-29) does not apply to shot-noise-limited systems. We will determine P_e for such systems later in this section.

In Section 11-2 we found signal-to-noise ratios of 10.8 and 22.8 dB for two thermal-noise-limited systems. The corresponding error rates, from direct calculation of Eq. (11-29) or from Fig. 11-11, are 4.2×10^{-2} and 2.6×10^{-12}, respectively.

TABLE 11-1.   The Error Function

x	erf x	x	erf x
0.00	0.00000	1.05	0.86244
0.05	0.05637	1.10	0.88021
0.10	0.11246	1.15	0.89612
0.15	0.16800	1.20	0.91031
0.20	0.22270	1.25	0.92290
0.25	0.27633	1.30	0.93401
0.30	0.32863	1.35	0.94376
0.35	0.37938	1.40	0.95229
0.40	0.42839	1.45	0.95970
0.45	0.47548	1.50	0.96611
0.50	0.52050	1.55	0.97162
0.55	0.56332	1.60	0.97635
0.60	0.60386	1.65	0.98038
0.65	0.64203	1.70	0.98379
0.70	0.67780	1.75	0.98667
0.75	0.71116	1.80	0.98909
0.80	0.74210	1.85	0.99111
0.85	0.77067	1.90	0.99279
0.90	0.79691	1.95	0.99418
0.95	0.82089	2.00	0.99532
1.00	0.84270	2.50	0.99959
		3.00	0.99998

For $x \geq 3$, $1 - \text{erf } x \cong \dfrac{e^{-x^2}}{x\sqrt{\pi}}$

Example 11-5

A 1-Mbps NRZ link uses a 100-Ω load at 300 K. The wavelength is 0.82 μm, and the desired error rate is 10^{-4}. The PIN detector quantum efficiency is unity. Compute the optic power incident on the photodetector, the photocurrent, and the incident number of photons per bit.

Solution:

From Fig. 11-11, we find that an error rate of 10^{-4} requires that $S/N = 17.5$ dB. This converts to the ratio $S/N = 56.2$. Solving Eq. (11-11) for the incident power yields

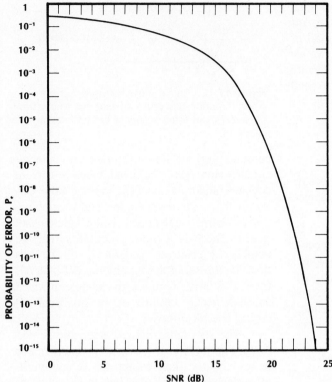

Figure 11-11 Probability of error for thermal-noise-limited systems.

$$P = \frac{hf}{\eta e}\sqrt{\frac{4kT\Delta f}{R_L}}\sqrt{\frac{S}{N}}$$

The optic frequency is $f = c/\lambda = 3.66 \times 10^{14}$ Hz, and the pulse duration is $\tau = 10^{-6}$ s. We will set the receiver's bandwidth conservatively at $\Delta f = 1/\tau = 10^6$ Hz. Then,

$$P = \frac{6.63 \times 10^{-34}(3.66 \times 10^{14})}{1.6 \times 10^{-19}}$$

$$\times \sqrt{\frac{4(1.38 \times 10^{-23})(300)(10^6)}{100}}\sqrt{(56.2)}$$

$$= 1.46 \times 10^{-7} \text{ W} = 146 \text{ nW}$$

This is the required incident optic power for a 10^{-4} BER. The detected current is $i = \eta eP/hf = 96.4$ nA. As developed in Section 7-2, the number of photons incident per second is P/hf. Thus the number of photons incident in the bit interval τ is $n_p = (P/hf)\tau$. In this example,

$$n_p = \frac{146 \times 10^{-9}(10^{-6})}{6.63 \times 10^{-34}(3.66 \times 10^{14})}$$

$$= 6 \times 10^5 \text{ photons/bit}$$

A large number of photons are required to achieve an error rate of 10^{-4} in a thermal-noise-limited system.

For signal-to-noise ratios better than 15 dB or so, Fig. 11-11 shows a large improvement in error rate with only a small increase in signal power. Consequently, we can improve the transmission quality significantly by lowering the system losses even a small amount.

Shot-Noise-Limited Error Rate

For a shot-noise-limited system, the postdetection processor counts the number of electrons produced during each bit interval and compares this number with a threshold. If the count exceeds the threshold, then the receiver assumes a 1 was transmitted. If the count is less than the threshold, then a 0 is assumed.

Errors occur when receiving 0s because the dark current occasionally contains enough electrons during a single bit interval to exceed the threshold. The dark currents found in detector manufacturers' literature are the average values. The instantaneous dark current varies randomly about this number. It can reach relatively large values for short periods of time.

When receiving 1s, errors occur if the number of electrons produced by the combination of the signal-plus-noise currents does not exceed the threshold. This happens if the noise current is large enough and if it adds out of phase with the signal current during most of a single bit interval. In this way, the total current occasionally falls below that needed to reach the threshold count. This type of error can even occur when there is no dark current. The signal-generated shot noise alone may decrease the total electron count. We can illustrate this last statement by referring to Fig. 11-12, showing the received current when the incident power is constant. (We can imagine this is the current when a series of 1s is received in a NRZ system.) On the average, a

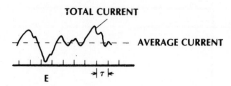

Figure 11-12 Signal plus shot-noise current when the optic power is constant due to a series of NRZ 1s. An error occurs in the bit interval labeled E.

constant current flows through the detector circuit. However, the instantaneous current deviates randomly about the average value owing to the random generation and recombination of charge carriers (this is the signal's shot noise). There is a finite probability that the number of electrons generated will be less than the threshold during any one bit interval. Interval E in the figure is an example in which an error occurs because of the small current during one bit interval.

The shot-noise-limited error rate is plotted in Fig. 11-13 for the case in which 1s and 0s are equally likely to appear. A few words of explanation are needed before you will understand this figure. The error probability depends on the average number of photoelectrons n_s generated by the signal during the bit interval τ when a 1 is received. In terms of the incident optic power, n_s is given by

$$n_s = \frac{\eta P \tau}{hf} = \frac{i_s \tau}{e} \qquad (11\text{-}30)$$

where η is the quantum efficiency, hf is the photon energy, and i_s is the signal current. The error rate also depends on the average number of electrons n_n produced by the dark current I_D. This is

$$n_n = \frac{I_D \tau}{e} \qquad (11\text{-}31)$$

The curves shown on the figure apply when

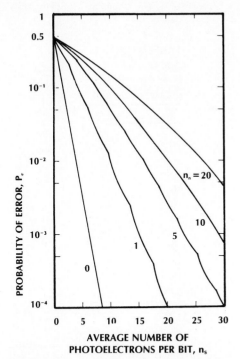

PROBABILITY OF ERROR, P_e

AVERAGE NUMBER OF
PHOTOELECTRONS PER BIT, n_s

Figure 11-13 Probability of error for a PCM
shot-noise-limited system. When a 1 is received,
n_s is the average number of photoelectrons
generated. The average number of electrons
generated by the dark current is n_n. (From
William K. Pratt, *Laser Communication Systems*,
John Wiley, New York, 1969, pp. 169–99.
Reproduced with permission.)

the detection threshold has been optimized.
When 1s and 0s are equally likely, the
threshold that minimizes P_e is

$$k_T = \frac{n_s}{ln(1 + n_s/n_n)} \qquad (11\text{-}32)$$

The actual threshold count k_D is an integer
that is set equal to k_T if k_T is itself an integer.
Otherwise k_D is set equal to the closest integer
that is greater than k_T. The breaks in the error
curves are caused by changes in k_D needed to
minimize P_e.

A few examples will illustrate threshold
optimization. Suppose that there is almost no
dark current ($n_n \cong 0$). Then, Eq. (11-32)

yields a threshold just barely above zero. We
set the actual threshold count to one electron
($k_D = 1$). Since there is virtually no dark current, the detected count will always be zero
and there will be no errors when the system
transmits 0s. Arrival of a 1 is assumed by the
detection of one or more electrons. The only
reason that errors occur at all in this situation
is that the incoming photon stream may not
generate any photoelectrons during a particular bit interval. When the incident power is
constant, we can determine the average number of photons per bit. However, the actual
number arriving during any one bit interval
varies randomly about this value. When the
average is low (say, just a few photons/bit), it
is entirely possible that no photons will actually strike the detector during some bit intervals. Additionally, the detector quantum efficiency is only an average value. For example,
if $\eta = 0.80$, then photons generate electrons
only 80% of the time. From another point of
view, a photon has an 80% probability of generating a free electron. It is possible that several incoming photons, on occasion, will not
free any electrons at all during the bit interval.
Of course, the larger the (average) number of
incident photons, the less the likelihood of
producing no electrons when transmitting 1s
and the lower the error rate. This discussion
explains the general nature of the zero-dark-
current curve in Fig. 11-13.

As noted previously, the random excitation of charge carriers is the source of shot-
noise current. Explanations of errors based directly on this probabilistic behavior or on the
resulting random currents are equivalent.

Here is another example. Suppose that
the dark current produces an average of $n_n =$
20 electrons per bit and there are an average
of $n_s = 10$ photoelectrons for each bit. The
threshold, from Eq. (11-32), is $k_T = 24.7$, so
we set the threshold count at $k_D = 25$. Note
that we must set the threshold above the aver-

age noise count. Errors can occur when the system transmits 1s or 0s. As explained earlier in this section, there is a finite probability that many more than the average number of dark current electrons (20 in this example) will be generated. If 25 or more electrons are produced when a 0 is being received, then an error results. When a 1 is received, on the average there will be $n_s + n_n = 30$ electrons per bit. This count will drop below 25 on occasion, causing errors. Raising the threshold closer to 30 makes it more likely that incident 1s will not produce enough electrons to equal, or exceed, the threshold. More 1 errors result and 0s are less likely to reach the new threshold. In general, increasing the threshold increases the 1 errors and reduces the 0 errors. Decreasing the threshold will decrease the 1 errors at the expense of the 0 errors. In any case, the optimum threshold provides for minimum errors.

A disadvantage of shot-noise-limited PCM now becomes apparent. The optic power and the noise must be known in order to set the threshold optimally. Since the error rate increases rapidly as the threshold moves away from the optimum, precise determination of the optimum threshold is critical.

The zero-dark-current curve in Fig. 11-13 can be approximated by

$$P_e = e^{-n_s}$$

when $n_s > 2$. Table 11-2 lists a few values obtained from this equation. This result can be used as a reference point from which to measure the quality of actual systems.

Example 11-6

A 1-Mbps NRZ pulse train is transmitted along a shot-noise-limited system at $\lambda = 0.82$ μm. The receiver has negligible dark current. How many photons per bit must be incident on the photodetector

TABLE 11-2
Approximate
PCM Error Rates
(No Dark Current)

P_e	n_s
10^{-1}	2.3
10^{-2}	4.6
10^{-3}	6.9
10^{-4}	9.2
10^{-5}	11.5
10^{-6}	13.8
10^{-7}	16.1
10^{-8}	18.4
10^{-9}	20.7
10^{-10}	23.0
10^{-11}	25.3
10^{-12}	27.6

if the desired error rate is 10^{-4}? Assume unity quantum efficiency. Compute the incident optic power. Compare the results with the thermal-noise-limited system investigated earlier in Example 11-5.

Solution:

From Table 11-2 we see that $n_s = 9.2 \cong 10$ photoelectrons per bit for $P_e = 10^{-4}$. Since the quantum efficiency is 1, 10 photons are required per bit. This is much less than the 600,000 needed in the comparable thermal-noise-limited system. The bit period is $\tau = 10^{-6}$ s. The optic power, from Eq. (11-30), is $P = hfn_s/\eta\tau = hcn_s/\eta\lambda\tau$, so that

$$P = \frac{6.63 \times 10^{-34}(3 \times 10^8)(10)}{0.82 \times 10^{-6}(10^{-6})}$$

$$= 2.4 \times 10^{-12} \text{ W} = 2.4 \text{ pW}$$

The corresponding power in the thermal system was 146 nW. The shot-noise-limited system is more sensitive by

$10 \log_{10}(146 \times 10^{-9}/2.4 \times 10^{-12}) =$
48 dB.

We should carefully note the steep slopes of the error curves in Fig. 11-13. Small changes in the power available to the receiver produce large changes in the probability of error. This result places a premium on obtaining efficient power transfer.

According to Fig. 11-13, even when the average number of photoelectrons is zero (maybe the fiber broke), the probability of error is not 1. The actual error rate is half. Why is this? The answer is simply that an observer at the receiving terminal can guess whether a 1 or 0 was transmitted and be correct 50% of the time, when (as usual) 1s and 0s are equally likely. Of course, no information is conveyed in this process.

11-4 MODAL NOISE, AMPLIFIER NOISE, LASER NOISE, AND JITTER

There are several important types of noise we have not yet described: modal noise, amplifier noise, and laser noise. We will see how they arise and how their effects can be minimized. In addition to noise, jitter in a digital link can increase the error rate. We will discuss jitter and its measurement.

Modal Noise

Modal noise[9] is a random variation in optic power occurring in multimode fibers. If the light source is highly coherent (say, a good laser diode), then the fiber modes interfere with one another and form a speckle pattern consisting of bright and dark spots. Figure 11-14 illustrates speckle. The spots are bright

FIBER END

Figure 11-14 Speckle pattern.

where the net mode interference is additive (in-phase modal fields) and dark where the net interference is subtractive (out-of-phase modal fields).

Because of its wide linewidth, a noncoherent source (such as an LED) will not form a speckle pattern. To explain this, we might consider the noncoherent spectrum to consist of a series of closely spaced wavelengths. Each of these wavelengths produces a slightly different speckle pattern. At locations where some patterns are dark, others are bright. The total pattern is the sum of the individual speckle intensities, because different wavelengths do not interfere with each other. Thus, the resulting pattern is uniform (or slowly varying) over the transverse cross section of the fiber.

Single-mode fibers do not contain speckle. More than one mode is needed, since speckle represents interference between two or more fields.

Speckle itself is not objectionable. However, consider what happens when the source shifts its output wavelength (e.g., owing to a temperature variation or modulation, as mentioned in Section 10-5). The interference pattern, whose shape critically depends on the wavelength, will change. The bright and dark spots will move to new positions. Similar shifting occurs if there is physical movement of the fiber, because this alters the paths (and the relative paths) of the many modes. Continuous random temperature variations or physical movements (vibration) create a continuous random shifting of the speckles. Even this effect is not detrimental in a perfect optic system. A photodetector placed at the fiber end can easily collect all the power in the fiber, irrespective of the particular pattern of the illuminating light. However, an imperfect fiber link has components whose losses are more selective. That is, their loss does depend on the pattern of illuminating light. Most connectors have this property. For example, a connector with a slight core misalignment (as shown in Fig. 11-15) has a mode-selective loss. Different speckle patterns couple differing amounts of light across the lossy junction simply because one pattern will concentrate more of the light within the overlapping portions of the two cores than another pattern. As the pattern changes, some of the speckles will move out of (or into) the overlap region. A randomly changing pattern will show up as a randomly time-varying connector loss. The resulting random variation in received power is modal noise.

Minimizing modal noise is simple in theory: use single-mode fibers, or low-loss connectors, or low-coherence sources. Single-mode fibers and low-loss connectors both add to system costs, and low-coherence sources increase material-dispersive pulse spreading (Section 3-2), lowering the fiber's capacity to handle data.

We conclude that modal noise must be considered in multimode system design. We realize, however, that the amount of modal noise is not easy to predict theoretically because we do not usually know the extent of connector misalignments and source wavelength shifts. Modal noise must be experimentally evaluated.

Figure 11-15 Misaligned fiber cores.

Electronic Amplifier Noise

An electronic amplifier normally follows the photodetector to boost the receiver signal to a useful level. In an ideal situation, both signal and noise powers would be multiplied by the amplifier's *power gain G*. Then, the signal-to-noise ratio at the amplifier output would equal that at the input. Unfortunately, real amplifiers not only multiply the input noise but also produce noise of their own. This reduces the SNR.

Let us represent the added noise by P_{out} watts. If we wish to include this power in our SNR calculations, then we can do so by assuming an ideal (noiseless) amplifier and adding a thermal noise source at its input that produces noise power $P_{in} = P_{out}/G$ watts. We now define the *amplifier noise temperature T_A* in such a way as to produce this power. That is, using Eq. (11-7),

$$P_{in} = \frac{P_{out}}{G} = 4kT_A\Delta f \qquad (11\text{-}33)$$

as illustrated in Fig. 11-16. Combining this with the load resistor's thermal noise yields the equivalent input thermal noise power

$$\bar{P}_N = 4k(T + T_A)\Delta f = 4kT_e\Delta f \qquad (11\text{-}34)$$

where T is the temperature of the resistor and

$$T_e = T + T_A \qquad (11\text{-}35)$$

is the *equivalent system noise temperature*. The actual thermal noise appears to come from a resistor operating at temperature T_e.

We can now compute signal-to-noise ratios (using all the equations previously derived) by simply replacing the actual system temperature T with the effective system noise temperature T_e. In other words, we assume that the amplifier is perfect and account for

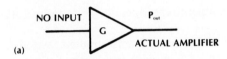

(a)

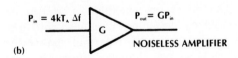

(b)

$P_n = 4k(T+T_A)\Delta f$ $GP_N = P_0$

(c)   NOISELESS AMPLIFIER

Figure 11-16 Amplifier noise. (a) Noise output owing to the amplifier alone. (b) Equivalent noise circuit that defines T_A. (c) Noise circuit including noise from the load resistor.

the noise it adds by increasing the apparent temperature of the load resistor.

Example 11-7

In Example 11-1 we found $S/N = 12$ (10.8 dB) for a thermal-limited constant-power system. The bandwidth was 10 MHz. The detected signal power was 2×10^{-12} W, and the thermal-noise power was 1.66×10^{-13} W at 300 K. Suppose that the photodetector is followed by an amplifier with a power gain of 10 dB and a noise temperature 454 K. Compute the SNR.

Solution:

The noise, referred to the amplifier's input terminals, is given by Eq. (11-34) with $T_e = T + T_A = 754$ K. $\bar{P}_N = 4kT_e\Delta f = 4(1.38) \times 10^{-23}(754)(10^7) = 4.2 \times 10^{-13}$ W. Then, $S/N = 2 \times 10^{-12}/4.2 \times 10^{-13} = 4.8$ (6.8 dB). Since the amplifier noise was included in T_e, this is the output signal-to-noise ratio. The 10-dB gain increases both the actual signal power and the apparent in-

put thermal noise power by a factor of 10. The amplifier noise reduces the SNR from 10.8 dB to 6.8 dB.

Sometimes an amplifier's *noise figure F* is given, rather than its noise temperature T_A. F is a property defined by

$$F = 1 + \frac{T_A}{T_s} \qquad (11\text{-}36)$$

where T_s is some reference temperature. In many applications a reference of 290 K has been agreed upon. Since T_A does not depend on the choice of reference, the value of the noise figure will. An interpretation of noise figure can be easily developed. The equivalent system noise temperature is

$$T_e = T + T_A = T + (F - 1)T_s \qquad (11\text{-}37)$$

where we eliminated T_A by using Eq. (11.36). Suppose that we choose the reference temperature equal to the system temperature, $(T_s = T)$. Then, $T_e = FT$ and the total output noise power becomes

$$\bar{P}_O = G\bar{P}_N = G4kT_e\Delta f$$
$$= G4kFT\Delta f \qquad (11\text{-}38)$$

Solving for the noise figure yields

$$F = \frac{\bar{P}_O}{G4kT\Delta f} = \frac{\bar{P}_O}{G\bar{P}_{NT}} \qquad (11\text{-}39)$$

where we identified the load resistor's thermal-noise power as $\bar{P}_{NT}$ from Eq. (11-7). This result permits us to define the noise figure as the thermal-noise power at the output divided by the product of the power gain and the input thermal noise. To use this definition, F must be measured (or calculated) at the temperature of the load resistor. For an ideal amplifier, $\bar{P}_O = G\bar{P}_{NT}$ and the noise figure is unity. In reality, all amplifiers add noise, making $\bar{P}_O > G\bar{P}_{NT}$ and $F > 1$.

Example 11-8

Compute the noise figure of the amplifier used in the preceding example.

Solution:

Assume that we want to know F at the actual system temperature of 300 K. Then, in Eq. (11-36), $F = 1 + 454/300 = 2.51$. The noise figure is often expressed in decibels. In this example, $F_{dB} = 10 \log_{10} 2.51 = 4$ dB.

Comparison of the last two worked examples shows that an amplifier having a 4-dB noise figure reduces the signal-to-noise ratio by this same amount. In fact, for thermal-noise-limited systems, the SNR (in decibels) will always be lowered by the amplifier's noise figure (expressed in decibels) if the noise figure is calculated at the actual system temperature. The equivalent statement in equation form is

$$\left(\frac{S}{N}\right)_{out} = \frac{1}{F}\left(\frac{S}{N}\right)_{in} \qquad (11\text{-}40)$$

This result can be obtained by noting that the amplifier output signal power is the gain G times the input signal power and the output noise power is given in terms of the input thermal noise power by Eq. (11-39). Rearranging this last equation yields

$$F = \frac{(S/N)_{in}}{(S/N)_{out}} \qquad (11\text{-}41)$$

This leads to the notion that, for thermal-noise-limited systems, the noise figure is the ratio of the signal-to-noise ratio at the amplifier's input to the signal-to-noise ratio at its output.

For nonthermal-noise-limited systems, the effects of amplifier noise on the SNR must be individually computed.

The effects of amplifier noise can be minimized by designing amplifiers with low noise figures. It is also true that shot-noise-limited systems are minimally affected by amplifier noise if the shot noise remains much larger than the thermal noise (when the amplifier's noise temperature is included in the calculation of the thermal noise power).

Shot-noise-limited fiber systems normally require APDs or heterodyne receivers, both of which may be looked upon as noiseless signal amplifiers (disregarding avalanche excess noise). The signal-to-noise ratio of a system containing a chain of amplifiers is determined primarily by the noise properties of the first amplifier. Thus, we again reach the conclusion that a receiver having gain (APD or heterodyne) will suffer less signal deterioration owing to the first electronic amplifier than a receiver without gain. We also conclude that the first amplifier (called the preamplifier) in a PIN diode receiver is a critical device in determining the system SNR.

Optical Amplifier Noise

It is now apparent that optical amplifiers, such as the erbium-doped amplifier discussed in Section 6-7, will be included in some networks where the transmission path is long or where the light power is distributed to many receiving terminals. In such applications cascaded amplifier chains may be required.

For the optical amplifier we will start with the definition of noise figure as given by Eq. (11-41). The amplifier chain illustrated in Fig. 11-17 consists of N amplifiers, each hav-

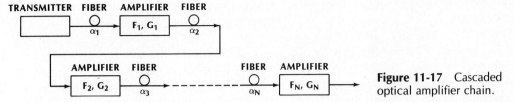

Figure 11-17   Cascaded optical amplifier chain.

ing a power gain of G_k and a noise figure of F_k. Thus, the noise figure for the kth amplifier is

$$F = \frac{(S/N)_{k,\text{in}}}{(S/N)_{k,\text{out}}} \qquad (11\text{-}42)$$

There is a loss of signal power between successive amplifiers given by the fractional transmission loss factor α_k. It results in a lowering of the signal-to-noise ratio between repeaters. The combined noise figure for the amplifier chain is[10]

$$F = \frac{F_1}{\alpha_1} + \frac{F_2}{\alpha_1 G_1 \alpha_2} + \frac{F_3}{\alpha_1 G_1 \alpha_2 G_2 \alpha_3}$$

$$+ \cdots + \frac{F_N}{\prod_{i=1}^{N-1} (\alpha_i G_i)\alpha_N} \qquad (11\text{-}43)$$

Example 11-9

Ideally, the gain of each amplifier will just equal the loss owing to fiber attenuation. Assume this to be the case (i.e., assume that $\alpha_i G_i = 1$). Also assume that each amplifier has a 3-dB noise figure, there are 10 amplifiers, the signal-to-noise ratio is 10^8 at the transmitter, and there is a 30-dB loss between amplifiers along the fiber. Compute the output SNR.

Solution:

For the assumed conditions, Eq. (11-43) reduces to $F = NF_iG_i$. A 30-dB loss

corresponds to a transmission loss of $\alpha_i = 10^{-3}$ and thus $G_i = 10^3$. Also, $N = 10$ and $F_i = 2$. Thus, $F = 10(2)10^3 = 20{,}000$. Finally, $(S/N)_{\text{out}} = (S/N)_{\text{in}}/F = 10^8/20{,}000 = 5000$ or 37 dB.

This problem could also have been solved by using decibels. In this case the input SNR is 80 dB and the system noise figure is $10 \log 20{,}000 = 43$ dB. The output SNR is then $80 - 43 = 37$ dB.

Laser Noise

Laser noise[11] is an undesirable random fluctuation in the output of a laser diode that occurs even when the driving current is constant. It is a characteristic associated with poor lasers but is present to some extent in all of them. Laser noise reaches a peak when modulating a diode at its resonant frequency (typically a few gigahertz). For this reason, laser noise is more significant for high-frequency links than for lower-frequency ones. Well-constructed laser diodes contribute only small amounts of noise to systems operating well below the diode's resonance.

For some lasers the relative noise reaches a peak at the oscillation threshold. When the driving current increases beyond threshold, the laser noise remains fixed while the output power rises rapidly. Thus, the relative noise falls, with a resultant improvement in the signal quality. The noise contribution is minimized by operating the diode well above threshold (say at currents more than 40% above threshold).

The *relative intensity noise* (RIN) describes the amount of noise emitted by the laser. We will introduce it in the following way: A laser emits an average power P. A photodetector having responsivity ρ, connected to a receiver whose bandwidth is Δf, measures the laser output. The average detected current is ρP, but the average value of the square of the noise current (i.e., the noise fluctuations) is[12]

$$\overline{i_{NL}^2} = \text{RIN} \, (\rho P)^2 \Delta f \qquad (11\text{-}44)$$

The average noise power generated by the laser must be

$$\sqrt{\overline{P_{NL}^2}} = \sqrt{\overline{i_{NL}^2}}/\rho \qquad (11\text{-}45)$$

Combining these last two equations yields the RIN,

$$\text{RIN} = \frac{\overline{P_{NL}^2}}{P^2 \Delta f} \qquad (11\text{-}46)$$

Notice that the RIN is a measure of the average noise power, normalized to the bandwidth. Its units are $(\text{Hz})^{-1}$. Often the RIN is expressed in dB/Hz, which is just,

$$(\text{RIN})_{\text{dB/Hz}} = 10 \, \log \left(\frac{\overline{P_{NL}^2}}{P^2 \Delta f} \right) \qquad (11\text{-}47)$$

To include the effects of laser noise in the signal-to-noise ratio equations previously developed, just add the laser noise current from Eq. (11-44) to the other noise currents (thermal and shot noise currents). As an example, Eq. (11-8) becomes

where the detector responsivity ρ has been used to replace the term $\eta e/hf$.

Example 11-10

A good laser diode has a RIN quoted as -140 dB/Hz. Compute the laser noise power detected by a receiver having a 100-MHz bandwidth if the average incident power is 10 μW. Also compute the average noise current. The detector responsivity is 0.5 μA/μW.

Solution:

Since $(\text{RIN})_{\text{dB/Hz}} = 10 \, \log \, \text{RIN} = -140$, we find $\text{RIN} = 10^{-14} \, (\text{Hz})^{-1}$. The laser noise power squared, from Eq. (11-46) is

$$\overline{P_{NL}^2} = \text{RIN} \, P^2 \Delta f$$
$$= 10^{-14}(10^{-5})^2 10^8 = 10^{-16}$$

so that the average laser noise power is 0.01 μW. The responsivity is 0.5, so that the average noise current is 0.005 μA or 5 nA.

Because analog systems (particularly ones transmitting video) require such high signal-to-noise ratios, the RIN should be included in analog S/N calculations. The RIN is less significant in digital transmission links.

Jitter and the Eye Diagram

As we have seen, digital light pulses are distorted in a number of ways. These include the distortion caused by noise and the distortion caused by pulse spreading. Pulse spreading is basically caused by the limited bandwidth of

$$\frac{S}{N} = \frac{(\rho P)^2 R_L}{2e \, R_L \, \Delta f \, (I_D + \rho P) + 4kT \, \Delta f + \text{RIN} \, (\rho P)^2 \, \Delta f R_L}$$

$$(11\text{-}48)$$

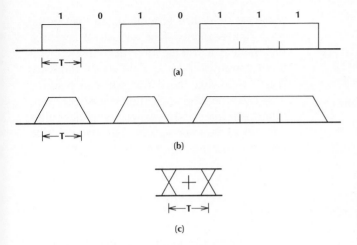

Figure 11-18 Idealized eye pattern for a NRZ signal. (a) Binary signal, (b) signal after bandwidth limiting, and (c) eye pattern formed by superposition of successive pulses.

the system, including the transmitter, the fiber, and the receiver. In addition the system introduces timing errors, a phenomenon referred to as *jitter*. All these distortions reduce the ability of the receiver to correctly identify the presence of binary ones and zeroes.

A convenient way to measure these distortions is the *eye diagram*. The eye diagram is an oscilloscope display that superposes digital waveforms. To appreciate the significance of the eye diagram, let us start with Fig. 11-18. Figure 11-18(a) illustrates a representative binary NRZ signal. The bit period is T seconds, correponding to a data rate of $1/T$ bps. Figure 11-18(b) shows this waveform after distortion (some pulse spreading and increase in the pulse rise and fall times) owing to transmission through the fiber and reception by the receiver. Figure 11-18(c) shows the superposition of successive pulses as would be observed on an oscilloscope whose time base was triggered by the receiver once every T seconds. A clear opening (the eye) exists on the pattern. This figure illustrates an ideal situation, since noise and jitter have not yet been introduced.

The cross at the center of Fig. 11-18(c) indicates the optimum time to sample the re-

ceived signal and the optimum level (the *threshold level*) at which to distinguish between zeroes and ones. The optimum sampling time is when the eye height is at its largest value.

Figure 11-19 illustrates the affects of jitter on the eye diagram. Jitter makes it more difficult for the receiver to sample the pulse at the optimum time. That is, there is less tolerance on the sampling time as the amount of jitter ($\Delta\tau$) increases. The sketch in Fig.11-20 shows the affects of noise together with jitter on the eye diagram. Noise tends to diminish the height of the eye, and jitter tends to narrow its width. Both conditions reduce the ability of the receiver to correctly identify individual pulses. Under the conditions indicated on the figure, slight errors in setting the threshold level and/or the sampling time will cause errors. In general, the error rate will increase as the eye closes.

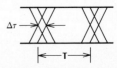

Figure 11-19 Jitter closes the eye, making it harder to sample the pulse at the optimum time.

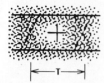

Figure 11-20   Noise and jitter close the eye.

11-5 OTHER SOURCES OF NOISE

Current noise and background noise may also contribute to system degradation.

Current Noise

Semiconductor devices produce a slowly fluctuating current called *current noise* or $1/f$ *noise*.[13] It is limited to low frequencies, varying as $1/f$ below 1 Hz. Current noise can be minimized by passing the amplified signals through filters that severely attenuate frequencies below about 10 Hz.

Background Noise

In an atmospheric optic communication system, light can enter the photodetector from sources other than the desired one. Energy from sunlight, street lamps, or car lights can be detected, increasing the receiver's dc current and, consequently, increasing the shot noise. This *background noise* is easily eliminated from fiber links because they are normally completely enclosed.

11-6 RECEIVER CIRCUIT DESIGN

In this section the receiver's front-end circuitry[14,15] (the photodetector and the first am-

plifier) is described in more detail than in the preceding accounts. One successful approach incorporates a voltage amplifier using either a bipolar transistor or a field-effect transistor (FET) following the detector's load resistor. Two other networks, the high-impedance amplifier and the transimpedance amplifier, offer advantages in some applications. We describe and compare the different circuits next.

Bipolar Transistor and FET Amplifiers

The simplest front-end arrangement consists of a reverse-biased photodiode terminated by a load resistor and followed by a conventional amplifier, as drawn in Fig. 11-21. Figure 11-21(a) illustrates a bipolar transistor amplifier and Fig. 11-21(b) a field-effect transistor amplifier. To simplify the sketch, the transistor's biasing networks are not shown.

The criteria for choosing the optimum load resistance were developed in Section 7-4 and summarized in Table 7-3. Briefly, we need a large R_L to obtain a high voltage and to reduce thermal noise, but require a small R_L for large bandwidth and wide dynamic range. The 3-dB bandwidth previously given by Eq. (7-16) must now include the capacitance and resistance associated with the amplifier. Then

$$f_{3\text{-dB}} = \frac{1}{2\pi R_T C_T} \qquad (11\text{-}49)$$

where R_T represents the combination of the load resistor R_L in parallel with the transistor's input resistance and biasing circuit resistance. C_T is the parallel combination of the diode's capacitance C_d and the transistor's input capacitance (typically, a few picofarads). The biasing resistances can be quite large, so that, being in parallel with R_L, they would not contribute much to the total resistance R_T. The

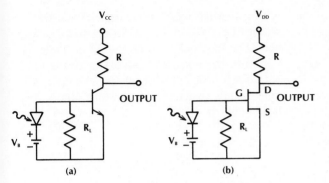

Figure 11-21 Simple receiver front end circuits. (a) Bipolar transistor amplifier. (b) FET amplifier.

FET's input resistance is quite high (typically 1–100 MΩ), so it can be neglected in determining R_T. For the FET front end, then, we can assume that $R_T = R_L$. The bipolar transistor's input resistance is moderate (a few kilohms). Its value should be included in calculating the total resistance, since it could significantly affect R_T and the resulting bandwidth, as calculated from Eq. (11-49).

The noise figure, introduced in the preceding section, accounts for the noise introduced by the transistor (and its biasing network). Contributors to the FET's noise figure include thermal noise generated by the drain-source channel conductance and by the biasing resistors. FET shot noise arises from the small leakage current between the gate G and the source S. Thermal noise in the bipolar amplifier comes from the transistor's base resistance and biasing resistance. Shot noise accompanies the base and collector currents in the bipolar transistor.

The noise power generated by a field-effect transistor increases as the cube of the system bandwidth, and the noise owing to a bipolar transistor increases only as the square of the bandwidth (if the base resistance is small, as is often the case). Thus, at high frequencies (corresponding to high data rates in a digital system) the bipolar transistor introduces less noise than the FET and is superior in this respect. At low frequencies, FETs pro-

duce less noise and are preferred. In addition, the gain of field-effect transistors drops considerably at high frequencies, greatly diminishing their usefulness for high-capacity systems. Generally, FETs provide the best results below about 25–50 MHz, and bipolar transistors perform better above this range.

High-Impedance Amplifier

If we make the load resistance R_L large for either the FET or bipolar front end, then the amplifier's input impedance (R_T in parallel with C_T) will be high. Hence the name *high-impedance* amplifier. As we know, the large resistance minimizes the thermal noise. Indeed, this is the reason for considering the high-impedance front end. However, large R_L reduces the receiver's bandwidth. For frequencies above the 3-dB value given by Eq. (11-49), the capacitance C_T mainly determines the input impedance. C_T tends to integrate the input waveform. The high-impedance amplifier operates above its 3-dB bandwidth, but amplifies higher frequencies much less than lower ones. Equalizers, placed in the receiver somewhere after the preamplifier, reverse this effect by attenuating lower frequencies more than higher ones. A differentiating network accomplishes this result. Equalizers ideally restore the signal waveform. The requirement for equalizers is the price paid for the improved noise characteristics of high-impedance front ends.

We should emphasize that the high-impedance front end does not have a wide dynamic range because of the large load resistance. This problem is solved by the transimpedance amplifier described next. When the application requires a sensitive (low-noise) receiver and only a narrow dynamic range, the high-impedance front end is appropriate.

Transimpedance Amplifier

The current-to-voltage converter, described in Section 7-4 and sketched in Fig. 7-11, is a *transimpedance amplifier*. The circuit diagram is repeated in Fig. 11-22 for convenience. The transimpedance amplifier operates over a wide dynamic range, linearly processing optic signals whose power levels differ by many decades because almost the entire bias voltage appears across the diode even when the incident power is large enough to produce high photocurrents. This behavior is illustrated in Fig. 7-12. It is unlike the situation illustrated by Eq. (7-13) and Fig. 7-8, where the maximum optic power for linear detection is limited to $V_B/\rho R_L$ and the maximum photocurrent is V_B/R_L.

The feedback resistor determines the thermal noise. R_F replaces R_L in all the pre-ceding thermal-noise calculations when computing the SNR of the transimpedance front end. The feedback resistor should be large to minimize the noise and to maximize the output voltage (iR_F). The feedback network contains shunt capacitance C_F, which limits the bandwidth to

$$f_{3\text{-dB}} = \frac{1}{2\pi R_F C_F} \qquad (11\text{-}50)$$

This is similar to Eq. (11-49) for circuits without feedback. However, the feedback capacitance can be much lower than the input capacitance C_T of circuits not containing feedback. Thus, R_F can be larger than R_L for a given bandwidth, increasing the receiver's sensitivity and decreasing the noise. Likewise, if R_F equals R_L, then the transimpedance front end will have a larger bandwidth than the nonfeedback amplifier.

The transimpedance front end has noise properties approaching that of the low-noise, high-impedance amplifier. It has a wider dynamic range and a larger bandwidth than the high-impedance network. If we also note that equalizing circuits are not generally required, then we see why the transimpedance amplifier is so popular in fiber optic receivers.

Table 11-3 summarizes the differences between the major front-end circuits. Recall that when an APD is used, the SNR is less dependent on the thermal noise than when a PIN detector is used. In other words, the APD determines the SNR, not the preamplifier. If this is the case, then the system designer should consider using the simplest front-end circuit.

Integrated Detector Preamplifier

Integrated detector amplifiers (mentioned briefly in Section 7-3) typically contain a pn photodetector, a transimpedance or a high-

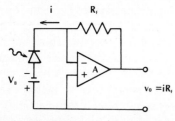

Figure 11-22 Transimpedance amplifier. A, operational amplifier; i, detector photocurrent plus dark current.

TABLE 11-3.   Receiver Front-End Comparisons

	Bipolar	FET	High Impedance	Transimpedance
Circuit complexity	Simple	Simple	Complex	Moderate
Equalizers required	No	No	Yes	No
Relative Noise	Moderate	Moderate	Very low	Low
Bandwidth	High	Low	Moderate	High
Dynamic range	Moderate	Moderate	Narrow	Wide

impedance amplifier, and one or more following amplifiers for additional gain and impedance matching. The IDP offers convenience. It provides the system designer with a completely designed, tested, and constructed front end. More important, IDPs reduce the possibility of electrical pickup of extraneous signals by the leads that run between the detector and the amplifier. The leads are short and easy to shield in the IDP. Noise introduced at the diode-amplifier connection could bury the tiny signal that was so carefully nurtured and protected as it traveled from the source, through the fiber input coupler, past several connectors and distribution networks, over several kilometers of fiber, and into a sensitive detector and low-noise receiver. The signal-to-noise ratio is much less affected by pickup later in the receiver where the signal is so much stronger.

A representative monolithic silicon IDP has a responsivity of 30 mV/μW and a rise time of 35 ns.[16] Note that the responsivity of an IDP relates the output voltage to the input optic power. The rise time of 35 ns corresponds to a 3-dB bandwidth, computed from Eq. (7-2), of 10 MHz.

Hybrid Receiver Modules

Hybrid receiver modules provide the same benefits as monolithic IDPs. These modules contain a photodiode and amplifier circuits, produced separately and subsequently connected within a small space. A hybrid PIN-FET receiver may have a PIN detector connected to either a transimpedance or high-impedance FET (or MESFET) amplifier using thick film circuit components and a ceramic substrate. The resulting devices are placed inside convenient structures, such as the dual in-line package popular for printed circuit boards. Optic connection to the photodiode may be through a connector or a fiber pigtail attached to the package.

11-7 SUMMARY

Because of shot noise, random fluctuations always accompany received signals. Thermal noise associated with the detector's load resistance and noise (shot and thermal) produced by the amplifiers add to the disturbance. To solve the noise problem, simply provide a strong signal to the receiver. This can often be done for short point-to-point links. When the signal must traverse a long path or be divided among several terminals, the losses incurred reduce the signal level to the point where noise must be considered.

The common measures of signal quality are signal-to-noise ratio and probability of error. Using specific numerical examples in this chapter, we illustrated that signals had acceptable quality even when the optic power was fairly small. We will generalize these results a bit to give a feel for the required power levels and allowed link losses. Some thermal-noise-limited systems can operate satisfactorily when about a microwatt (−30 dBm) of optic power arrives at the receiver. Similar shot-noise-limited systems can operate at nanowatt (−60 dBm) levels. If the source emits 10 mW (10 dBm), then a total system loss of 40 dB (10 + 30) can be tolerated in the thermally limited system and 70 dB (10 + 60) in the shot-noise-limited system. We should emphasize, however, that the ideal shot-noise-limited results are difficult to obtain in practice. APDs allow systems to approach within about 10 dB of this ideal. Even so, a receiver coming within 10 dB of the ideal will still be 20 dB or so more sensitive than the thermal-noise-limited receiver.

When noise is a problem, close attention must be paid to the receiver design. The design starts with the choice of photodetector. A PIN diode is chosen if thermal-noise-limited operation yields sufficient signal quality, and an APD is chosen if the signal must be improved. The cost and circuit complexity associated with an APD make it desirable to lower system losses as much as possible (perhaps by improving source coupling and connector efficiency) before employing an APD. The variety of preamplifier circuits available (e.g., the terminated detector plus amplifier and the high-impedance and transimpedance front ends) gives the designer flexibility for optimizing the receiver on the basis of cost, complexity, or performance.

Important receiver qualities are sensitivity, bandwidth, and dynamic range. Sensitive receivers, capable of detecting very weak signals, increase the permissible repeater spacings and path lengths and provide high-quality reception. They also allow the power to be divided among several terminals in distributed networks. Large-bandwidth receivers increase the system capacity. They permit reception of more information. Receivers with wide dynamic range operate satisfactorily even when the received optic power varies considerably.

This is a requirement in distributed communication networks, in which signals from nearby transmitters are much stronger than signals arriving from more distant ones.

The signal processing networks that follow the receiver front end include circuits that perform as integrators, differentiators, equalizers, comparators, peak detectors, and power amplifiers. These are conventional electronic devices and, not being peculiar to fiber optic links, are not covered in detail in this book.

PROBLEMS

11-1. The temperature of a receiver is 35°C, its bandwidth is 6 MHz, and the load resistance is 50 Ω.
   (a) Compute the rms (root-mean-square) thermal noise current.
   (b) Compute the rms thermal noise voltage that appears across the resistor and the thermal noise power generated.
   (c) Repeat this problem if the resistance is changed to 50,000 Ω.

Problems 11-2 through 11-6 are for binary PCM systems:

11-2. The receiver's bandwidth is 6 MHz and the average photocurrent is 1 nA. $T = 300$ K and the dark current is zero. The load resistance is 50 Ω.
   (a) Compute the rms shot-noise current.
   (b) Compute the rms signal voltage and the rms shot-noise voltage across the resistor.
   (c) Compute the signal-to-noise ratio neglecting thermal noise.
   (d) Repeat without neglecting thermal noise.

11-3. A photodetector's dark current is 5 nA and its responsivity is 0.5 A/W. At what value of optic power is the signal-generated shot noise equal to the dark-current-generated shot noise?

11-4. A PIN photodetector has responsivity 0.5 A/W and 2 nA of dark current. The load resistance is 2000 Ω, and the system bandwidth is 50 MHz. The temperature is 40°C.
   (a) At what value of received optic power is the thermal noise equal to the shot noise?
   (b) What is the signal-to-noise ratio at this power level?
   (c) What is the value of the shot-noise power at this value of the received optic power?

11-5. Repeat Problem 11-4 if the photodetector is an APD having a gain of 100 and an unamplified dark current of 2 nA.

11-6. The optic power reaching the receiver is 1 μW. The detector's responsivity is 0.5 A/W, and its dark current is 4 nA. The temperature is 27°C, the receiver's bandwidth is 500 MHz, and the load resistance is 50 Ω.
   (a) Compute the signal-to-noise ratio.
   (b) Compute the thermal-noise-limited SNR.
   (c) Compute the shot-noise-limited SNR.
   (d) What value of photodetector gain is required to make the actual SNR just 5 dB less than the quantum limit? Assume that the photodetector has negligible excess noise.

11-7. A PIN photodiode has responsivity 0.3 A/W and a dark current of 10 nA. The temperature is 300 K, and the bandwidth is 100 MHz.

   (a) Produce a plot of NEP versus load resistance like that in Fig. 11-9.
   (b) Compute the minimum detectable power when the load resistance is 50 Ω.
   (c) Repeat for a load resistance of 5000 Ω.
   (d) Repeat for a load resistance of 50,000 Ω.

11-8. An analog system has a bandwidth of 10 MHz, a photodetector whose responsivity is 0.5 A/W (it has no gain) and whose dark current is 1 nA, and a SNR of 50 dB when the optic modulation factor is 0.4. The receiver's temperature is 27°C, and the load resistance is 50 Ω. How much optic power (average) must reach the receiver?

11-9. Consider a heterodyne receiver for a digital system. The photodetector has 2 nA of dark current and 0.5 A/W of responsivity. The temperature is 27°C, the load resistance is 100 Ω, the IF bandwidth is 500 MHz, and the received optic signal power is constant at 5 nW when a binary 1 is received.
   (a) How much local-oscillator power is required to make the SNR just 1 dB less than the quantum limit?
   (b) If this were not a heterodyne system, then the receiver's bandwidth could be as small as 250 MHz. For this case determine the signal power required to achieve a SNR equal to that found in part (a).

11-10. Extend the error function table (Table 11-1) from $x = 3$ to $x = 6$ in steps of 0.5.

11-11. A thermal-noise-limited PCM system must operate with an error probability better than 10^{-9}. The load resistance is 50 Ω, and the temperature is 300 K.

The data rate is 500 Mbps (NRZ), the wavelength is 1.3 μm, and the photodetector's quantum efficiency is 0.9.

(a) What is the required minimum SNR?

(b) How much optic power must reach the receiver?

(c) Compute the number of incident photons per bit (that is, the number of photons when a binary 1 is received) at this power level.

11-12. A shot-noise-limited 500-Mbps PCM system operates with an error probability better than 10^{-9}. The wavelength is 1.3 μm, and the photodetector's quantum efficiency is 0.9. The dark current is negligible.

(a) How much optic power must reach the receiver?

(b) Compute the number of incident photons per bit at this power level.

(c) Compare the results of this problem with those of Problem 11-11.

(d) You should have found that the quantum-limited system required much less power than the thermal-limited one. How can the system be designed to approach the quantum-limited result?

11-13. An amplifier has a power gain of 8 and a noise figure of 3 dB. This amplifier follows a photodetector having a responsivity of 0.5 A/W. The load resistance is 100 Ω, and the received optic power is 0.5 μW. The temperature is 300 K, and the receiver's bandwidth is 1 MHz.

(a) Compute the signal power flowing through the load resistor.

(b) Compute the signal power exiting the amplifier.

(c) Compute the thermal noise power generated by the load resistor.

(d) Compute the thermal noise power exiting the amplifier.

(e) Compute the amplifier's noise temperature.

(f) Compute the equivalent input noise power.

(g) Compute the signal-to-noise ratio at the input to the amplifier.

(h) Compute the SNR at the output of the amplifier.

11-14. An optical receiver consists of a PIN photodiode and a FET amplifier as in Fig. 11-21. The load resistance is 2000 Ω. The diode's capacitance is 3 pF, and the transistor's capacitance is 6 pF.

(a) Compute the 3-dB bandwidth of this receiver.

(b) Compute the approximate rise time of this receiver.

11-15. Repeat Problem 11-14 if the FET is replaced by a bipolar transistor having input capacitance 6 pF and input resistance 2000 Ω. Compare the results of this problem with those of Problem 11-14.

11-16. Draw a bipolar transistor receiver circuit, like that in Fig. 11-21(a), and include a biasing network. Explain (with words and/or drawings) how your circuit works.

11-17. Repeat Problem 11-16 for the FET receiver in Fig. 11-21(b).

11-18. Consider the transimpedance amplifier optical receiver as drawn in Fig. 11-22. The feedback resistance is 10 kΩ. The feedback capacitance is 0.2 pF. The diode's capacitance is 5 pF, and its responsivity is 0.5 A/W. The incident optical power is 0.5 μW.

(a) Compute the receiver's output voltage.

(b) Compute the receiver's 3-dB bandwidth.

(c) Compute the rms thermal noise current generated in the feedback resistor assuming a temperature of 300 K.

(d) Compute the signal current.

(e) Assuming no dark current and an amplifier noise figure of 4 dB, compute the output SNR.

11-19. Prove Eq. (11-40).

11-20. Using Eq. (11-40), show that the signal-to-noise ratio (expressed in decibels) is decreased by the noise figure (expressed in decibels) of the amplifier in a thermal-noise limited system.

11-21. Repeat Example 11-9 if the loss between repeaters is increased to 32 dB and all other parameters of the system are unchanged. The amplifier gain remains the same as in Example 11-9.

11-22. A laser diode has a RIN = −135 dB/Hz, a receiver bandwidth of 1 GHz, and a received average power of 20 μW.

(a) Compute the laser noise power at the receiver.

(b) Compute the average laser noise current if the detector's responsivity is 0.3 μA/μW.

11-23. Rewrite Eq. (11-24) for analog modulated systems to include the effects of laser noise (i.e., to include the RIN).

11-24. An avalanche photodiode has an excess noise factor of 5 when its gain is 100. Compute the value of the factor n in the avalanche photodiode excess noise factor expression.

11-25. The receiver in Problem 11-6 uses an APD whose value of n in the excess noise factor is 2.35.

(a) Compute and plot the SNR as a function of the APD gain for gains from 0 to 1000.

(b) What is the value of the optimum gain?

REFERENCES

1. Amnon Yariv. *Introduction to Optical Electronics,* 2nd ed. New York: Holt, Rinehart and Winston, 1976, pp. 282–87.

2. Ibid, pp. 280-282.

3. Tien Pei Lee and Tingye Li. "Photodetectors." In *Optical Fiber Telecommunications,* edited by Stewart E. Miller and Alan G. Chynoweth. New York: Academic Press, 1979, pp. 608–21.

4. R. J. McIntyre. "Multiplication Noise in Uniform Avalanche Diodes," *IEEE Trans. Electron Devices* 13, no. 1 (Jan. 1966): 164–68.

5. Gerd Keiser. *Optical Fiber Communications,* 2nd ed. New York: McGraw-Hill, 1991, pp. 253–55.

6. Peter K. Cheo. *Fiber Optics.* Englewood Cliffs, N.J.: Prentice Hall, 1985, p. 253.

7. Yariv. *Introduction to Optical Electronics.* pp. 292–95.

8. Milton Abramovitz and Irene A. Stegun (eds.). *Handbook of Mathematical Functions.* Washington, D.C.: United States Department of Commerce, 1964, pp. 295–329.

9. M. Chown, A. W. Davis, R. E. Epworth, and J. G. Farrington. "System Design." In *Optical Fibre Communication Systems,* edited by C. P. Sandbank. New York: John Wiley, 1980, pp. 249–65.

10. Keiser. *Optical Fiber Communications.* pp. 423–424.

11. P. A. Kirkby. "Semiconductor Laser Sources for Optical Communication," *Inst. Electron. Radio Eng.* 51, no. 7/8 (July/ Aug. 1981): 362–76.

12. Keiser. *Optical Fiber Communications.* p. 361.

13. William K. Pratt. *Laser Communication Systems.* New York: John Wiley, 1969, pp. 152–53.

14. R. G. Smith and S. D. Personick. "Receiver Design for Optical Fiber Communication Systems." In *Semiconductor Devices for Optical Communication,* edited by H. Kressel. New York: Springer-Verlag, 1980, pp. 89–160.

15. Stewart D. Personick. *Optical Fiber Transmission Systems.* New York: Plenum, 1981, pp. 57–98.

16. *Motorola Optoelectronics Device Data.* Phoenix, Ariz.: Motorola, 1989.

Chapter 12

System Design

We started with a discussion of fiber optics from a broad, systems point of view. The block diagram in Fig. 1-3 identified the major components and their positions within the system. The accompanying description provided the purpose of each component in general terms. Following chapters dealt with details of the theory, design, operation, and characteristics of the individual components. Now we put the components back together and see how their individual behaviors affect total system performance. Thus we have come full circle in our treatment of fiber systems.

12-1 ANALOG SYSTEM DESIGN

In a fiber system the combined component losses must be low enough to ensure that sufficient power reaches the receiver. For an analog system, sufficient power means the amount of power that produces a specified signal-to-noise ratio. As an additional requirement, the combined components must have sufficient bandwidth to pass the highest modulation frequencies contained in the optic signal. Up to this point we have discussed individual device losses and bandwidths. Now we investigate how they work together. We do this by working through a sample problem, which illustrates computation of the power and bandwidth budgets.

System Specification

We will design a relatively simple point-to-point video system. This link could deliver signals from a TV studio to a remote transmitter. The link could serve just as well as part of a closed-circuit security monitor in a building or on a campus. Path lengths on the order of half a kilometer or so are required.

For simplicity, we use the signals generated by the TV camera to intensity modulate the light source. The signals cover a bandwidth of nearly 6 MHz. To obtain a clear picture, a signal-to-noise ratio of 50 dB ($S/N = 10^5$) is specified.

The simplest systems use multimode fibers (either SI or GRIN) together with LEDs emitting in the range 0.8–0.9 μm and silicon PIN photodetectors. If these components do not have enough bandwidth or provide enough power, then we would have to consider using laser diodes, avalanche photodiodes, single-mode fibers, and the longer-wavelength, second-window region.

The SNR given by Eq. (11-24) applies. We will assume 100% modulation. To evaluate this equation, we need a value for R_L, the detector's load resistance. We will assume that the PIN diode has a capacitance of 5 pF and a responsivity of 0.5 A/W at 0.85 μm. The maximum value of R_L, assuming a 6-MHz cutoff frequency, is determined from Eq. (7-16).

$$R_L = (2\pi C_d f_{3\text{-dB}})^{-1}$$
$$= [2\pi (5 \times 10^{-12})(6 \times 10^6)]^{-1} = 5305 \ \Omega$$

We choose $R_L = 5100 \ \Omega$ for the succeeding calculations. It would be unwise to set $R_L = 5305 \ \Omega$ because then the photodetector would use up the entire bandwidth budget. We also must allow some signal bandwidth degradation due to the source and fiber. We will evaluate the combined bandwidth effects shortly.

The bandwidth of the receiver, using the 5100-Ω load resistor in Eq. (7-16), is 6.24 MHz. This is the figure we must use in the following calculations.

Power Budget

Since we are using a PIN diode, we expect a thermal-noise-limited system. We will proceed on this assumption and check it after the received power has been calculated. With this assumption, Eq. (11-24) reduces to

$$\frac{S}{N} = \frac{0.5R_L(\rho P)^2}{4kT_e \Delta f} \qquad (12\text{-}1)$$

where $\rho = \eta e/hf$ is the PIN diode's responsivity and we have replaced the actual temperature with the equivalent-system noise temperature to account for amplifier noise. Let us assume an ambient temperature of 300 K and a preamplifier noise figure of 2 (3 dB). The equivalent temperature, $T_e = FT = 600$ K, is used in Eq. (12-1). Solving for the average optic power required at the receiver, we obtain

$$P = \sqrt{\frac{4(1.38 \times 10^{-23})(600)(6.24 \times 10^6)(10^5)}{0.5(0.5)^2(5100)}}$$
$$= 5.7 \ \mu\text{W}$$

For simplicity, round this off to 6 μW. At this power level, the PIN diode generates average current $I = \rho P = 3 \ \mu$A. This is much larger than typical PIN diode dark currents, which are a few nanoamperes. Therefore, the dark current can be ignored in this system. Evaluation of Eqs. (11-1) and (11-2) shows that the thermal-noise power is nearly seven times larger than the shot-noise power, confirming our initial suspicion that we would have a thermal-noise-limited system.

We should check that the expected current of 3 μA does not drive the detector into nonlinear operation. As indicated in Fig. 7-8, the maximum current before saturation equals the ratio of the bias voltage to the load resistance. Using a 5-V bias, we obtain a maximum allowed current of $5/5100 = 980 \ \mu A$, far greater than our operating value of just $3 \ \mu A$. Saturation is no problem in this system.

We will proceed with the design assuming components are available having the following characteristics:

1. *Light source.* A surface-emitting LED operating with average power (denoted by P_{dc} in Fig. 6-7) of 1 mW at 0.85 μm. Its rise time is 12 ns, its spectral width is 35 nm, and its emitting surface has a diameter less than 50 μm.
2. *Multimode SI fiber.* NA = 0.24, optic bandwidth $f_{3\text{-dB}} \times L = 33$ MHz $\times$ km, loss of 5 dB/km, and a 50-μm core diameter.
3. *Multimode GRIN fiber.* Axial NA = 0.24, optic bandwidth $f_{3\text{-dB}} \times L = 500$ MHz $\times$ km (if the source is a laser diode), loss of 5 dB/km, and a 50-μm core diameter.

The power budget is easily handled by writing the power levels in dBm. The source emits 0 dBm (1 mW) and the receiver requires -22.2 dBm (6 μW). Thus, the combined component losses must be no more than 22.2 dB. The SI fiber-source coupling loss is $\eta = NA^2 = 0.0576$ (12.4 dB). According to Eq. (8-12), the coupling loss into the GRIN fiber is 3 dB worse. The loss in this case is 15.4 dB. There is a 0.2-dB reflection loss at the entrance to the fiber and at the exit. Assuming the need for only two connectors (one at the transmitter and one at the receiver), each having 1 dB of loss, adds 2 dB of loss. This leaves $22.2 - 12.4 - 0.4 - 2 = 7.4$ dB for the permissible SI fiber loss and 4.4 dB for the GRIN fiber loss. At 5 dB/km, attenuation restricts the SI fiber to lengths less

than $7.4/5 = 1.48$ km. A 1-km SI link will leave a 2.4-dB margin. For the GRIN fiber, the maximum link length is $4.4/5 = 0.88$ km = 880 m.

Bandwidth Budget

Next we should examine the bandwidth restrictions when combining the source, fiber, and detector. At this point you may realize that some of the response data have been given in terms of rise time (e.g., the 12-ns LED rise time) and some in terms of bandwidth (e.g., the 6-MHz system bandwidth). We will convert all the data to equivalent rise times. Approximations will have to be made because neither the rise time nor the bandwidth completely characterizes a component. (The impulse response does completely characterize a component. However, the impulse response is usually not known, it is difficult to obtain experimentally and, once obtained, is difficult to use.) Rise time and bandwidth give sufficient information for initial system design. One or the other is usually given on data sheets, and they are easy to use.

The rise times t_S, t_{LS}, t_F, and t_{PD} of the system, light source, fiber, and photodetector, respectively, are related by

$$t_S^2 = t_{LS}^2 + t_F^2 + t_{PD}^2 \qquad (12\text{-}2)$$

We will assume that Eq. (7-2) correctly converts bandwidth to rise time for the system and for the fiber. We will apply it with some care in the following, however. The system rise time is then $t_s = 0.35/6 \times 10^6 = 58.3$ ns. From Eq. (7-15), $t_{PD} = 2.19 R_L C_D = 2.19(5100)(5 \times 10^{-12}) = 55.8$ ns. This is much larger than the typical transit-time limit of about 1 ns, so the detector is circuit limited. In this example the receiver uses up most of the rise-time budget. This could be changed by lowering R_L (which would subsequently lower the receiver's sensitivity, requiring more power). The LED's rise time is 12 ns. According to Eq. (12-2), the fiber's rise time (in nanoseconds) must be no more than that given by solving

$$t_F^2 = t_S^2 - t_{LS}^2 - t_{PD}^2 = 58.3^2 - 12^2 - 55.8^2$$
$$= 141$$

Thus we require that $t_F \le 11.9$ ns.

We will digress a moment before proceeding with the calculation of the allowed fiber length based on its bandwidth. The 3-dB bandwidth used in Eq. (7-2) is the frequency at which the electrical power in a circuit drops to half of its maximum value. However, when calculating the bandwidth of fibers, we have been using the reduction in optic power. For example, the 3-dB bandwidth of a fiber corresponds to a reduction in modulated optic power of half. Consider the measurement of the 3-dB optic bandwidth. The modulation frequency starts out low and a photodetector measures the amplitude of the received sinusoidal current. This amplitude is recorded and used as the reference. The modulation frequency is now increased while monitoring the detected current. Because the current is proportional to the optic power, we know that the 3-dB frequency has been reached when the current drops to half of the reference value. However, because the power in the detector's load resistor is proportional to the square of the current, a reduction by half in the current corresponds to a drop to a fourth of the electrical power. This is a 6-dB electri-

cal power loss. We have just found out that the 3-dB optic bandwidth corresponds to the 6-dB electrical bandwidth. In general, the electrical loss is double the optic loss, both measured in decibels. Table 12-1 emphasizes this point.

Now, to use Eq. (7-2) for the fiber we must find the fiber's 3-dB *electrical* bandwidth. As indicated in Table 12-1, this corresponds to the frequency at which the optic power is reduced by only 1.5 dB. From the loss characteristic given by Eq. (3-17), we find that $f_{1.5\text{-dB}} = 0.71 f_{3\text{-dB}}$. Thus the 3-dB electrical and optic bandwidths are related by

$$f_{3\text{-dB}}(\text{electrical}) = 0.71 f_{3\text{-dB}}(\text{optic}) \quad (12\text{-}3)$$

For the SI fiber in our TV system, the electrical bandwidth-length product is $0.71(33) = 23.4$ MHz × km. The corresponding rise time is $0.35/23.4 \times 10^6 = 15$ ns/km. Recalling that the rise-time budget for the fiber was 11.9 ns, the allowed SI fiber length is $11.9/15 = 0.793$ km = 793 m. In this case, although the power budget permits nearly 1.5 km, the rise-time budget (or bandwidth budget) restricts the link to just under 800 m. This system is *bandwidth limited* rather than *power limited*. If the required length is less than 800 m, then the design suffices. To lengthen the path, a variety of adjustments are possible. Decreasing the load resistance is the simplest one. This will reduce the receiver's rise time and allot more of the rise-time budget to the fiber.

Now let us substitute the GRIN fiber for the SI fiber while retaining the LED source. The fiber's 3-dB electrical bandwidth-length product is $0.71 (500 \times 10^6) = 355$ MHz ×

TABLE 12-1. Electrical Versus Optic Loss

Optic loss (dB)	0	0.5	1.0	1.5	2.0	2.5	3.0
Electrical loss (dB)	0	1	2	3	4	5	6

km and its rise time is $t_{\text{mod}}/L = 0.35/355 \times 10^6 = 1$ ns/km. This result accounts only for the modal distortion. Material dispersion must also be computed. The dispersive pulse spread by using the LED is $M\Delta\lambda = 90(35) = 3150$ ps/km $\cong 3.2$ ns/km, where we obtained $M = 90$ ps/(nm $\times$ km) at 0.85 μm from Fig. 3-8. Converting the pulse spread to bandwidth [by using Eq. (3-19)] and then to the equivalent rise time [from Eq. (7-2)] shows that the pulse spread and rise time are nearly equal. The rise time owing to material dispersion is thus $t_{\text{dis}}/L = 3.2$ ns/km. The total fiber rise time is found from

$$t_F^2 = t_{\text{dis}}^2 + t_{\text{mod}}^2 \qquad (12\text{-}4)$$

which in this case yields $t_F/L = 3.4$ ns/km. The allowed length is then $11.9/3.4 = 3.5$ km, much longer than the 880 m permitted by the power budget. The GRIN system in this example is power limited to lengths less than 880 m. More efficient source-fiber coupling would dramatically improve this system.

The example treated in this section outlined a general analog design procedure and illustrated some of the possible compromises and component choices. Keep in mind that the results obtained were approximate because characteristics such as rise time, bandwidth, and pulse spread do not completely characterize components. These parameters are simply good (and convenient) measures of component response. However, we have been conservative in the rise-time and power-budget calculations, so our solutions should produce workable results.

The system specifications in this section were rather moderate. Relatively simple devices sufficed. Longer links carrying more information (e.g., several multiplexed video channels transmitted over a few kilometers) would require more sophisticated components. Shorter links carrying less information (e.g.,

a short, single-channel telephone connection) would permit use of cheaper, lower-quality components.

12-2 DIGITAL SYSTEM DESIGN

In Section 12-1 we worked through a specific example to illustrate the general procedures involved in analog system design. Similarly, in this section we start with a set of specifications for a digital link and show how the requirements might be met. The methods used can be applied quite generally to fiber optic digital design.

System Specification

The analog system developed in Section 12-1 had rather moderate requirements placed on it. We will now consider a digital system that must meet fairly difficult standards. The link must transmit a 400-Mbps NRZ pulse train over a 100-km path with an error rate of 10^{-9} or better. This must be accomplished without repeaters. From the outset it is apparent that a fiber having a very high rate-length product and very low attentuation will be needed. Also, a very fast light emitter and photodetector will be necessary to accommodate the 400-Mbps data rate. We might also foresee that the signal levels reaching the receiver will be quite low, in which case a very sensitive receiver will be needed. The analyses to follow will indicate to what extent these predictions are correct. Computations of the rise-time (that is, bandwidth) and power budgets constitute the bulk of the analyses.

Rise-Time Budget

A sample pulse appears in Fig. 12-1. For the NRZ code the pulse duration (τ) and the repe-

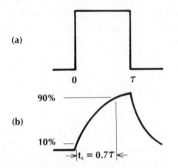

(a)

(b)

90%

10%

$t_s = 0.7\tau$

0　　　　　τ

Figure 12-1　System rise-time requirement. (a) Ideal input pulse. (b) Minimum system pulse response.

tition period (T) are both equal to $1/R$, where R is the data rate. A reasonable estimate of the required total system rise time t_S is that it be no more than 70% of the pulse duration, as illustrated in the figure. That is, the rise time must be limited to

$$t_S = 0.7\tau = \frac{0.7}{R_{\mathrm{NRZ}}} \qquad (12\text{-}5)$$

A similar argument for a return-to-zero signal, in which the pulse duration is half the repetition period T, leads to

$$t_S = 0.7\tau = \frac{0.7T}{2} = \frac{0.35}{R_{\mathrm{RZ}}} \qquad (12\text{-}6)$$

Thus, for the 400-Mbps NRZ signal, the allowable rise time is $t_S = 0.7/4 \times 10^8 = 1.75$ ns. This time must be apportioned between the light source, the fiber, and the photodetector (including its load circuit) in the manner indicated by Eq. (12-2). That is,

$$t_S^2 = t_{\mathrm{LS}}^2 + t_F^2 + t_{\mathrm{PD}}^2$$

Before determining the effect that a 1.75-ns rise time has on the choice of the fiber, we must first develop the relationship between a fiber's rise time and its pulse

spread. Assuming the validity of Eq. (7-2) for our fiber, and using $f_{3\text{-}\mathrm{dB}}(\text{electrical}) = 0.35/\Delta\tau$ from Eq. (3-19), we conclude that the fiber's rise time obeys

$$t_F = \frac{0.35}{f_{3\text{-}\mathrm{dB}}(\text{electrical})} = \Delta\tau \qquad (12\text{-}7)$$

The fiber's electrical rise time and its full-duration half-maximum pulse spread are equal. Although not exact, this relationship can be useful for initial design calcualtions.

From this result we see that the chosen fiber must have a pulse spread of less than 1.75 ns for 100 km (a spread per unit length less than 17.5 ps/km). This is unachievable with multi-mode SI or GRIN fibers, whose typical pulse spreads are closer to 15 ns/km and 1 ns/km, respectively (as indicated in Table 5-2). Even single-mode fibers have nearly 500 ps/km pulse spreads at operating wavelengths near 0.8 μm. The choice is now limited to single-mode fibers operating at 1.3 or 1.55 μm. Because of the extreme fiber length, the loss per kilometer must be quite small. Even a loss as small as 0.5 dB/km (contributing a total loss of 50 dB to our 100-km link) would be intolerable. Later in this section we show that the total system loss, including all couplers and connectors, must be less than about 37 dB. Thus we will use a single-mode fiber operating at 1.55 μm, the region of minimal attenuation. We will assume a loss of 0.25 dB/km, achievable at 1.55 μm, for the remainder of this example.

The pulse spread of the single-mode fiber is due to material and waveguide dispersion. From Figs. 3-8 and 5-23 we note that the dispersion factors are $M = -20$ ps/(nm $\times$ km) and $M_g = 4.5$ ps/(nm $\times$ km) for the material and waveguide effects, respectively, at 1.55 μm. Being of opposite sign, the two dispersion factors tend to cancel, leaving a net dispersion having magnitude $M_t = 20 - 4.5 = 15.5$ ps/(nm × km).

A single-mode (both longitudinal and transverse modes) 1.55 μm InGaAsP laser diode is required because its radiation pattern closely resembles the pattern of the fiber's single propagating mode. Additionally, the linewidth is sufficiently narrow to minimize pulse spreading. The laser diode we will use has a 0.15-nm spectral width and a rise time of 1 ns, so that the fiber's total pulse spread is $\Delta\tau = LM_t\Delta\lambda = 100(15.5)(0.15) = 233$ ps $= 0.23$ ns. According to Eq. (12-7), this is the fiber's rise time as well. Fortunately, $t_F = 0.23$ ns is a small fraction of the 1.75-ns rise-time budget.

Note that an LED would not suffice for several reasons. First, good LEDs emitting in the range 1.3–1.55 μm have spectral widths around 50 nm. The fiber pulse spread would then be $100(15.5)(50) = 77.5 \times 10^3$ ps $= 77.5$ ns, much too large for the proposed system. Second, LEDs radiate over such a wide angle that coupling into a small, low-numerical-aperture, single-mode fiber is very inefficient. For a long system, we must launch as much power into the fiber as possible.

Now we can compute the photodetector's rise-time allotment. From Eq. (12-2),

$$t_{PD}^2 = t_s^2 - t_{LS}^2 - t_F^2 = 1.75^2 - 1 - 0.23^2$$

$$= 2, \text{ or } t_{PD} = 1.4 \text{ ns.}$$

At high frequencies it is important that the photodiode's capacitance be as small as possible. This allows a larger load resistance, which increases the receiver's sensitivity whenever thermal noise is a factor. The photodetector's surface collects the light emerging from the fiber's end. Since the single-mode fiber is small, the active surface can be small, minimizing the detector's capacitance C_d. Suppose that we find a photode-

tector with $C_d = 1$ pF and a transit-time-limited rise time of $t_{TR} = 0.5$ ns. The circuit-limited rise time, from Eq. (7-15), can be written as $t_{RC} = 2.19R_LC_d$. The total rise time of the photodiode is computed from

$$t_{PD}^2 = t_{TR}^2 + t_{RC}^2 \qquad (12\text{-}8)$$

Using $t_{TR} = 0.5$ ns and $t_{PD} = 1.4$ ns, we find that $t_{RC} = 1.3$ ns. The resulting maximum value of the load resistance is then $R_L = t_{RC}/2.19C_d = 1.3 \times 10^{-9}/2.19 \times 10^{-12} = 594$ Ω. If a high-impedance or transimpedance receiver front end is used, then the load resistance can be increased. Table 12-2 summarizes the rise-time calculations.

We have now completed computation of the rise-time budget. By using a narrow line width LD and operating at 1.55 μm, we were able to design a system in which the fiber contributed very little to the bandwidth limitation. Next we will investigate the power budget.

Power Budget

At this point we will assume a few things about the components and techniques available for use in our system. The assumptions made are all reasonable, although they represent very high-quality characteristics rather than

TABLE 12-2. Rise-Time Budget Calculations

Component		Rise Time (ns)	
System budget, $t_s = 0.7/R_{NRZ}$			1.75
Light source, t_{LS}		1.0	
Fiber, $t_F = \Delta\tau$		0.23	
Photodetector			
Transit time, t_{TR}	0.5		
Circuit, $t_{RC} = 2.19R_LC_d$	1.3		
Total, $t_{PD} = (t_{TR}^2 + t_{RC}^2)^{1/2}$		1.4	
System rise time, $(t_{LS}^2 + t_F^2 + t_{PD}^2)^{1/2}$		1.75	1.75

typical ones. Our assumptions include 5 dBm (about 3.2 mW) source output power, 3-dB source-fiber coupling loss, two connectors of 1 dB loss each, and 50 splices of 0.1 dB loss each. The splices are placed, on the average, every 2 km along the 100-km path. We need them to simplify construction and installation of the fiber cable. It is also wise to allow for losses that might occur if the fiber unexpectedly breaks and needs to be spliced.

The power budget calculations are summarized in Table 12-3. The total system loss, including 25 dB attenuation in the fiber, comes to 35 dB. Since we started with a power level of +5 dBm, this leaves $5 - 35 = -30$ dBm as the optic power level at the receiver. As we shall see later in this section, a receiver using an APD photodetector requires about -40 dBm to achieve a 10^{-9} error rate at 400 Mbps. The signal delivered is 10 dB higher. The *power margin,* the power available minus the receiver sensitivity, is $-30 - (-40) = 10$ dB for the APD receiver. A margin of 6 dB or so is generally desirable to account for reduced power caused by aging of the laser diode and for other incidental losses not easily determined beforehand. Results obtained with a high-impedance PINFET receiver are also included in Table 12-3. This

receiver is 8 dB less sensitive than the avalanche device, providing only a 2-dB power margin.

The design of the specified link is now complete. The combination of chosen components operate within the rise-time and power budgets of the system. In this example the receiver sensitivity was assumed to be known. Many times this will be the case, particularly when the receiver has been constructed and tested previously. If this is not the situation, the receiver sensitivity must be determined. We will show how this might be accomplished in the following paragraphs.

Quantum-Noise-Limited Receiver Sensitivity

A quantum-limited receiver provides the ultimate in detection. We will compute its sensitivity to provide a benchmark against which other receivers can be measured. As noted in Section 11-3, the error rate for a quantum-limited system with negligible dark current is given by $P_e = \exp(-n_s)$, where n_s is the average number of signal photoelectrons generated when a binary 1 is received. As shown in Table 11-2, $n_s = 20.7$ corresponds to an error

TABLE 12-3. Power Budget Calculations

Laser diode output power		5 dBm
Source coupling loss	3 dB	
Connector loss (2 connectors)	2 dB	
Splice loss (50 splices)	5 dB	
Fiber attenuation (100 km)	25 dB	
Total loss	35 dB	
Power available at receiver (5 − 35)		−30 dBm
APD receiver		
Sensitivity	−40 dBm	
Loss margin (40 − 30)		10 dB
Hybrid PINFET high-impedance receiver		
Sensitivity	−32 dBm	
Loss margin (32 − 30)		2 dB

rate of 10^{-9}. Since fractional electrons are not meaningful, we will take $n_s = 21$ for our system.

The number of incident photons necessary to produce n_s electrons is n_s/η, where η is the quantum efficiency. As developed in Example 11-6, the peak optic power in a rectangular pulse is related to n_s by

$$P = \frac{hfn_s}{\eta\tau} = \frac{hcn_s}{\eta\lambda\tau} \qquad (12\text{-}9)$$

where τ is the pulse duration. For NRZ signals, $\tau = 1/R$. For RZ signals, $\tau = 1/2R$. For the NRZ system, then

$$P = \frac{hcn_sR}{\eta\lambda} \qquad (12\text{-}10)$$

while for RZ pulses the peak power is twice this value. Essentially, the energy in a pulse (the peak power times the pulse duration, for a rectangular pulse) must exceed the level $hcn_s/\eta\lambda$, regardless of the coding scheme.

Using $n_s = 21$, $\eta = 1$, $\lambda = 1.55$ μm, and $R = 400$ Mbps in Eq. (12-10) yields a sensitivity of $P = 1.08$ nW. This equals 1.08×10^{-6} mW or -59.7 dBm. Under ideal

quantum-limited conditions, the system must deliver at least -59.7 dBm of optic power to the photodetector to ensure a 10^{-9} BER. Assuming a 70% quantum efficiency increases the required power level to -58.1 dBm. The APD receiver referred to in Table 12-3 has a sensitivity (-40 dBm) about 20 dB poorer than the ideal quantum limit. This occurs when operating in the 1.3–1.6-μm region because of the high excess noise of the InGaAs avalanche photodiodes. The noise is so great that optimum gains are quite moderate (i.e., 10 or so). The small gain helps overcome thermal noise only a bit. Later in this section we will calculate the thermal-noise-limited sensitivity and see that the low-gain APD in this example provides about a 12-dB improvement over that case.

Figure 12-2 is a plot of Eq. (12-10) with 100% quantum efficiency assumed and power levels converted to dBm. In practice, efficiencies range between 55 and 80%, as mentioned in Section 7-4.

The true sensitivity of a detector can be computed directly from Eq. (12-10). Also, it can be found from Fig. 12-2 by dividing the power obtained from the curve by the actual quantum efficiency. Detector manufacturers

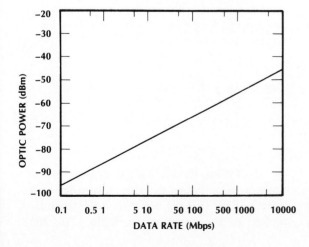

Figure 12-2 Quantum-noise-limited receiver sensitivity at 1.55 μm for a 10^{-9} BER in a NRZ system.

more often list the responsivity ρ than the quantum efficiency η. The two are related by Eq. (7-7) which, combined with Eq. (12-10), yields

$$P = en_sR/\rho \qquad (12\text{-}11)$$

This form of the quantum-limited sensitivity is easier to use than Eq. (12-10) if the detector's responsivity is known.

Assuming equal quantum efficiencies at all wavelengths, Eq. (12-10) predicts that the sensitivity improves as the operating wavelength increases. The ratio of the longest wavelength of major interest (1.55 μm) to the shortest (0.8 μm) is almost two. Thus, the shorter wavelengths require almost twice as much power as the longer ones. This represents nearly a 3-dB advantage for long-wavelength systems.

Another point regarding the sensitivity calculations is worth mentioning. The symbol P in Eqs. (12-10) and (12-11) and in Fig. 12-2 represents the peak optic power. It is the peak power delivered when receiving a 1. Since typical messages contain equal numbers of 1s and 0s, the average power in NRZ pulse trains is just half the peak value. Then,

$$P_{\text{AVE}} = \frac{hcn_sR}{2\eta\lambda} \qquad (12\text{-}12)$$

A plot of the average power sensitivity would be 3 dB *below* the curve in Fig. 12-2.

tector's responsivity. In addition, if we account for amplifier noise by using the equivalent system noise temperature, then the result is

$$\frac{S}{N} = \frac{R_L(\rho P)^2}{4kT_e\Delta f} \qquad (12\text{-}13)$$

The thermal-limited SNR corresponding to a given error rate is determined from Eq. (11-29) or from the plot of that equation in Fig. 11-11. From these data we find that a 10^{-9} BER requires a 21.5-dB SNR (i.e., $S/N = 142$). Solving Eq. (12-13) for the required power yields

$$P = \frac{\sqrt{4kT_e(\Delta f/R_L)(S/N)}}{\rho} \qquad (12\text{-}14)$$

Earlier in this section we chose $R_L = 594\ \Omega$ based on rise-time restrictions. This was the maximum resistance allowed when using the simple terminated-photodiode receiver front end. Using this value of resistance and the 1-pF photodiode capacitance in Eq. (7-16), we find the receiver bandwidth to be 268 MHz. A receiver with a larger bandwidth could be chosen, but its sensitivity would be poorer as predicted by Eq. (12-14). Assuming a temperature of 300 K, a noise figure of 2, and a detector responsivity of 1 A/W yields a receiver sensitivity of

$$P = \sqrt{\frac{4(1.38 \times 10^{-23})(2)(300)(2.68 \times 10^8)(142)}{594}}$$

$$= 1.46\ \mu W$$

Thermal-Noise-Limited Receiver Sensitivity

The thermal-noise-limited SNR, given by Eq. (11-11), can be rewritten in terms of the de-

or -28.4 dBm. Recall that the ideal quantum-limited sensitivity in this example was -59.7 dBm, 31.3 dB better than the thermal-limited result. As noted in Table 12-3, the power level at the receiver is only -30 dBm, 1.6 dB *lower* than that required by the thermal-limited receiver. In this system a resistor-terminated PIN diode receiver will not work. It is not sensitive enough.

We have several options at this point for improving the sensitivity. We can use high-impedance or transimpedance front ends, which operate at lower levels than the terminated diode. Equalizers must be included in these receivers. For operation at 400 Mbps, hybrid InGaAs high-impedance and trans-impedance PINFET receivers have been constructed having sensitivities around -32 dBm.[1] This is more than 3 dB better than the resistor-terminated circuit and provides a 2-dB margin for our system.

InGaAs avalanche photodiodes are somewhat more sensitive than high-impedance PINFET receivers in the long-wavelength regions.[2] Assuming an 8-dB improvement yields a -40-dBm sensitivity and provides a 10-dB power margin. Table 12-3 summarizes the power budget results.

Generalized Thermal-Noise-Limited Receiver Sensitivities

We will now develop some general results for thermal-noise-limited receivers, applicable to a range of data rates. At first we restrict the discussion to NRZ pulse trains. Later in this section we show how the results apply to RZ systems. For the analysis we must assume some value for the receiver's bandwidth. Let us choose it equal to half the data rate. This is the minimum bandwidth required to transmit NRZ pulses, as indicated by Fig. 3-12. This choice does not allow for bandlimiting in the transmitter and fiber. In other words, the receiver uses up the entire system bandwidth budget. Although this cannot always be al-

lowed (e.g., in the system treated earlier in this section), it does produce the most sensitive thermal-limited receiver. In this sense the results we obtain are idealized. The sensitivities of receivers having other bandwidths can be obtained by following the procedures outlined in this example. To make the results as general as possible, we will normalize them so that they can be used with arbitrary values of photodetector responsivity and amplifier noise figure. In addition to this normalization, we will assume that the receiver bandwidth is entirely determined by the load resistor and the photodiode's capacitance. These assumptions will give us the best sensitivity that can be obtained with the simple terminated front-end receiver.

Now, combining Eqs. (12-5) and (7-15) yields $t_{RC} = 0.7/R = 2.19R_L C_d$, or

$$R_L = \frac{1}{\pi R C_d} \qquad (12\text{-}15)$$

Equation (12-14) now becomes

$$\frac{\rho P}{F^{1/2}} = R\sqrt{2kT\pi C_d(S/N)} \qquad (12\text{-}16)$$

Evaluating the right side of this equation for $T = 300$ K, $C_d = 1$ pF, and $S/N = 142$ (as appropriate for a 10^{-9} error rate) leaves $\rho P/F^{1/2} = 1.92 \times 10^{-15} R$. With R written in Mbps and P in nW, this simplifies to

$$\frac{\rho P}{F^{1/2}} = 1.92R \qquad (12\text{-}17)$$

This result is plotted in Fig. 12-3. If $\rho = 1$ A/W and $F = 1$, then the sensitivity can be read directly in dBm. Otherwise, the vertical scale must be converted to milliwatts, multiplied by the square root of the noise figure, and divided by the responsivity. As mentioned in Section 7-4, responsivities are near

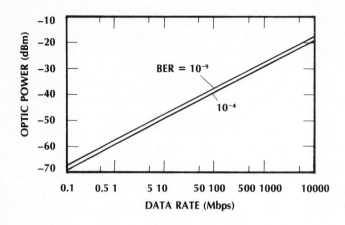

Figure 12-3 Thermal-noise-limited receiver sensitivity in a NRZ system.

0.5 for silicon in the region 0.8–0.9 μm, 0.7 for germanium in the region 1.0–1.8 μm, and 0.7–1.1 for InGaAs in the region 1.0–1.7 μm.

If the error rate is only 10^{-4}, then Fig. 11-11 indicates that the SNR should be 17.5 dB or better. Then, $S/N = 56$ in Eq. (12-16), leaving

$$\frac{\rho P}{F^{1/2}} = 1.21R \qquad (12\text{-}18)$$

where we still have $C_d = 1$ pF, $T = 300$ K, R in Mbps, and P in nanowatts. This result also appears in Fig. 12-3. As noted from the figure, the required power levels for 10^{-4} and 10^{-9} error rates differ only by 2 dB. Once again we see the remarkable sensitivity of the error rate to the optic power level.

The curves in Fig. 12-3 depend on the operating wavelength only through the wavelength dependence of the responsivity. A 1.55-μm InGaAs photodiode with $\rho = 1$ requires only half as much power as a 0.8-μm silicon photodiode with $\rho = 0.5$. The long-wavelength detector in this comparison is 3 dB more sensitive.

Figures 12-4 and 12-5 compare the sensitivities of different receivers. Figure 12-4 applies in the region 0.8–0.9 μm and Fig. 12-5 in the region 1.3–1.6 μm.

In the short-wavelength first window, optimum receivers using APDs approach the quantum limit within about 10–13 dB.[3] Excess noise and dark current prevent APD receivers from actually achieving the ideal quantum-noise-limited sensitivity. As noted from Fig. 12-4, thermal-noise-limited, resistor-terminated, PIN-diode receivers require about 25–30 dB more power than the ideal quantum receiver. As a final comparison, the APD receiver represents about a 15-dB improvement over the PIN receiver. This is a substantial advantage. For example, if the fiber attenuation is 3 dB/km, then the APD link can be 5 km longer than the one using the PIN diode.

At the longer wavelengths, InGaAs APD receivers require about 20 dB more power than the ideal.[4] Resistor-terminated PIN-diode receivers are another 10–12 dB poorer. High-impedance (and transimpedance) PINFET front ends, with sensitivities lying between those of terminated PIN photodiodes and APD receivers, are attractive for long transmission links.[5] A hybrid PINFET receiver consists of an InGaAs photodiode followed by a GaAs MESFET preamplifier.[6] Its sensitivity, plotted in Fig. 12-5, begins to degrade above

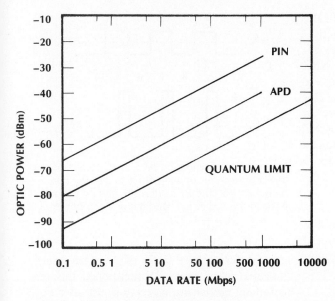

Figure 12-4 Receiver sensitivities.
BER = 10^{-9}, λ = 0.82 μm. The
photodiodes are silicon devices.

100 Mbps owing to the limited frequency response of the MESFET preamplifier.

When using Figs. 12-4 and 12-5, remember the assumptions made. Corrections must be applied to account for the actual preamplifier noise figure and the photodetector's responsivity and capacitance. We hope that these figures were developed in enough detail to enable you to construct similar curves for specific situations. Nonetheless, the parameters for which Figs. 12-4 and 12-5 apply were chosen to be representative of practical devices. These figures can be used directly as a guide for initial system investigation.

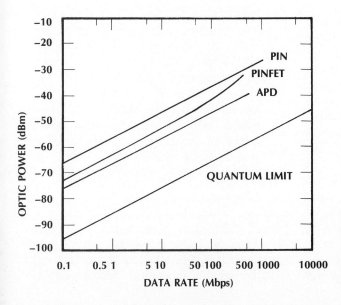

Figure 12-5 Receiver sensitivities.
BER = 10^{-9}, λ = 1.55 μm. The
photodiodes are InGaAs devices.
The PINFET receiver includes a
high-impedance (or transimpedance)
preamplifier.

Figures 12-2–12-5, derived specifically for NRZ pulse trains, easily yield the sensitivities of return-to-zero systems. RZ pulses require twice as much peak power as NRZ pulses for the same data rate because RZ pulses last half as long as NRZ pulses. This statement applies whether thermal noise or shot noise limits the receiver. Thus, for RZ pulse trains, simply double the power obtained from Figs. 12-2–12-5. That is, *add* 3 dB to the power levels found in the figures.

Example 12-1

Compute the quantum-noise-limited sensitivity and the thermal-noise-limited sensitivity for a 100-Mbps RZ system with a 10^{-9} error rate. The system operates at 0.82 μm.

Solution:

From Fig. 12-4 we find required NRZ power levels of -63 and -36 dBm for the quantum and thermal limits, respectively. *Adding* 3 dB yields -60 and -33 dBm as the corresponding RZ peak power levels.

So far we have been dealing mostly with peak powers. When taking data on real systems, the average power in a pulse train is easier to measure than the peak power. We can easily deduce the relationship between peak and average power by considering a series of alternating 1s and 0s. This strategy is valid because typical messages contain equal numbers of 1s and 0s. As illustrated in Fig. 12-6, the average NRZ power is half the peak power. In this system the power is *on* half the time and *off* half the time. However, for RZ pulse trains (also sketched in the figure) a series of alternating 1s and 0s is *on* only one-fourth of the time and *off* three-fourths of the time. Thus the average power is one-fourth of the peak power. The figure shows the peak power

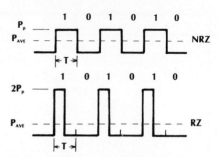

Figure 12-6 Comparing peak and average power levels for a series of alternating 1s and 0s. For both the NRZ and RZ codes, the energy in each pulse is the same ($P_p T$) and the average power is the same ($P_{AVE} = P_p / 2$).

levels differing by a factor of 2, as necessary for equal error rates. We mentioned this requirement earlier in this section just after Eq. (12-10). We conclude, as indicated in the figure, that the *average* powers are equal for the two coding schemes.

Example 12-2

Compute the average power levels for Example 12-1.

Solution:

The quantum and thermal limits are -63 and -36 dBm, respectively, for the NRZ system. The average power, being half as much, is 3 dB *less,* that is, -66 and -39 dBm. We could arrive at the same result starting with the RZ sensitivities of -60 and -33 dBm determined in Example 12-1. The average power, being one-fourth as much, is 6 dB less, that is, -66 and -39 dBm.

12-3 SUMMARY

Experienced engineers realize that a gap may exist between a theoretical design and its physical implementation. This occurs when analytical models only approximate the actual behavior of the system and its components. Nonetheless, an extensive theoretical analysis pays dividends by ultimately leading more quickly and economically to the desired result than a purely experimental approach. Even approximate developments clearly show trends that suggest the proper design changes. As an example, we found how sensitive the bit-error rate was to small changes in the power level. This information tells us the added power needed to achieve a desired BER if the present one is known. A good design strategy is to complete as much theoretical analysis as practical before proceeding with construction or purchase of the system hardware.

In most cases, design is an iterative process. Several different approaches are often considered, compared, and retained or discarded. Next, working models are constructed and tested. The tests indicate any differences between the theoretical predictions and the results actually obtained. If the original design is conservative, then the system might still work satisfactorily. If the specifications are not met, then corrections and improvements can now be made by using the theory as a guide. Usually, the final design is based on a combination of theoretical and experimental work.

Many manufacturers produce extensive applications notes, which they distribute free. These notes can be quite helpful because they are usually written clearly and simply and because they are based on devices actually built and sold by the manufacturer. Component and subsystem manufacturers are willing and able to help. Use their services.

A large volume of literature can be consulted to increase understanding of fiber optics. Details on specific subjects addressed in this book can be found in the materials refer-enced at the end of each chapter. The bibliography directs your attention to useful books. These books range from introductory to advanced. Some cover practical matters and others cover more theoretical ones.

To keep abreast of current topics, regularly read one or more of the periodicals listed in the bibliography.

PROBLEMS

In the problems for this chapter, you are asked to design certain fiber optic systems. These are open-ended problems. There are no unique solutions. This state of affairs corresponds to real-life situations. Find a solution based on components having reasonable properties (as suggested in this text or as determined from other reliable sources). In any case, make assumptions which are reasonable and state them clearly. As implied by the examples given in this chapter, design includes selection of components (source, detector, fiber, couplers, connectors, etc.), specifying their operating characteristics, setting up a power and a bandwidth budget, and evaluating system performance. It also includes selection of the network topology (including the use of repeaters or optical amplifiers, if necessary) and the form of modulation. Both direct detection and heterodyne detection should be considered.

Design a system that will satisfy the requirements stated in each of the following problems.

12-1. Transmit a video signal having a bandwidth of 4.5 MHz over a 10-km path. The SNR at the receiver must be 48 dB or more. Use analog modulation.

12-2. Repeat Problem 12-1, using digital modulation with an error rate of 10^{-9} or better.

12-3. Transmit a 2-Gbps NRZ signal over a 100-km path without the use of repeaters. The error rate must be 10^{-9} or better.

12-4. Suppose that the system in Problem 12-3 has been constructed, but now the error rate is allowed to degrade to 10^{-4}. How much farther than 100 km can the signal be transmitted based only on power considerations? Make the necessary system changes to allow transmission over this longer path.

12-5. Transmit a voice signal with a 4 kHz bandwidth over 100 m with a SNR = 30 dB.

12-6. Repeat Problem 12-5, using digital modulation. The error rate must be 10^{-5} or better.

12-7. Transmit a 2-Gbps Manchester coded signal over a 100-km path without the use of repeaters. The error rate must be 10^{-9} or better.

12-8. Transmit at a rate of 10 Mbps over a system having five terminals. The terminals are situated along a straight path and are 200 m from each other. The error rate must be better than 10^{-9}.

12-9. Transmit at a rate of 250 Mbps over a system having 25 terminals. The terminals are located uniformly on the perimeter of a circle whose diameter is 1 km. The error rate must be 10^{-9} or better.

12-10. Transmit simultaneously over three channels: one for voice, one for video, and one for data. The path length is 10 km. The voice band width is 4 kHz, the video bandwidth is 4.5 MHz, and the data rate is 10 Mbps (NRZ). The SNR for the voice channel is 25 dB and for the video channel is 40 dB. The error rate for the data channel is 10^{-9} or better.

12-11. Transmit at a rate of 2 Gbps (RZ) over a 5000-km path. The error rate must be 10^{-9} or better.

12-12. Transmit three channels simultaneously on the same fiber. The channels have optical carriers near 1550 nm and are spaced 2 GHz apart. The data rate is 100 Mbps (NRZ). The path length is 100 km, and the data error rate must be better than 10^{-9}.

12-13. Design a 20 Gbps digital system operating over 10,000 km without repeaters. A 10^{-9} error rate is required. (*Hint:* Solitons and optical amplifiers must be used.)

REFERENCES

1. Manufacturer's literature. Burlington, Mass.: Lasertron.

2. S. R. Forrest. "Photodiodes for Long-Wavelength Communication Systems," *Laser Focus* 18, no. 12 (Dec. 1982): 81–90.

3. Tien Pei Lee and Tingye Li. "Photodetectors." In *Optical Fiber Telecommunications,* edited by Stewart E. Miller and Alan G. Chynoweth. New York: Academic Press, 1979, pp. 622–23.

4. Forrest. "Photodiodes for Long-Wavelength Communication Systems." pp. 84–85.

5. Michael Ettenberg and Gregory H. Olsen. "Diode Lasers for the 1.2 to 1.7 Micrometer Region," *Laser Focus* 18, no. 3 (March 1982): 61–66.

6. Manufacturer's literature. Burlington, Mass.: Lasertron.

Bibliography

BOOKS

Adams, M. J. *An Introduction to Optical Waveguides*. New York: John Wiley, 1981.

Adams, M. J., and I. D. Henning, *Optical Fibres and Sources for Communications*. New York: Plenum, 1991.

Agrawal, G. P., and N. K. Dutta. *Long-Wavelength Semiconductor Lasers*. New York: Van Nostrand Reinhold, 1986.

Agrawal, G. P., *Nonlinear Fiber Optics*, New York: Academic, 1989.

Agrawal, I.D., and G. Lu, eds., *Flouride Glass Fiber Optics*, New York: Academic, 1991.

Allard, F. C., *Fiber Optics Handbook*, New York: McGraw-Hill, 1990.

Arnaud, J. A. *Beam and Fiber Optics*. New York: Academic, 1976.

Baack, C., *Optical Wideband Transmission Systems*, Boca Raton, Fla.: CRC Press, 1986.

Baker, D. G. *Local-Area Networks with Fiber Optic Applications*. Englewood Cliffs, N.J.: Prentice Hall, 1986.

Barnoski, M. K., ed. *Fundamentals of Optical Fiber Communications,* 2nd ed. New York: Academic Press, 1981.

Basch, E. E., ed. *Optical-Fiber Transmission*. Indianapolis, Ind.: Howard W. Sams, 1986.

Bendow, B., and S. Mitra, eds. *Fiber Optics: Advances in Research and Development*. New York: Plenum, 1979.

Cancellieri, G., and U. Ravaioli. *Measurement of Optical Fibers and Devices: Theory and Experiments*. Norwood, Mass.: Artech House, 1984.

Cancellieri, G. *Single-Mode Optical Fibres*. New York: Pergamon Press, 1991.

Chaffee, C. D., *The Rewiring of America: The Fiber Optics Revolution*, New York: Academic, 1987.

Chaimowicz, J. C. A., *Lightwave Technology: An Introduction.* Stoneham, Mass.: Butterworth, 1989.

Chamberlain, G. E., G. W. Day, D. L. Franzen, R. L. Gallawa, E. M. Kim, and M. Young. *Optical Fiber Characterization.* Washington, D. C.: U. S. Government Printing Office, 1983.

Chang, K., ed., *Handbook of Microwave and Optical Components*, Vol. 3: *Optical Components*, Vol. 4: *Fiber Optical Components*, New York: John Wiley, 1989.

Cheo, P. K. *Fiber Optics*. 2nd ed. Englewood Cliffs, N.J.: Prentice Hall, 1990.

Cherin, A. H. *An Introduction to Optical Fibers*. New York: McGraw-Hill, 1983.

Clarricoats, P. J. B., ed. *Optical Fibre Waveguides*. Stevenage, Herts., England: Peter Peregrinus, 1975.

Clarricoats, P. J. B. "Optical Fibre Waveguides: A Review." In *Progress in Optics.,* Vol. 14. Amsterdam: North-Holland, 1976.

Comyns, A. E., ed., *Flouride Glasses*, New York: John Wiley, 1989.

CSELT. *Optical Fibre Communication*. New York: McGraw-Hill, 1980.

CSELT. *Fiber Optic Communications Handbook*. 2nd ed. Blue Ridge Summit, Penn.: TAB, 1991.

Culshaw, B. *Optical Fibre Sensing and Signal Processing*. Stevenage, Herts., England: Peter Peregrinus, 1984.

Dakin, J. P. and B. Culshaw, eds., *Optical Fiber Sensors*, Vol. 1: *Principles and Components*, Norwood, Mass.: Artech House, 1988.

Daly, J. C., ed. *Fiber Optics*. Boca Raton, Fla.: CRC Press, 1984.

Diament, P., *Wave Transmission and Fiber Optics*. New York: Macmillan, 1990.

Elion, G. R., and H. A. Elion. *Fiber Optics in Communications Systems*. New York: Marcel Dekker, 1978.

Ezekiel, S., and H. J. Arditty, eds. *Fiber-Optic Rotation Sensors and Related Technologies*. New York: Springer-Verlag, 1982.

France, P. W., ed., *Flouride Glass Optical Fibres*, Boca Raton, Fla.: CRC Press, 1990.

Gallawa, R. L. *A User's Manual for Optical Waveguide Communications*. Springfield, Va.: National Technical Information Service, U.S. Department of Commerce, 1976.

Geckeler, S., *Optical Fiber Transmission Systems*, Norwood, Mass.: Artech House, 1987.

Ghatak, A. K., and K. Thyagarajan. *Contemporary Optics*. New York: Plenum, 1978.

Ghatak, A. K., and K. Thyagarajan, *Optical Electronics*. New York: Cambridge University Press, 1989.

Gloge, D., ed. *Optical Fiber Technology*. New York: IEEE Press, 1976.

Gowar, J. *Optical Communications Systems*. Englewood Cliffs, N.J.: Prentice Hall, 1984.

Halley, P., *Fiber Optic Systems*. John Wiley, New York, 1987.

Hanson, A. G., L. R. Bloom, A. H. Cherin, G. W. Day, R. L. Gallawa, E. M. Gray, C. Kao, F. P. Kapron, B. S. Kawasaki, P. Reitz, and M. Young. *Optical Waveguide Communications Glossary*. NBS Handbook 140. Washington, D.C.: National Telecommunications and Information Administration, 1982.

Hoss, R., *Fiber Optic Communications Design Handbook,* Englewood Cliffs, N.J.: Prentice Hall, 1991.

Howes, M. J. and D. V. Morgan, eds. *Optical Fiber Communications*. New York: John Wiley, 1980.

Hunsberger, R. G. *Integrated Optics: Theory and Technology*, 2nd ed. New York: Springer-Verlag, 1984.

Jeunhomme, L. B. *Single-Mode Fiber Optics*. 2nd ed. New York: Marcel Dekker, 1989.

Jones, W. B., *Introduction to Optical Fiber Communication Systems*. New York: Holt, Rinehart and Winston, 1988.

Kao, C. K. *Optical Fiber Systems: Technology, Design and Applications*. New York: McGraw-Hill, 1982.

Kao, C. K., ed. *Optical Fiber Technology*, Vol. 2. New York: McGraw-Hill, 1981.

Karim, M. *Electro-Optical Devices and Systems*. Boston, Mass.: PWS-Kent Publishing, 1990.

Katsuytama, T., and H. Matsumura. *Infrared Optical Fibers*. New York: Taylor and Francis, 1989.

Keiser, G. E. *Optical Fiber Communications*. 2nd ed. New York: McGraw-Hill, 1991.

Killen, H. B., *Digital Communications with Fiber Optics and Satellite Applications*, Englewood Cliffs, N.J.: Prentice Hall, 1988.

Killen, H. B., *Fiber Optic Communications*. Englewood Cliffs, N.J.: Prentice Hall, 1991.

Kressel, H., ed. *Semiconductor Devices for Optical Communications*. New York: Springer-Verlag, 1980.

Kressel, H., and J. K. Butler, eds. *Semiconductor Lasers and Heterojunction LEDs*. New York: Academic Press, 1977.

Kuecken, J. A., *Fiberoptics: A Revolution in Communications*. Blue Ridge Summit, Penn.: TAB, 1987.

Lacy, E. A. *Fiber Optics*. Englewood Cliffs, N.J.: Prentice Hall, 1982.

Lee, D. L. *Electromagnetic Principles of Integrated Optics*. New York: John Wiley, 1986.

Li, T., ed. *Optical Fiber Communications*. Vol. 1: *Fiber Fabrication*. New York: Academic Press, 1985.

Lin, C., *Optoelectronic Technology and Lightwave Communications Systems*. New York: Van Nostrand Reinhold, 1989.

Mahlke, G. and P. Gossing, *Fiber Optic Cables*. New York: John Wiley, 1987.

Marcuse, D. *Light Transmission Optics*. 2nd ed. New York: Van Nostrand Reinhold, 1982.

Marcuse, D. *Principles of Optical Fiber Measurements*. New York: Academic Press, 1981.

Marcuse, D. *Theory of Dielectric Optical Waveguides*. New York: Academic Press, 1974.

Midwinter, J. E. *Optical Fibers for Transmission*. New York: John Wiley, 1979.

Miller, C., S. Mettler, and I. White. *Optical Fiber Splices and Connectors*. New York: Marcel Dekker, 1986.

Miller, S. E., and A. G. Chynoweth, eds. *Optical Fiber Telecommunications*. New York: Academic Press, 1979.

Miller, S. E., and I. P. Kaminow, eds. *Optical Fiber Telecommunications*, Vol. 2. New York: Academic Press, 1988.

Morris, D. J. *Pulse Code Formats for Fiber Optical Data Communications*. New York: Marcel Dekker, 1983.

Murata H. *Handbook of Optical Fibers and Cables*. New York: Marcel Dekker, 1988.

Neumann, E. G., *Single-Mode Fibers*. New York: Springer-Verlag, 1988.

Nishihara, H., M. Haruna, and T. Suhara.

Optical Integrated Circuits. New York: McGraw-Hill, 1988.

Noda, K., ed. *Optical Fiber Transmission.* Amsterdam: North-Holland, 1986.

Okoshi, T. *Optical Fibers.* New York: Academic Press, 1982.

O'Shea, D. S., W. R. Callen, and W. T. Rhodes. *Introduction to Lasers and Their Applications.* Reading, Mass.: Addison-Wesley, 1977.

Ostrowsky, D. B., ed. *Fiber and Integrated Optics.* New York: Plenum, 1979.

Owyang, G. H. *Foundations of Optical Waveguides.* New York: Elsevier North-Holland, 1981.

Personick, S. D. *Fiber Optics: Technology and Applications.* New York: Plenum, 1985.

Personick, S. D. *Optical Fiber Transmission Systems.* New York: Plenum, 1981.

Pratt, W. K. *Laser Communications Systems.* New York: John Wiley, 1969.

Runge, P. K., and P. R. Trischitta, eds. *Undersea Lightwave Communications.* New York: IEEE Press, 1986.

Sandbank, C. P., ed. *Optical Fiber Communications.* New York: John Wiley, 1980.

Senior, J. *Optical Fiber Communications: Principles and Practice.* 2nd ed. Englewood Cliffs, N.J.: Prentice Hall, 1991.

Sharma, A. B., S. J. Halme, and M. M. Butusov. *Optical Fiber Systems and Their Components.* New York: Springer-Verlag, 1981.

Simons, R. *Optical Control of Microwave Devices,* Norwood, Mass.: Artech House, 1990.

Snyder, A. W., and J. D. Love. *Optical Waveguide Theory.* London: Chapman and Hall, 1983.

Sodha, M. S. and A. K. Ghatak. *Inhomogeneous Optical Waveguides.* New York: Plenum, 1977.

Sterling, D. J. *Technician's Guide to Fiber Optics.* Albany, N.Y.: Delmar, 1987.

Suematsu, Y., ed. *Optical Devices and Fibers* (several volumes in a continuing series). Amsterdam: North-Holland, 1982, 1983, 1984.

Suematsu, Y., and K. Iga. *Introduction to Optical Fiber Communications.* New York: John Wiley, 1982.

Tamir, T., ed. *Integrated Optics,* 2nd ed. New York: Springer-Verlag, 1982.

Tamir, T., ed. *Guided-Wave Optoelectronics,* New York: Springer-Verlag, 1988.

Taylor, H. F., ed. *Fiber Optics Communications.* Norwood, Mass.: Artech House, 1983.

Tsang, W. T., ed. *Lightwave Communications Technology,* Part A: *Material Growth Technologies.* New York: Academic Press, 1985.

Tsang, W. T., ed. *Lightwave Communications Technology,* Part B: *Semiconductor Injection Lasers,* Vol. 1. New York: Academic Press, 1985.

Tsang, W. T., ed. *Lightwave Communications Technology,* Part C: *Semiconductor Injection Lasers,* Vol. 2., *Light-Emitting Diodes.* New York: Academic Press, 1985.

Tsang, W. T., ed. *Lightwave Communications Technology,* Part D: *Photodetectors.* New York: Academic Press, 1985.

Tsang, W. T., ed. *Lightwave Communications Technology,* Part E: *Integrated Optoelectronics.* New York: Academic Press, 1985.

Ungar, S, *Fiber Optics Theory and Applications.* New York: John Wiley, 1991.

van Etten, W., and J. van der Plaats. *Principles of Optical Fiber Communications.*

Englewood Cliffs, N. J.: Prentice Hall, 1991.

Verdeyen, J. T. *Laser Electronics*. Englewood Cliffs, N.J.: Prentice Hall, 1981.

Weik, M. H. *Fiber Optics and Lightwave Communications Standard Dictionary*. New York: Van Nostrand Reinhold, 1981.

Weik, M. H. *Fiber Optics Standard Dictionary*. New York: Van Nostrand Reinhold, 1988.

Wickersham, A., *Microwave and Fiber Optics Communications*, Englewood Cliffs, N.J.: Prentice Hall, 1988.

Willardson, R. K., and A. C. Beer, eds. *Lightwave Communications Technology (Semiconductors and Semimetals)* (5 volumes). New York: Academic Press, 1985.

Wolf, H. F., ed. *Handbook of Fiber Optics: Theory and Applications*. New York: Garland, 1979.

Yariv, A. *Introduction to Optical Electronics*, 4th ed. New York: Holt, Rinehart and Winston, 1991.

Yeh, C., *Handbook of Fiber Optics*. New York: Academic Press, 1990.

Young, M. *Optics and Lasers*, 2nd ed. New York: Springer-Verlag, 1984.

Zanger, H., and C. Zanger, *Fiber Optics: Communications and Other Applications*, New York: Macmillan, 1990.

PERIODICALS

Applied Optics. New York: Optical Society of America.

Bell System Technical Journal. New York: American Telephone and Telegraph.

Electronics Letters. Stevenage, Herts., England: The Institution of Electrical Engineers.

Fiber and Integrated Optics. New York: Taylor and Francis.

International Journal of Optoelectronics. Philadelphia, Penn.: Taylor and Francis.

Journal of Quantum Electronics. New York: The Institute of Electrical and Electronics Engineers.

Journal of Lightwave Technology. New York: The Institute of Electrical and Electronics Engineers.

Journal of Optical Communications. Berlin: Fachverlag Schiele and Schon.

Journal of the Optical Society of America. New York: Optical Society of America.

Journal of Optics (formerly *Nouvelle Revue D'Optique*). Paris: Masson.

Lasers and Optronics. Torrance, Calif.: High Tech Publications.

Laser Focus World. Tulsa, Okla.: PennWell Publishing Company.

Lightwave. Tulsa, Okla.: PennWell Publishing Company.

Lightwave Communications Systems Magazine. New York: The Institute of Electrical and Electronics Engineers.

Optica Acta. London: Taylor and Francis.

Optical Engineering. Bellingham, Wash.: Society of Photo-Optical Instrumentation Engineers.

Optical and Quantum Electronics (formerly *Opto-Electronics*). London: Chapman and Hall.

Optics and Laser Technology. Guildford, Surrey, England: Butterworth.

Optics and Photonics News. New York: Optical Society of America.

Optics Communications. Amsterdam: North-Holland.

Optics Letters. New York: Optical Society of America.

Optik.   Stuttgart: Wissenchaftliche Verlags-
gesellschaft.

Photonics Spectra.   Pittsfield, Mass.: Laurin
Publishing.

Photonics Technology Letters.   New York:

The Institute of Electrical and Electronics
Engineers.

Soviet Journal of Optical Technology (English
translation). New York: American Institute
of Physics.

Answers

CHAPTER 1

CHAPTER 1

1-1. $dB = 10 \log(P_2/P_1)$.

1-2. $P = 0.001 \times 10^{dB/10}$.

1-3. 0.16 mW.

1-4. 1 mW.

1-5. 3920 lb.

1-6. 1.8 km (coax), 8 km (fiber).

1-7. 698.

1-8. 2 or 3 pulses per second.

1-9. 4.5 of the copper cables will equal the capacity of the 672 channel per fiber cable, and 27 copper cables yield the same capacity as a single DS-4 fiber cable.

1-10. 30.

1-11. Use wavelength $= c/f$.

1-12. 4.28×10^{14} Hz, 7.5×10^{14} Hz, bandwidth $= 3.2 \times 10^{14}$ Hz.

1-13. 3.3×10^{-19} J, 2.4×10^{-19} J, 1.5×10^{-19} J. Visible photons have more energy than infrared photons.

1-14. $P = 2.48 \times 10^{-9}$ W, $I = 1.6$ nA.

1-15. 6.54×10^9 photons/s.

1-16. 0.1 kbps, 10 kbps, 1 Mbps, 100 Mbps, 3×10^{12} bps.

1-17. The carrier oscillations do not have enough time to build up.

1-18. 7×10^8 channels.

1-19. (a) Monitoring of rocket launch. (b) Live video distribution of classroom lecture to other rooms in the same building.

1-20. For 10 billion homes the bandwidth

needed is 4×10^{13} Hz. An optical carrier whose frequency is 3×10^{14} Hz might carry one-tenth of this traffic when modulated at about 1% of its center frequency. Ten fibers, or 10 different carrier frequencies, would probably be needed.

1-21. $R = 6.4 \times 10^{14}$ bps. An optical carrier could not be turned on and off fast enough.

1-22. (a) 4×10^{11} photons/s. (b) 7.8×10^{11} photons/s. (c) The longer wavelength requires more photons.

1.23. 0.5 mW.

1.24. 2.7 errors/min.

1-25. 5.175 million.

CHAPTER 2

2-1. $\alpha_i = 8°$.

2-2. NA = sin (acceptance angle).

2-3. Plot.

2-4. Plot.

2-5. $d = 3.9 \, \mu$m.

2-6. $w_o = 5.09 \, \mu$m.

2-7. $I/I_o = e^{-2r^2}$

2-8. Divergence $= 5.09 \times 10^{-4}$ rad, $w_o = 96.5$ km (60 miles) on moon, 0.255 m at 1 km, and 2.55 m at 10 km.

2-9. (a) 30 ms. (b) 236 ms. (c) Only the satellite delay is noticeable.

2-10. The incident medium has the highest index.

CHAPTER 3

3-1. For wavelengths $< 1.3 \, \mu$m, the longer wavelength will reach the receiver first.

For wavelengths $> 1.3 \, \mu$m, the shorter wavelength will reach the receiver first.

3-2. 2.7 ns/km, 0.18 ns/km.

3-3. 0.6 ns/km, 0.04 ns/km.

3-4. Sample results (0.85 μm, 30 nm spectral width, path length 1 km): Optical bandwidth = 185 MHz, electrical bandwidth = 130 MHz, data rate (RZ) = 130 Mbps, data rate (NRZ) = 260 Mbps.

3-5. $k = 7.77 \times 10^6$ rad/m (air), $k = 1.149 \times 10^7$ rad/m (glass).

3-6. 0.12%, 2.4%, 4.39×10^{11} Hz, 8.78×10^{12} Hz.

3-7. $R = 0.319$, loss = 1.67 dB.

3-8. Plot.

3-9. Proof.

3-10. Plot.

3-11. Proof.

3-12. The total optical power is $P = 4 + 2 \cos(d/2) \cos (\omega_m t + \phi_1 + d/2)$, where $d = \phi_2 - \phi_1$.

3-13. (a) magnitude = 1, angle = 74.9°. (b) 2.87 μm.

3-14. Proof.

3-15. 0.0183, -17.4 dB.

3-16. 0.023 km^{-1}.

3-17. 1 GHz.

3-18. 0.967 ps/(nm x km).

3-19. (a) -0.095×10^3 s/m^3. (b) -0.095×10^{-3} ns/(nm^2 x km).

3-20. 15 nm.

3-21. (a) 50 Gbps. (b) System losses.

CHAPTER 4

4-1. $d = 0.847 \, \mu$m, $n_{\text{eff}} = 3.586$.

4-2. Plot.

4-3. $d = 1.689 \, \mu$m.

4-4. Number of modes: 6, 12, 120.

4-5. TE_0, $d = 0$; TE_1, $d = 1.69$ μm; TE_2, $d = 3.38$ μm; TE_3, $d = 5.06$ μm.

4-6. $1.69 < d < 2.68$ μm.

4-7. Proof.

4-8. Proof.

4-9. $\theta > 80.6°$ and $\theta < 42.5°$.

4-10. Use a lens with focal length in the range: 0.67 mm < f < 7.85 mm.

4-11. Proof.

4-12. Proof.

4-13. (a) 3.2 μm. (b) 13 μm.

CHAPTER 5

5-1. $V = 12.227$ cm^3, $D = 20.06$ cm.

5-2. $D = 23$ cm.

5-3. 2321 reflections per meter.

5-4. Plot.

5-5. Proof.

5-6. Plot.

5-7. Sketch. The ray turns back just before $r/a = 0.7$.

5-8. $\theta = 8.11°$.

5-9. $R = 1$ cm.

5-10. 3286 modes, 1531 modes.

5-11. Proof.

5-12. Plot.

5-13. $a/\lambda = 2.48/[n_1(n_1 - n_2)]^{1/2}$, $N = 132$ modes, $N = 121$ modes.

5-14. $a/\lambda = (p + q + 1)/\{3.14[2n_1(n_1 - n_2)]^{1/2}\}$.

5-15. (a) Single mode, $f \times L = 0.227$ GHz $\times$ km; multimode, $f \times L = 0.222$ GHz $\times$ km.
(b) Single mode, $f \times L = 4.55$ GHz $\times$ km; multimode, $f \times L = 1.08$ GHz $\times$ km.

(c) Single mode, $f \times L = 0.667$ GHz $\times$ km; multimode, $f \times L = 0.574$ GHz $\times$ km.
(d) Single mode, $f \times L = 33.3$ GHz $\times$ km; multimode, $f \times L = 1.11$ GHz $\times$ km.

5-16. $a = 4.43$ μm, $N = 14, 12, 8$, modes.

5-17. Plot.

5-18. Plot.

5-19. 92.2 Mbps $\times$ km.

5-20. Discussion.

5-21. Discussion.

5-22. 14.

5-23. Plot, $r = 6.287$ μm.

5-24. 5.95 MHz.

5-25. 50% (3 dB).

CHAPTER 6

6-1. $v(t) = 1 - \exp(-t/RC)$, rise time $= 2.19RC$.

6-2. V (magnitude) $= 1/[1 + (\omega RC)^2]^{1/2}$.

6-3. Proof.

6-4. Proof.

6-5. (a) Sketch. (b) $I_{peak} = 500$ mA, $I_{dc} = 400$ mA, $P_{ave} = 8$ mW, $m = 0.25$.
(c) $I_{peak} = 500$ mA, $I_{dc} = 250$ mA, $P_{AVE} = 5$ mW, $m = 1$. (d) Note the signal clipping.

6-6. 1%.

6-7. Carrier lifetime $= 1.6$ ns.

6-8. Bandgap energy $= 2.24 \times 10^{-19}$ J.

6-9. (a) $P = 20 + 2 \sin \omega t$ (mW). (b) Sketch.

6-10. Discussion.

6-11. 151°.

6-12. Sketch. 6 modes centered at 900 nm.

6-13. 0 dBm, 22.2 dB.

6-14. 253.

6-15. 0.06 mW.

6-16. 0.8 mA/°C.

6-17. 120 GHz × km.

6-18. 150 GHz/°C.

6-19. 0.185 μm, 0.37 μm.

CHAPTER 7

7-1. $i = 25$ nA.

7-2. $t_r = 700$ ps.

7-3. Cutoff wavelength = 0.99 μm, $f = 3 \times 10^{14}$ Hz.

7-4. Wavelength = 1.98×10^{-25}/work function.

7-5. Responsivity = 0.4, 1.05, and 1.37 at 0.5, 1.3, and 1.7 μm, respectively.

7-6. $i = 188$ nA, $v = 9.4$ μV, 188 μV, and 188 mV.

7-7. $i = 48$ μA, $v = 2.4$ mV, 48 mV, and 48 V.

7-8. Wavelength = 1.98×10^{-25}/bandgap energy.

7-9. (a) $W_g = 1.76 \times 10^{-19}$ J. (b) $W_g = 1.07 \times 10^{-19}$ J.

7-10. $I_D = 0.06$, 0.24, 0.96, and 3.84 nA at $T = 25°$, 45°, 65°, and 85°C, respectively.

7-11. $P = 286$ nW or -35.4 dBm.

7-12. (a) Plot. (b) Load-line equation: $10 + v_d + 2 \times 10^6 \ i_d = 0$. (c) $v = 0.5 \times 10^6 \ P$. (d) $P_{\text{sat}} = 20$ μW.

7-13. (a) $f = 583$ MHz. (b) $R_L = 11.4$ Ω. (c) $f = 2.7$ MHz.

7-14. Responsivity = 53 A/W, $P = 0.38$ nW = -64 dBm.

7-15. Responsivity = 8.7 A/W, $P = 2.3$ nW = -56 dBm.

7-16. Discussion.

7-17. Discussion.

7-18. 0.25 μA/μW.

7-19. 159 kHz.

CHAPTER 8

8-1. Derivation.

8-2. Answers given on Fig. 8-3.

8-3. $d/2a = 0.8054$.

8-4. $d/w = 1.52$, $d/2a = 0.836$.

8-5. $d = 1.7$ μm.

8-6. $I/I_o = 0.19$.

8-7. Plot. Sample results [angle(degrees), loss(dB)]: [1, 0.248], [5, 1.41], [10, 3.52].

8-8. Same plot for both wavelengths. Sample results [angle(degrees), loss(dB)]: [1, 0.845], [2, 3.38], [4, 13.53].

8-9. Derivation.

8-10. Plot. At 1.4 μm, $w = 5.74$ μm, loss = 0.13 dB; at 1.6 μm, $w = 6.46$ μm, loss = 0.1 dB.

8-11. Plot. Sample results [spacing(micrometers), loss(dB)]: 0.8 μm [50, 0.215], [200, 2.59], [500, 7.84]; 1.3 μm [50, 0.083], [200, 1.17], [500, 4.66].

8-12. Focal length 5.1 mm; $d = 398$ μm.

8-13. NA = 0.2, $P = 0.2$ mW, 0.08 mW, 0.00002 mW; NA = 0.5, $P = 1.1$ mW, 1.25×10^{-50} mW, 1.25×10^{-500} mW.

8-14. $(m - 1)\text{NA}^2 \leq 0.4$.

8-15. 9.7% (10.1 dB).

8-16. 5.94% (12.3 dB).

CHAPTER 9

9-1. (a) 4/5, 1/5, 0. (b) $L_{\text{THP}} = 0.97$ dB, $L_{\text{TAP}} = 6.99$ dB, $L_E = 0$, $L_D = $ infinity.

9-2. (a) 0.504, 0.126, 0.0001. (b) $L_{\text{THP}} = $

2.97 dB, $L_{TAP} = 8.99$ dB. (c) $L_E = 2$ dB.

9-3. 6 dB.

9-4. 17.8 dB.

9-5. (a) Sketch. (b) $L = 9$ dB, 12 dB, 15 dB, 15 dB.

9-6. $L = 10.9$ dB, 11.38 dB, 11.84 dB, 11.84 dB. The "best" coupler depends on the application.

9-7. One solution, using a combination of connectors and splices, yields the following losses: $L = 19.2$ dB, 27.6 dB, 36 dB, 39.7 dB.

9-8. (a) Sketch. (b) $L = 6.99$ dB.

9-9. $L = 17.99$ dB to terminals 2, 3, and 4, $L = 15.2$ dB to terminal 5.

9-10. Plot. Sample results of [numbers of terminals, loss (dB)]: [3, 9.37], [10, 14.6],[15, 16.4], [20, 17.6].

9-11. $d = 23.1$ μm.

9-12. $d = 231$ μm.

9-13. Sketch.

9-14. $L = 9$ dB, 12 dB, 15 dB, 15 dB.

9-15. $L = 30$ dB, 20.92 dB, 21.38 dB, 21.38 dB.

9-16. (a) 30. (b) 6.

9-17. 22 dB.

9-18. 17 dB.

9-19. -11.14 dB.

9-20. (a) 1.678×10^3 A/m. (b) 1.678×10^5 A/m.

9-21. Proof.

9-22. (a) 0.785 mm. (b) 0.306 mm.

9-23. (a) 2.355 mm. (b) 1.33 rad/mm. (c) 1.18 mm.

CHAPTER 10

10-1. (a) Sketch. (b) $m' = 0.6$. (c) $P = 5 + 2.68 \cos \omega t$ (mW). (d) $m = 0.54$.

10-2. (a) Sketch, $m' = 0.6$, $P = 5 + 2.12 \cos \omega t$ (mW), $m = 0.42$. (b) Sketch, $m' = 0.6$, $P = 5 + 1.66 \cos \omega t$ (mW), $m = 0.33$.

10-3. (a) $R_e = 87$ Ω. (b) $v_{CE} = 3.9$ V. (c) $m' = 0.84$.

10-4. $a_2 = 0.000158$ mW/(mA)^2.

10-5. $I_C = 96.7$ mA, $I_B = 1.93$ mA.

10-6. Bandwidth = 1.07 MHz.

10-7. Bandwidth = 0.8 MHz for an 8-bit code.

10-8. Bandwidth = 20 MHz.

10-9. Bandwidth = 72 MHz.

10-10. Bandwidth = 1152 MHz for an RZ 8-bit code.

10-11. Input current $i = 50 + 16.7$ [(1 + $0.5 \cos \omega_m t) \cos \omega_{sc1} t + (1 + 0.5 \cos \omega_m t) \cos \omega_{sc2} t$], ω_m = modulation frequency, ω_{sc1} and ω_{sc2} = subcarrier frequencies.

10-12. Drawing.

10-13. If the supply voltage is 5 V, then a suitable circuit has load resistance 45 Ω, and transistor drive currents greater than 1.6 mA.

10-14. (a) In Eq. (10-9) set $i_s = I_s(1 + m \cos \omega_{m1} t) \cos \omega_{sc1} t + I_s(1 + m \cos \omega_{m2} t) \cos \omega_{sc2} t$ and multiply out all terms. (b) Sketch. (c) Reduce the nonlinearity coefficient a_2 and separate the two subcarriers so no overlapping of spectra will occur.

10-15. 9677.

CHAPTER 11

11-1. (a) i_{NT}(rms) = 45.2 nA. (b) v_{NT}(rms) = 2.26 μV, P_{NT} = 0.1 pW. (c) i_{NT}(rms) = 1.43 nA, v_{NT}(rms) = 71.4 μV, P_{NT} = 0.1 pW.

11-2. (a) i_{NS}(rms) = 43.8 pA. (b) v_s = 50 nV, v_{NS}(rms) = 2.19 nV. (c) SNR = 520 = 27 dB. (d) SNR = 5×10^{-4} = −33 dB.

11-3. P = 10 nW.

11-4. (a) 54 μW. (b) SNR = 8.43 × 10^5 = 59.3 dB. (c) P_{NS} = 864 fW.

11-5. (a) P = 1.4 nW. (b) SNR = 5.67 = 7.5 dB. (c) P_{NS} = 864 fW.

11-6. (a) SNR = 1.51 = 1.79 dB. (b) SNR = 1.79 dB. (c) SNR = 3101 = 34.9 dB. (b) M = 31.

11-7. (a) Plot. (b) P = 607 nW. (c) P = 60.7 nW. (d) P = 19.3 nW.

11-8. P = 132.7 μW.

264 fW. (e) T_A = 300 K. (f) P = 33.2 fW. (g) SNR = 376 = 25.8 dB. (h) SNR = 188 = 22.8 dB.

11-14. (a) f = 8.84 MHz. (b) t_r = 39.6 ns.

11-15. (a) f = 17.7 MHz. (b) t_r = 19.8 ns.

11-16. Sketch.

11-17. Sketch.

11-18. (a) 2.5 mV. (b) f = 79.6 MHz. (c) i_{NT}(rms) = 11.5 nA. (d) i = 0.25 μA. (e) SNR = 185 = 22.7 dB.

11-19. Proof.

11-20. Proof.

11-21. 22.7 dB.

11-22. (a) 0.112 μW. (b) 0.034 μA.

11-23. $$\frac{S}{N} = \frac{(m^2/2)(M\rho P)^2 R_L}{M^n 2e\, R_L \Delta f (I_D + \rho P) + 4kT\, \Delta f + M^n \text{RIN}(\rho P)^2 \Delta f\, R_L}$$

11-9. (a) P_{LO} = 4 mW. (b) P = 2 μW.

11-10. Table.

11-11. (a) SNR = 21.6 dB. (b) P = 5.2 μW. (c) 67,800 photons/bit.

11-12. (a) 1.8 nW. (b) 23 photons per bit. (c) The shot-noise-limited system requires much less power. (d) Use an APD.

11-13. (a) P = 6.25 pW. (b) P = 50 pW. (c) P_{NT} = 16.6 fW. (d) P_{NT} =

11-24. n = 2.35.

11-25. (a) Plot. (b) M = 55.

CHAPTER 12

The design problems in this chapter have no unique solutions.

Index

A

Absorption (*see* Attenuation in fibers)
Acceptance angle, 46, 90–91, 103 (*see also* Numerical aperture)
AlGaAs, 156
Amplifier noise:
  noise figure, 163, 285–87
  noise temperature, 284
Amplifiers:
  electronic, 256–59
  erbium-doped fiber, 162–63
  Fabry-Perot laser, 162
  optical, 161–63
  semiconductor, 162
Amplitude modulation (AM), 12, 246–47
Analog modulation, 4–5, 145, 239–43, 245, 246–47
Analog system design:
  bandwidth budget, 301–3

power budget, 300–1
Antireflection coating, 73
APD (*see* Avalanche photodiode)
Asymmetric slab waveguide:
  mode chart, 88
  modes, 87–88
Attenuation coefficient (*see* Electromagnetic waves)
Attenuation factor, 75, 81
Attenuation in fibers:
  absorption, 107–14
  atomic defects, 109–10
  bends, 111–12
  Rayleigh scattering, 110–11
Avalanche photodiode:
  advantages, 179–80, 271–72
  amplification, 179
  construction, 180
  excess noise, 272